Primer of Ecological Theory

Primer of Ecological Theory

JONATHAN ROUGHGARDEN
Stanford University

Prentice Hall, Upper Saddle River, New Jersey 07458

Library of Congress Cataloging-in-Publication Data

Roughgarden, Jonathan.
Primer of ecological theory / Jonathan Roughgarden.
p. cm.
Includes bibliographical references and index.
ISBN 0-13-442062-4
1. Ecology—Mathematical models. I. Title.
QH541.15.M3R68 1998
577'.01'5118—DC21 97-5917
CIP

ACQUISITION EDITOR: ***Teresa Ryu***
EXECUTIVE EDITOR: ***Sheri L. Snavely***
PRODUCTION EDITOR: ***Joanne E. Jimenez***
MANUFACTURING MANAGER: ***Trudy Pisciotti***
COVER DESIGNER: ***Bruce Kenselaar***
COVER ART: *Seascape with Heavenly Body* by Paul Klee (1879–1940)

Trademark Information
MATLAB is a registered trademark of The MathWorks, Inc.

Printed in the United States of America
10 9 8 7 6 5 4 3

ISBN 0-13-442062-4

PRENTICE-HALL INTERNATIONAL (UK) LIMITED, LONDON
PRENTICE-HALL OF AUSTRALIA PTY. LIMITED, SYDNEY
PRENTICE-HALL CANADA INC. TORONTO
PRENTICE-HALL HISPANOAMERICANA, S.A., MEXICO
PRENTICE-HALL OF INDIA PRIVATE LIMITED, NEW DELHI
PRENTICE-HALL OF JAPAN, INC., TOKYO
SIMON & SCHUSTER ASIA PTE. LTD., SINGAPORE
EDITORA PRENTICE-HALL DO BRASIL, LTDA., RIO DE JANEIRO

Contents

Preface

Ecology and evolution differ from other areas of biology in having a mathematical theory. A glance at texts and journals for ecology and evolution reveals a level of math that provokes culture shock in those accustomed to pictures of mitochondria, the circulatory system, and the double helix. But that's the way it is, and here's why. Most ecological phenomena are produced by several mechanisms acting simultaneously, and mathematical models are needed to splice what's going on into a whole picture. At their most basic level, almost all ecological states are brought about by combining some sort of input with an output. How many people there are in a population depends on how fast births *and* deaths happen, so it's impossible to understand how many people there are without some model that combines both births and deaths. Similarly, how hot or cold you feel depends on how fast you're gaining *and* losing heat, so again, to understand how you feel, or how a plant or animal "feels," a model with both inputs and outputs of heat is needed. So, ecological states are almost always dynamic; that is, they have the potential for changing if either the inputs or outputs change. And if a state is staying relatively constant for a while, then the inputs happen to be balancing the outputs during that time.

OK, if there is a basic need for models to keep track of ecological inputs and outputs, is that all that models do? Well no. Once you get used to working with models you have in effect acquired a new language. When you look at an equation you should think of it as a sentence. The equal sign is a verb, the left hand side is the subject, and the right hand side is the object. "Lollapalooza sounds good" is no more a sentence than "$dN/dt = rN$" which translated, means the speed of a population's growth is proportional to its present size. If you can get used to saying interesting things with equations, then you've become in a sense multilingual, and can enjoy your readings in ecology a lot more than you can if you're unilingual—it's obviously more fun to visit Paris if you know French.

People new to modeling often don't realize that a model doesn't have to be totally correct to be useful. Think of cooking. In most dishes the ingredients don't have to be measured to a milligram, nor the baking timed to the millisecond. A model too doesn't have to get every thing exactly right, because it may still account for what is going on pretty well. Modelers use the phrase "structural stability" for a model whose predictions change only a little bit if the model itself is changed only a little bit. The recipe for baking bread is structurally stable; the loaf is still a loaf, but just a little darker, if the baking is increased 5 minutes. Of course, you can't push a model too far and get away with it, but most good models don't fall out of the sky if pushed a little bit. We will look into how to tell whether a model is sensitive to being pushed around. If a model is very sensitive to pushing, then either the model itself has been put together with poor craftmanship, or else the phenomenon the model represents is in truth very unstable, like a field of dry grass just waiting for a match.

Most models in ecology fall into three categories. A minimal model for an idea tries to illuminate a hypothesis. For example, sexual reproduction befuddles biologists to this day. Reproduction is certainly possible asexually, that is without mating, by mitosis of single cells, and by fission or fragmentation of multicellular organisms. So why is there sexual reproduction when asexual reproduction might be, in some sense, easier? Well, this kind of question invites models that offer ideas for why sexual reproduction may be somehow better than asexual reproduction. A minimal model for an idea is not intended to be tested literally, any more than one would test whether the models for a frictionless pulley or frictionless inclined plane are wrong. Of course models of ideal machines are wrong because there is always some friction in the real world, but they still correctly convey the idea of mechanical advantage, and they are structurally stable because they are approximately correct if the friction is small. In this book we will look at many models for key ecological ideas, models that should not be tested as such, but are important to think about.

A minimal model for a system tries to illuminate how some natural grouping of real organisms and processes work together. A forest might be idealized as a collection of stems positioned at initially random locations on a plane. Then as each stem grows it intersects a neighbor, and the forest as a whole stops growing when the average tree is shaded enough by its neighbors that photosynthesis balances respiration. This rather simple picture could lead to a minimal model for how a forest grows. Such a model would obviously not apply to a population of mobile fish, nor to birds or butterflies, and so forth. The point of such a model is to refer in simplified terms to a specific category of ecological systems. How much detail to include and still count as "minimal" is usually determined by what might be called academic market forces. A model of a forest that includes only photosynthesis and respiration while ignoring seeds and tree falls would not generate any readership and would gradually be forgotten. On the other hand, a model that added seeds, tree falls, weevils, woodpeckers, earth worms, and a kitchen sink's worth of everything else would also fail to gain attention because no one can afford to find out enough about so many factors to represent them in equations. So, somehow through peer review and readership, minimal models for systems emerge that express a compromise between simplicity and detail, and the job of the modeler is as much to negotiate the compromise as it is to solve the model's mathematics. This book will explore many minimal models of systems.

The third type of model in ecology is a synthetic model of a system. The goal of a synthetic model is to coordinate the activities and integrate the findings of many investigators on a large project. Models of global change divide the globe into a large grid and predict the temperature around the world according to various scenarios of emission controls. There is a lot of money involved in CO_2 emissions and global warming, and it's important to get this model accurate. Global-change models are examples of synthetic models, as are ecological models designed to be coupled to global-change models. Synthetic models usually exist only as large computer programs, and are often called "systems models" or "simulations," although simulation may also be used for smaller computer calculations too. Synthetic and systems models do not aim to be minimal, they aim instead to include everything relevant that is known about a system, to be numerically accurate, and to incorporate, or to "assimilate," new information into the model as it becomes available to improve accuracy. We will illustrate how systems and minimalist approaches may be taken to the same problem.

If you're a physicist, talk about kinds of models, and negotiations between simplicity and detail, seems strange indeed. One doesn't negotiate about $F = ma$ or the laws of thermodynamics. So ecology is different and it's a good idea to accept this now. For the most part, mathematical models in ecology are not foundational, they are supplemental, and judged by their utility. And if you're a field biologist, other aspects of this book may seem strange. We will not be concerned very much with testing whether the assumptions or predictions of a model are true. Instead, we will test, so to speak, whether the predictions of a model follow logically from the assumptions. Should someone say, "If two species compete for resources then P," we will focus on whether P really can be derived from the assumption of competition for resources, and not on whether the two species in fact compete for resources in nature. We do want to know whether a model is well motivated biologically, whether it is about a biological issue or phenomenon, but this is not the same as constructing a field test of a model. Constructing a field test requires thoroughly understanding the model, possibly retooling the model so that it is more appropriate to the test system, knowing the natural history of the test system and its components, and knowing about experimental design and the power of the statistical tests that are anticipated. Hopefully though, if you're interested in the models in this book, you'll press on to see if some of them really are true representations of nature.

One might digress here to speculate philosophically on why ecology, and not other areas of biology, has a strong tradition of mathematical modeling. After all, physiology too concerns inputs and outputs, and one might imagine a different biology from what we have today, one in which models were used through out. Also a different medicine—imagine using your next physical exam to parameterize a model of your body, so that your drug dosages and other therapies could be predicted in advance. I guess two reasons why ecology leads biology with modeling are that ecological experiments are relatively expensive and time-consuming, and that ecological situations are always unique if looked at closely enough. Today, it's faster for a doctor to determine a patient's drug dosage by trial and error than to develop a model of a person's physiology, and there are many patients around to experiment with. It's hard to do large-scale ecological experiments—try rerouting the Gulf Stream to test a hypothesis about how global warming affects ocean circulation, and there's only one Gulf Stream anyway. Small ecological systems can be studied with experiments, but the comparative advantage of models to experiments favors models relatively more in ecology than in the rest of biology. Of course, models and experiments can help each other, but in most of biology one doesn't bother with mathematical models at all, whereas in ecology, models are needed in addition to experiments. The other distinguishing feature of ecology is the near absence of universal facts and theories. Instead, there is a general *approach* to ecological phenomena. Think of building a bridge at the mouth of a river. There's no universal bridge—one size doesn't fit all, but civil engineering offers a general approach to building bridges. Similarly, no two lakes are the same, nor are the tropical forests of different continents the same. Through modeling one can present the information about different systems in a common format, and see general features emerge.

Ecological knowledge is organized along a progression of scales from small to large, from individual to population to ecosystem. The small scales are contained in the large scales, and phenomena at small scales usually, but not necessarily, change faster than phenomena at large scales. The chapters in this book thus follow each other in this traditional sequence.

If you're a student, here's what I hope you will come away with. First, a knowledge of the basic models presently in circulation in ecology. This is like knowing that the Arc de Triumph and the Seine are in Paris, and there's no point in going there if you don't know this. Second, a knowledge of how to frame some simple models of your own, and how to get answers from them. This is like composing your own sentences in French. This is an elementary book though, so don't expect to end up as a great French novelist overnight. I assume you know high school algebra well, and are comfortable rearranging equations and making graphs. You should know things like the shape of an exponential function, a parabola and a hyperbola. You should also know basic calculus—that the derivative of x^2 is $2x$ should definitely not come as a surprise. Concerning computers, you should know how to turn one on and get it to operate, how to back up diskettes, how to operate your printer, and so forth. You should also be willing to speak in computerese, to think with "if-then" statements and "for-loops." You don't need to know any particular computer language before you begin, it's the willingness to think this way that's important.

If you're an instructor, here's what I hope to have provided you with. First, a book that you can assign as a supplement for texts or collected readings, as a laboratory manual in ecology, or as a standalone text for a course in ecological theory and modeling. Second, a set of generic topics that you can present without cramping your own style. I'm counting on *you* to make a course out of this—to present examples, to offer alternative models, to discuss the strengths and shortcomings of the present state of theory in ecology. Ecology, and theoretical ecology in particular, are thriving academic disciplines, slowly but surely developing reliable and interesting models, and enjoying unprecedented public attention. I hope this primer will assist you in getting that message across.

A word on software. In the past a primer was organized somewhat like a workbook, with some textual material interspersed with set problems. Today though, a primer probably should offer an opportunity to explore the models with a computer. This book uses the program MATLAB (plus the optional SIMULINK add-on) to do the derivations "on line" so the reader can participate in the derivations, and thereby learn how to tailor existing models for particular situations. MATLAB is primarily used for numerical analysis and it has acquired limited capability for symbolic analysis through an add-on called a toolbox. The Symbolic Math Toolbox, which is automatically part of the student edition of MATLAB, was originally developed as an independent program called Maple, but is now grafted onto the MATLAB product[1]. An alternative to MATLAB is Mathematica. Mathematica is primarily

[1]This book was developed using the version of MATLAB generally available during the 1996–97 academic year. The version of MATLAB and any tool kits you're using can be determined by typing `ver` at the command line. The version I have used is found in this way to be: MATLAB Version 4.2c; MATLAB

used for symbolic analysis and its numerical capabilities are relatively cumbersome. At present, marketing favors MATLAB because the student edition is about half the price of the student edition of Mathematica, and it also retains almost all the functions of the professional edition. On the whole, MATLAB seems the best for general purposes because it's easier to do simple things with it, whereas in Mathematica simple things can be surprisingly hard to do. You may want to switch to Mathematica or to the stand-alone version of Maple if most of your work is with the derivation of formulas and not with numbers. Also, the program, Splus, is best if most of your work is with statistical data analysis. Still, I bet you'll find MATLAB satisfactory most of the time.

What will you do with the knowledge you gain from this book? If you're going into an environmental career, I hope you'll integrate it into your future work. Which is more dangerous to the environment, a greedy developer or an uninformed conservation biologist? At this time, it's probably an uninformed conservation biologist, because, to the credit of environmentalism, the public increasingly prevents the environmental destruction wrought by greedy developers. But as an environmentalist, after you've won your political battles, will you really know what to do to save the environment? Or will you blow it? *The San Francisco Chronicle* reported on November 12, 1996 that the spotted owl of the Pacific Northwest is being threatened with competitive exclusion by another owl, the barred owl, from the eastern US. Do you know how to calculate how much area should be conserved to protect an owl? Do you know how much more area would be needed if that owl were surrounded by a competing species of owl? If you don't know how to answer these questions, how can you bargain for the amount of old-growth forest to conserve? Don't count on someone having already done experiments that completely answer these questions—you're going to need a model. To illustrate how ecological models are used in conservation biology, example applications are found as special sections at the end of each chapter. I'm grateful to Professor Peter Kareiva of the Department of Zoology at the University of Washington for contributing these.

Prof. Kareiva's applications jump-start the larger venture of gathering teachable examples, vignettes, of applied ecological theory for everyone to use. I am maintaining a WWW site at `rough.stanford.edu`, and invite you to contribute examples you have developed in teaching, research, consulting, or other work. With a web-browser, connect to `rough.stanford.edu` and take it from there[2]. All kinds of applications of ecological models are appreciated, including epidemiology, resource management, pest control, ecological economics, conservation, and so forth, from as many locales as possible. I've also placed complete MATLAB scripts for each chapter on this web site. They may be downloaded

Toolbox Version 4.2a, 25-Jul-94; Symbolic Math Toolbox, Version 1.1, 1-Mar-95; SIMULINK model analysis and construction functions, Version 1.3c, 15-August-94; SIMULINK block library, Version 1.3c, 15-August-94. I have also tested the software with Version 4 of the Student Edition of MATLAB. As the book was going to press, I received a beta copy of the new release of MATLAB, Version 5. The new release introduces major changes, especially to the Symbolic Toolbox. Dr. Cleve Moler, one of the founders of MATLAB, has worked on the design of MATLAB itself to ensure that the programs for this book will continue to work with Version 5 of MATLAB when it becomes available.

[2] You may also connect to the home page of The Math Works at `www.mathworks.com` and of Prentice Hall at `www.prenhall.com` and follow the links from there.

with your web browser, or by anonymous FTP from the pub directory. This will allow you to download all the code presented in the book without having to type it in from scratch[3].

I thank many people who have contributed to this project, especially Sean Connolly who served as head teaching assistant both for a graduate seminar in which this material was first developed and for an undergraduate course in which the book was first used; my colleagues at Stanford who have reviewed sections of the manuscript: Hans Andersen and Steven Schneider; colleagues from other universities who reviewed the manuscript at Prentice Hall's request: Leslie S. Bowker, California Polytechnic State University; Martin L. Cody, University of California, Los Angeles; Robert Holt, University of Kansas; Peter Kareiva, University of Washington; Kate Lajtha, Oregon State University; Matthew A. Leibold, University of Chicago; John Pastor, University of Minnesota; Donald C. Potts, University of California, Santa Cruz; Daniel Simberloff, Florida State University; Alan E. Stiven, University of North Carolina; Fannie Toldi for her gentle touch with the copy editing; and above all, Teresa Ryu, Biology Editor of Prentice Hall, whose support, skill and enthusiasm have been invaluable.

J. R.
Palo Alto, California

[3]If you're migrating from Version 4 to Version 5 of MATLAB, be sure to check in for the few touch-ups that will be needed.

Primer of Ecological Theory

Chapter 1

Getting Comfortable Outdoors

If an organism can't survive in a particular environment there's no point in trying to discuss its ecology there, so let's begin by seeing what's needed for an organism to stay alive in a habitat without either shivering to death or overheating.

1.1 Equilibrium Body Temperature

Body temperature results from combining the flows of heat into and out of an organism. Heat enters an organism directly from the sun's rays or indirectly, as reflections from other surfaces. Let's focus on a small animal. Heat also enters if a wind blows that is hotter than the animal's present temperature (convection), or if it is sitting on a hot rock (conduction). Heat may be lost if the wind is cooler than the body temperature, or if the surface the animal contacts is cooler than it is. Solar radiation is unidirectional, and convection and conduction are bidirectional. Furthermore, if the animal evaporates water it loses heat—through breathing where water is released from the lungs, and also from exterior surfaces including the skin and eyes. This is a unidirectional path, but requires rather dry air because evaporation is slow in high humidity. Finally, in birds and mammals, metabolism generates a lot of heat. To find the body temperature that results from all this, we simply add up all the flows and see what we get.

Because combining heat flows is relatively simple, it's a good place to start using MATLAB. You can probably figure out what we're going to do more quickly with pencil and paper, but later on MATLAB will prove faster. Also, if you can figure out a problem by hand, you can check whether the computer is right. Computers are often wrong and you have to watch them like a hawk.

Let's restrict ourselves to just two of the many pathways, solar radiation and convection, because these are the two most important for small lizards, which are creatures dear to my heart. Our objective is to find the body temperature at which the animal stabilizes if it remains in one spot. If the sun is shining on it, the animal tends to heat up, and if the wind is blowing over it, the animal tends to cool off. Therefore it will stabilize at the temperature for which the solar input balances the convective loss. So, to figure all this out, fire up MATLAB and you should see the prompt[1]

```
EDU>>
```

or if you're not using the student version, simply

```
>>
```

Let the solar radiation on the animal measured in cal/h be q. This quantity depends on the surface area of the animal, the color (or reflectivity) of the skin, and of course, on how sunny it is. A typical q might be 1000 cal/h. Next, the convective heat loss is proportional to the difference between the body temperature and the air temperature. Let b be the body temperature, let a be the air temperature (both in degrees Celsius), and the constant of proportionality be h. The h is called the convective heat-transfer coefficient, and a typical value for it might be 50 cal/(h °C). The h depends on the surface-to-volume ratio of the animal, its total size, and the wind speed. The total heat flux, f, in cal/h, into the animal is then the solar radiation minus the convective heat loss,

$$f = q - h(b - a)$$

Now look carefully at this equation. If the animal starts out with a body temperature equal to the air temperature, then the first term is positive and the second term is zero, so f is positive, and the animal begins to heat up. Then, as its body temperature rises, convection begins to remove more and more heat because the convection depends on b minus a, and this difference gets bigger as b gets bigger, provided a says the same. So, as b continues to rise, eventually the convection comes to balance the solar input, and the body temperature will then stay constant as long as q, a, and h remain unchanged. The body temperature at which the solar input balances the convective heat loss is called the *equilibrium* or *steady-state* body temperature, and is found by setting the equation for the total flux equal to zero and solving for b. Now let's do this with MATLAB.

MATLAB can manipulate both formulas and numbers with different commands. We'll start with formulas, and when we get the formula we want, we'll switch to numbers. I suggest that you first make sure your working directory or folder is correct, and that the diary feature is on,

```
EDU>> cd c:\primer
EDU>> diary on
```

In the example above, my working directory on a Windows PC was `c:\primer` and the diary, which is a transcript of all input and output, will be saved in that directory. So, we're ready with the first formula,

[1]To fire up MATLAB, on a Windows PC or Macintosh you should click on its icon, and the prompt will appear in a new window. In UNIX type the command `matlab` in a terminal window and the prompt should appear in that window where the command was issued. For all computers, before the prompt appears, a signon message, including a graph or other picture, will appear for a few moments.

```
EDU>> f = 'q - h * (b - a)'
```

MATLAB then replies with

```
f =
q - h * (b - a)
```

which confirms the definition of `f`. Now the important point here is that one initially defines a formula by placing it between single quotation marks. In MATLAB a formula is an instance of what is called a string, which is a vector of characters. In MATLAB, manipulating formulas basically amounts to converting one string into another. Now we want to solve the flux, `f`, set equal to zero for the body temperature, `b`, and we will call the result `be` which will stand for equilibrium body-temperature. Type the command

```
EDU>> be = solve(f,'b')
```

and MATLAB replies with

```
be =
(q+h*a)/h
```

The `solve()` command[2] solves `'f=0'` for the variable `'b'` and returns a formula, which is itself a string, that is assigned to `be`. Specifically, `be` is the formula for the equilibrium body-temperature given `q`, `a`, and `h`. To further rearrange the formula by dividing the `h` into both terms in the parentheses, type

```
EDU>> be = expand(be)
```

and MATLAB replies with

```
be =
1/h*q+a
```

The `expand()` command replaces `be` with a formula that has the individual terms separated out. Still, the way the formula appears remains computerish and hard to read, so you can use two commands to improve on the visual presentation. For a better appearance at the computer screen, type

```
EDU>> pretty(be)
```

and MATLAB replies with

```
q/h + a
```

and for a version of the formula that can be plugged directly into a LaTeX manuscript[3] use

[2]The exact syntax for this and other commands is found online using the help facility. For example, typing **`help solve`** displays information about **`solve()`**.

[3]LaTeX is a word processing system especially useful for mathematical documents.

```
EDU>> latex(be)
```

to which MATLAB replies with

```
{\frac {q}{h}}+a
```

In a LaTeX formated document, as this book is, the formula will appear as

$$\frac{q}{h} + a$$

The `latex` command will also write the formula directly to a file.

Now that we have the formula for the equilibrium body-temperature, `be`, we want to try it out with some numbers. So, set the heat-transfer coefficient, air temperature, and solar radiation, as follows:

```
EDU>> h = 50
h =
    50
EDU>> a = 18
a =
    18
EDU>> q = 1000
q =
    1000
```

Notice that no quote marks are used when variables are defined as numbers, and that MATLAB confirms the values after they are defined. Now to get a numerical result from the formula, we evaluate it by typing

```
EDU>> eval(be)
ans =
    38
```

So, the equilibrium body-temperature of the animal in these circumstances is 38°C. The `eval()` command is the way to convert a formula to a number just as though you had typed the formula at the command line[4].

MATLAB is famous for computing with tables and matrices and for graphics. To illustrate, we plot a graph of how the equilibrium body-temperature depends on the amount of sunlight. Let the solar radiation, `q`, now be defined as a vector of three possible values, 0, 1000, and 1500. This is done in MATLAB using square brackets,

```
EDU>> q = [0 1000 1500]
q =
           0        1000        1500
```

[4]If `s='2+2'` then typing `eval(s)` at the command line is the same as typing `2+2` itself at the command line.

See how MATLAB responded by confirming that `q` is a vector of three numbers. Now here's something neat. MATLAB will automatically generate a vector of results on the left-hand side if handed a vector on the right-hand side. Let `y` be a vector of equilibrium body-temperatures corresponding to the vector of solar-radiation values. With one command, `y` is found as a whole:

```
EDU>> y = eval(be)
```

and MATLAB responds with

```
y =
    18    38    48
```

Next, `q` and `y` can be viewed as a vector of x and y coordinates on a graph. The way graphs are done in MATLAB is to open a window that becomes, in effect, a blackboard on which lines, curves, dots, dashes, letters, labels and so forth can be drawn. The command **`figure`** opens a window, and the command **`hold on`** states that the blackboard should not be erased as information is written on it—this makes the blackboard accumulate successive writes.

```
EDU>> figure
EDU>> hold on
```

Now that the blackboard is available, one command is sufficient to draw the curve:

```
EDU>> plot(q,y)
```

The `plot()` command plots two vectors against each other. The first vector contains all the horizontal coordinates, the second vector contains all the vertical coordinates.

A straight line now appears on the blackboard; this is the graph of the equilibrium body-temperature vs. solar radiation. The figure can then be spiffed up with a title and labels:

```
EDU>> title('Equilibrium Body Temperature')
EDU>> xlabel('Solar Radiation')
EDU>> ylabel('Body Temperature')
```

Notice that the title and labels are strings. The overall result is shown as Figure 1.1.

Finally, you may want to print the graph on the printer. The commands

```
EDU>> print -dpsc fg1
EDU>> print -depsc fg1
```

produce the files **`fg1.ps`** and **`fg1.eps`** on the disk. **`fg1.ps`** is a full page PostScript file that can be sent directly to a PostScript printer. **`fg1.eps`** is an encapsulated PostScript file that can be incorporated into a document, such as a LaTeX document, by use of the **`psfig.sty`** style option and printed with the **`dvips`** program, or into a Microsoft Word document as a picture object. Alternatively, you may also be able to print out the graph by clicking menu

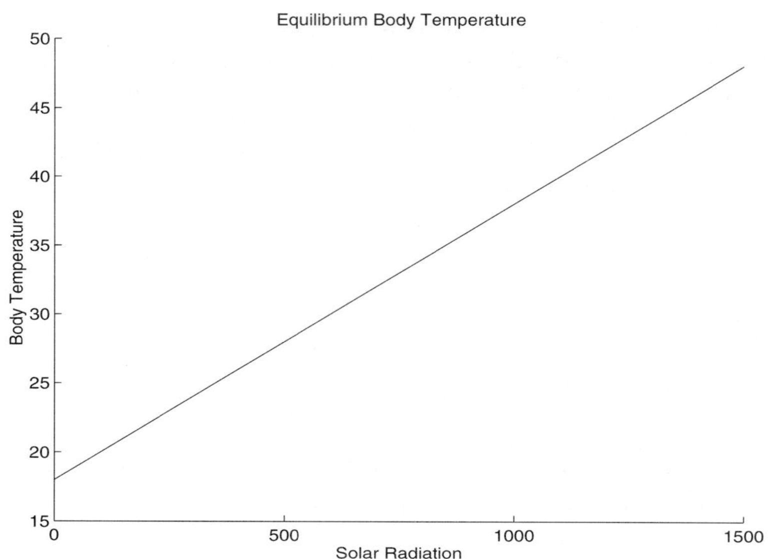

Figure 1.1: Equilibrium body-temperature as a function of solar radiation, with the assumption that the convective heat-transfer coefficient, h is 50 cal/(h °C), and air temperature, a, is 18°C. Solar radiation is in cal/h and body temperature in °C.

items on the graph's window, or by copying it to the window's clipboard system, depending on the kind of computer you're using.

We have presented the analysis of the equilibrium body-temperature in full, with both MATLAB commands and responses. To save space, in the remainder of the book we'll present only the commands, and after you type them in, MATLAB will respond to each as appropriate. Also, we'll indicate the prompt as `>>` rather than `EDU>>` and edit the MATLAB responses so that they fit on one line if possible. Now that you've seen the complete procedure from derivation to numerical example to drawing a graph, it's a good time to study the MATLAB manual. Be sure to check out sections 5.1–5.7, 5.19, 6.1–6.3, and 6.6 of the *User's Guide to the Student Edition of MATLAB.*

1.2 Climate Space

Animals have upper and lower lethal limits to body temperature. We can use the concept of an equilibrium body-temperature to pick out parts of the environment where it is safe to be—these are spots where the equilibrium body temperature remains between the lethal limits. The spot in the environment where it is safe is called the "climate space."

A spot in the environment is characterized by the amount of sun shining there together with the air temperature, for a given wind speed. The climate space, i.e., the spots where the combination of sunlight and air temperature are safe, can be plotted as a region on a graph of air temperature, `a`, vs. solar radiation, `q`. Let `b_h` be the upper lethal limit for the body temperature. Then the upper boundary to the climate space is where the equilibrium body temperature equals `b_h`. This boundary is found by setting the heat flux equal to zero, to indicate an equilibrium, and solving for `a` as a function of `q` so that the curve can be plotted on a graph of `a` vs. `q`. The formula for this curve is found in MATLAB with three commands as

```
>> f_h = 'q - h * (b_h - a)'
>> a_h = expand(solve(f_h,'a'))
```

and MATLAB yields

```
a_h = -1/h*q+b_h
```

This is a straight line with slope `-1/h` and intercept `b_h`. Similarly, if `b_l` is the lower lethal-limit body-temperature, and

```
>> f_l = 'q - h * (b_l - a)'
>> a_l = expand(solve(f_l,'a'))
```

yields the curve for the lower boundary of the climate space

```
a_l = -1/h*q+b_l
```

This is a straight line with the same slope as the upper boundary but with a different intercept.

Now let's switch to numbers so we can produce a graph. Plausible upper and lower limits are 36°C and 24°C respectively, and a reasonable `h` is 50 cal/(h °C).

```
>> b_h = 36
>> b_l = 24
>> h = 50
```

Because two points determine a line, we can use a vector for `q`; the vector would consist of two values

```
>> q = [0 1500]
```

for 0 and 1500 cal/h. The corresponding vectors for this `q` are found from

```
>> y_h = eval(a_h)
>> y_l = eval(a_l)
```

Now to plot the upper and lower limits of the climate space, type

```
>> figure
>> hold on
>> plot(q,y_h)
>> plot(q,y_l)
```

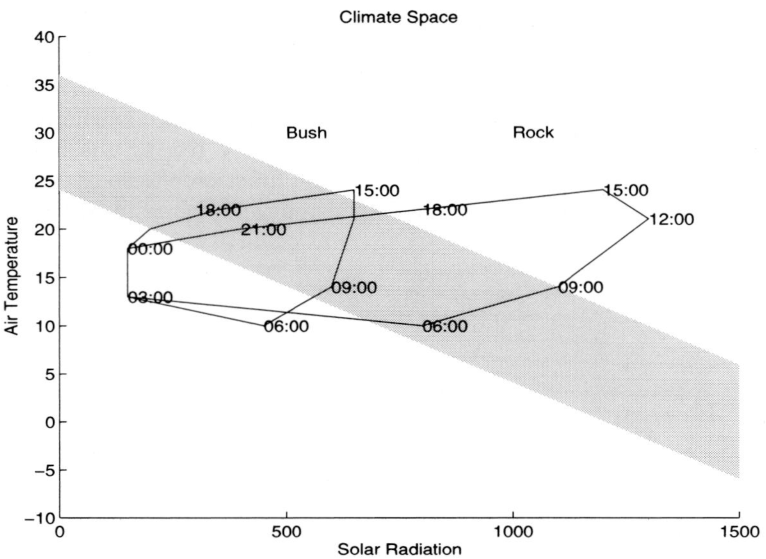

Figure 1.2: Climate space is shaded, with an upper safe limit of 36°C and a lower safe limit of 24°C. Convective heat-transfer coefficient is assumed to be 50 cal/(h °C). The loops illustrate possible combinations of air temperature and solar radiation available at two types of sites in the habitat, large exposed rocks and protected bushes.

These appear as two parallel lines with negative slopes. The area between them is the climate space, where it is safe to live. It would be nice to shade the climate space somehow, and we can color the region between the lines green with the `fill` command. This command uses the corners of the region, and we go around them clockwise, with one vector for the x-coordinates of the corners and another vector with the y-coordinates. The color green is specified with the character `'g'`.

```
>> fill([q(1) q(1) q(2) q(2)],[y_l(1) y_h(1) y_h(2) y_l(2)],'g')
```

Notice that particular elements of the vectors are specified by their positions. `q(1)` is the the first element of `q`, which is 0, and `q(2)` is 1500. So at this point we have the climate space as a green shaded region on the graph, and the picture is finished by the addition of a title and labels:

```
>> title('Climate Space')
>> xlabel('Solar Radiation')
>> ylabel('Air Temperature')
```

To apply the climate space to a particular habitat, we need to know what pairs of air temperature and solar radiation actually occur there throughout the day at various spots.

Beginning and ending at midnight, and moving in steps of 3h, the solar radiation on a large exposed rock in the habitat might be the following values:

```
>> rockq = [150 150 800 1100 1300 1200 800 400 150]
```

and the air temperature at these times might be

```
>> rocka = [18 13 10 14 21 24 22 20 18]
```

These values can be plotted in red as a closed loop:

```
>> plot(rockq,rocka,'r')
```

Similarly, a more protected bush in the habitat might experience the following solar-radiation and air temperatures:

```
>> bushq = [150 150 450 600 650 650 350 200 150]
>> busha = [18 13 10 14 21 24 22 20 18]
```

which can be plotted in yellow as another closed loop with

```
>> plot(bushq,busha,'y')
```

The times for these coordinates and other labels can be annotated on the graph with

```
>> text(150,18,'00:00')
>> text(150,13,'03:00')
>> text(800,10,'06:00')
>> text(1100,14,'09:00')
>> text(1300,21,'12:00')
>> text(1200,24,'15:00')
>> text(800,22,'18:00')
>> text(400,20,'21:00')
>> text(450,10,'06:00')
>> text(600,14,'09:00')
>> text(650,24,'15:00')
>> text(300,22,'18:00')
>> text(1000,30,'Rock','Color','r')
>> text(500,30,'Bush','Color','y')
```

The completed graph appears as Figure 1.2.

This figure illustrates how the animal must remain in its burrow from midnight until one of the loops enters the climate space. It can inhabit the rock sometime around dawn, about 5:00 A.M., but as the rock gets too hot by about 9:00 A.M., the animal must move to the bush til about 2:30 P.M. Even the bush is outside the climate space at 3:00 P.M., so the animal must return into its burrow from about 2:30 P.M. to 3:30 P.M.; this pattern implies a bimodal activity schedule. It can re-emerge from the burrow at about 3:30 P.M.

and reoccupy the bush. Then it can move back to the rock and remain active to about 10:00 P.M., when it must retire to the burrow for the remainder of the night.

A climate space reveals that much of an apparently empty habitat is really not suitable to organisms. One may stroll through a park seeing lots of unoccupied space, on green leaves, tree trunks, and so forth, and wonder why some animal isn't sitting there. The answer is often that the thermal characteristics of the space make it unsuitable for living. A caterpillar cannot digest a leaf that is sufficiently shaded and a butterfly would die of overheating if it remained too long in the sun.

The climate space is a valuable tool in predicting environmental impact on wildlife. If the habitat is thinned by logging or clearing, or as global warming proceeds, then the microclimate in the habitat will get hotter, and the activity patterns of the animals in the habitat will change in response. Perhaps some will benefit and others suffer, but there is certain to be some impact nonetheless, and the climate space can predict who gets to do what in the new regime.

1.3 Living at the Edge

The climate space is not a rigid cage imprisoning an organism to a fixed set of spots in the habitat. A careful animal can dash outside its climate space and return to tell about it. The trick is for the animal to start out well within the climate space, to go outside its boundaries for a little while, and then return before its body temperature has exceeded the lethal limit. If the animal were at the edge of the climate space to begin with, then it couldn't leave. But by starting from well inside the climate space, the animal can buy some time outside the climate space. So, let's see how much time an animal can buy in this way. To answer this question we have to go beyond the concept of an equilibrium or steady-state temperature and consider how quickly the body temperature can change. The way the body temperature changes as it moves from its initial condition to the equilibrium is called a "transient response."

We will illustrate how to solve for the transient response of a organism's body temperature in three ways that illustrate some basic methods in modeling. First, we derive a formula for how the body temperature varies through time, and simply plot it a graph. This approach is the most informative intellectually, is exact, but isn't often possible. Second, we will bypass the formula and generate a graph straight away. This approach is purely numerical and works when a formula can't be derived, but may be inexact. Third, we will develop a simulation model for the body temperature. This approach can be extended to include substantial realism in a model, but because it relies on the numerical approach, it may also be inexact.

1.3.1 Symbolic Analysis

Recall that the flux in energy into an organism is $q - h(b - a)$ where q is solar radiation, h is the convective heat transfer coefficient, b is the body temperature, and a is the air temperature. To find out how this flux is changing the animal's body temperature, we need to know its weight and the specific heat of its tissue. Because an organism is mostly water, we approximate the specific heat as 1 calorie per gram; i.e., one calorie will raise one gram of the organism by 1 degree C. Let w be the animal's weight. The heat flux then will change the body temperature according to

$$\frac{db}{dt} = \frac{q - h(b - a)}{w}$$

This says that the speed with which the body temperature is changing equals the heat flux divided by the animal's weight. The equation is a differential equation, which means that what we have to solve for is a function, $b(t)$. Contrast this with solving for the equilibrium body-temperature, which is a number; here the unknown is an entire function that has the property that its slope at any time equals the heat flux at that time divided by the body weight. Well, in the old days, finding the formula for the unknown function took a bit of work, but now MATLAB can generate the formula for $b(t)$ in one command,

```
>> b = dsolve('Db = (q - h * (b - a)) / w','b(0) = c','t')
```

The differential equation is the first string, the initial condition is the second string, and the independent variable is the third string. MATLAB replies that

```
b = 1/h*q+a+exp(-1/w*h*t)*(-1/h*q-a+c)
```

The formula for $b(t)$, when formated for LaTeX, is

$$b(t) = \frac{q}{h} + a + e^{-\frac{ht}{w}} \left(-\frac{q}{h} - a + c\right)$$

Now that we have the formula, let's look at it closely. When t is zero, $b(t) = c$, according to the initial condition that we stipulated. If $t \to \infty$ then $b \to q/h + a$, which is the equilibrium value for b. The approach to equilibrium is governed by an exponential function, so the distance to equilibrium will decrease exponentially with time.

To see if $b(t)$ looks like what we think it does, let's graph it with some numbers and see. Again we let

```
>> a = 18
>> h = 50
>> q = 1000
```

The equilibrium b is then 38°C as before. We also suppose the animal is 10g in weight and its initial body temperature is 30°C, which is between its lethal limits of 24°C and 36°C.

```
>> w = 10
>> c = 30
```

Thus, if the animal sat at this location forever, it would be outside its climate space, but if, when it arrives at this location, it has an initial temperature of 30°C, it can stay there for a little while before its temperature hits the edge of the climate space at 36°C. To see how long it can stay there, let's plot $b(t)$ over the course of an hour. First, generate about 30 time points with

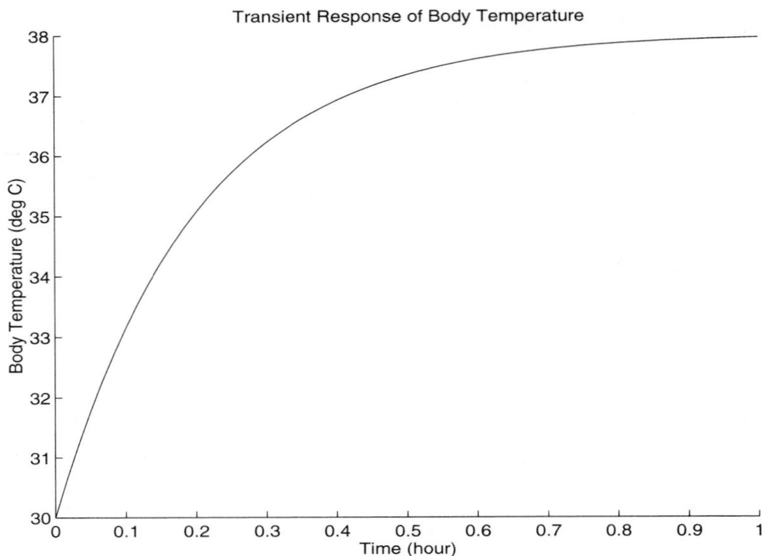

Figure 1.3: Transient response of body temperature to heating together with convection starting at 30°C and approaching an equilibrium at 38°C. The body weight is 10g.

```
>> t = 0:(1/30):1
```

and second, evaluate $b(t)$ at these points with

```
>> y = eval(b)
```

and third, plot the result with

```
>> plot(t,y)
```

Upon adding the labels

```
>> xlabel('Time (hour)')
>> ylabel('Body Temperature (deg C)')
>> title('Transient Response of Body Temperature')
```

we obtain Figure 1.3. The figure shows that the body temperature crosses the lethal limit of 36°C in about 15-20 minutes.

1.3.2 Numerical Analysis

If you're not interested in the formula for $b(t)$ or if, in more complicated situations, the `dsolve` command doesn't supply an answer, you can generate a graph of $b(t)$ directly from the differential equation without bothering with explicit formulas. Because we will be dealing only with numerical values for $b(t)$, and not with its formula, we'll call this a numerical analysis. A numerical analysis is not exact though, and rounding error from the computer's calculations can lead to incorrect answers. Here we can compare the numerical solution with the exact mathematical solution that we just derived above.

To carry out the numerical analysis of a differential equation in MATLAB, you must first prepare an external file on the disk. You should use a text editor and save the file as a plain text file (an ASCII file) named `heatup.m`. Here is the file `heatup.m`:

```
function bdot = heatup(t,b)
global a h q w;
bdot = (q - h * (b - a)) / w;
```

This file describes the differential equation. It says that the function named `heatup` takes as independent variables the time, `t`, and the body temperature, `b`, and returns the value of db/dt, which is named `bdot` within the function. The parameters needed to compute `bdot` are declared as `global` so that they can be set and changed from the command line within MATLAB. The name of the file must be the same as the name of the function, except that the file has the extension `.m`. Once this file resides on the disk in the current working directory, the parameters are assigned within MATLAB with the commands

```
>> global a h q w
>> a = 18
>> h = 50
>> q = 1000
>> w = 10
```

The solution consists of the numerical values for time and body temperature, and the solution is placed into the vectors `time` and `bt` by the command

```
>> [time,bt] = ode23('heatup',0,1,30)
```

This command means that a particular numerical technique[5], which is contained in `ode23`, will be used to generate the solution, that the independent variable (time) should run from 0 to 1 (hour), and that the initial condition of the body temperature is 30 (°C). Now that we have these numbers, they can be immediately plotted with

```
>> plot(time,bt)
```

which, when labeled, yields a figure that appears identical to Figure 1.3. Thus, the numerical analysis evidently did not produce any noticeable rounding errors in this case.

[5]The details are revealed, if you're interested, by typing **help ode23**.

1.3.3 Systems Analysis

It doesn't take a weatherman to know that wind comes and goes in gusts, that sunlight flickers as leaves tremble in the breeze and clouds blow by. That is the real world, and models with constants to describe solar radiation and convective heat transfer perhaps idealize too far. Or, perhaps they don't. So let's see how to include environmental variation into the model for how an animal heats up, and how this consideration will change what we have learned so far. The approach that lends itself to an almost limitless degree of realism and detail is called systems analysis. The type of model that results is often called a systems model, or simulation model, or, as I prefer, a synthetic model.

The systems approach originates with electrical engineering in which networks of electrical components are analyzed. From this tradition, systems analysis inherits a terminology and notation that speaks of signals, components, and connections. In our case, the varying sunlight and wind speed will be viewed as "signals" to which the "system" responds. The system is the organism and its response is its body temperature. The components of the system are little boxes that do algebraic operations such as addition and multiplication on the signals. By stringing together a bunch of boxes any algebraic formula can be duplicated.

Systems analysis is carried out in MATLAB through an extra toolbox called SIMULINK. This toolbox has two capabilities. First, it is a graphical frontend to MATLAB and lets you assemble a model without ever explicitly writing any equations. Instead, using the computer's mouse, you make what looks like a wiring or circuit diagram that embodies your idea of how the system works, what kinds of components are in it, and how they're connected to one another. Second, SIMULINK is what is called a code generator, which means it translates the wiring diagram that you've composed into differential equations that are then solved numerically with `ode23` or similar commands. Mathematically, systems analysis boils down to the numerical solution of differential equations. In practice though, as an approach to model formulation, systems analysis leads to much more complete models than might otherwise be developed. Although the wiring diagram for a simple model is more trouble than it's worth, even a medium-sized model may be more easy to visualize in a system diagram than as several coupled differential equations.

We will first develop a systems formulation for a constant environment, and then add the environmental variation. We proceed in two steps because we want to confirm that this approach is correct for a case we have already solved before we extend it into new territory.

So let's start drawing the system diagram. To the MATLAB prompt, type

```
>> simulink
```

and a window appears with folders for `Sources`, `Sinks`, `Linear`, and `Nonlinear` among others. The folders each contain the little boxes we need for the wiring diagram. On the SIMULINK window click `File` and select `New`. This will open the window in which the drawing will be done. Open each of the folders from `Sources` through `Nonlinear` and look over the kinds of boxes there are. Now check out Figure 1.4. This is the diagram we have to assemble. From the `Sources` folder locate the little box for a `Constant` and drag four of them into the drawing window with the mouse (click on the `Constant` box and move it, with the mouse button pressed down, into the graphing window—do this four times in

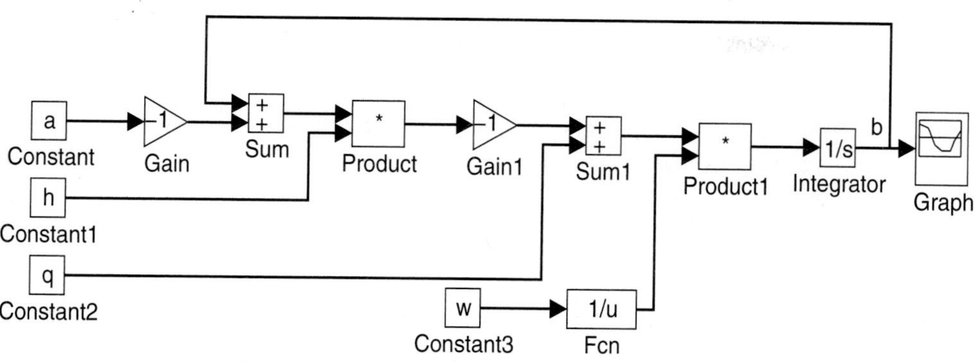

Figure 1.4: System diagram for simulating the transient response of body temperature to heating.

all). Similarly, drag in copies of the other boxes, `Gain`, `Sum`, `Product`, `Integrator`, `Fcn`, and `Graph` from their folders. Next, with the mouse, draw the wires between the boxes as specified in Figure 1.4. See how this is leading to an equation. For example, the `Constant` box at the top left will stand for the air temperature. To initialize this constant to a, click on the box, and fill in the menu with `a`. Then draw a line to a `Gain` box. Then click on this box and initialize its value to -1. Therefore, the signal leaving this `Gain` box is $-a$. We next need to add $-a$ to b, so a wire from this `Gain` box is connected to the input of a `Sum` box together with another wire carrying the signal for b. The output from that box represents $b-a$ and this is fed into a box which multiplies it by h, which is another constant. And so forth. Note the `Integrator` box, which takes the signal for db/dt (which by this time equals $h(b-a)/w$) and integrates it to yield b, which is then fed back to the `Sum` box where it was combined with the $-a$. The b that emerges from the `Integrator` box is also fed into a `Graph` block in which the answer is displayed. Thus, to make the system diagram, you drag in copies of all the boxes from their folders into the drawing window, hook up the boxes to represent a formula, and initialize each box as necessary. The `Constant` boxes are initialized to the variables `a`, `h`, `q`, and `w`. The `Fcn` box is initialized to the expression `1/u`, both `Gain` boxes are initialized to `-1`, and the `Integrator` box is initialized to `30`, which is the initial condition of the animal's body temperature. Finally, the `Graph` box is initialized to the x and y range for the coordinates of the graph, which are 0-1 for time and 30-38 for y. To save the assembled model, click within the `File` menu of the graph's window. I've called this model `heatsim.m`.

To run the model, first assign numerical values to the parameters in the MATLAB command window

```
>> global a h q w
```

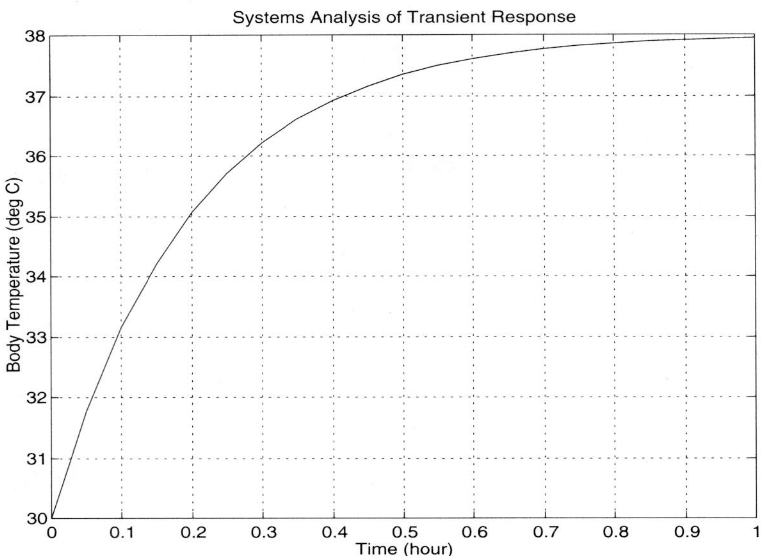

Figure 1.5: Output from simulation model for the transient response of body temperature to heating.

```
>> a = 18
>> h = 50
>> q = 1000
>> w = 10
```

(You could, incidentally, initialize the `Constant` boxes to these values to begin with, but I think it's more flexible to have the constants declared as `global` variables so that they can be used in other contexts as well.) Now we can run the model. Click on the `Simulation` menu of the graph's window. Select `Parameters` and set the start time at 0 and the stop time at 1 and the maximum step size at 0.05. Then select `OK` and return to the `Simulation` menu and select `Start`. Then a graph appears, as illustrated in Figure 1.5. (But ignore the label on the x axis for a moment.) A tiny bug in SIMULINK always labels the x axis as `Time (second)`, whether the time is really in seconds or not. (MATLAB actually has no way of knowing the units.) So, in the MATLAB command window, we spiff up the graph with

```
>> xlabel('Time (hour)')
>> ylabel('Body Temperature (deg C)')
>> title('Systems Analysis of Transient Response')
```

and the correct `xlabel` overwrites the incorrect one. The result appears as Figure 1.5. To save both the system diagram and the output of the simulation in encapsulated PostScript within the command window, type

```
>> print -sheatsim -depsc heatsim
>> print -depsc heatout
```

and the first command produces `heatsim.eps`, which is the graph of the system wiring diagram, and the second produces `heatout.eps`, which is the graph of the output. The model itself, including all the initializations in the little boxes, but not including the parameters, `a`, `h`, `q`, and `w` that are set on the MATLAB command line, still resides on the disk as `heatsim.m` because this was saved earlier.

At this point you may be thinking we've been through a lot and we're no farther along than we were before. Well, you're right. Figure 1.5 is a dead ringer for Figure 1.3, and all this wiring diagram stuff has been superfluous. But matters are about to improve.

Consider the fate of an organism living near the edge of its climate space. Suppose the animal is living at a spot where the equilibrium body temperature is 35°C. Suppose also that the upper lethal limit is 36°C, so the animal has a 1-degree margin of safety between where it is presently sitting and the edge of its climate space. The question is, is this 1-degree margin safe enough, given the natural fluctuation of wind and sunlight?

The virtue of the systems-analysis approach is that additional considerations are very easy to add to a systems diagram. Open up a SIMULINK graphing window again and load in `heatsim.m`. Now also open up the `Sources` folder and drag in two copies of the `Band-Limited White Noise` box. These provide sources of random variation that will be added to `h` and `q` to simulate fluctuating wind and sunlight. Also drag in two copies of the `Sum` box from the `Linear` folder to do the addition. Now connect these components into the wires from `h` and `q` as shown in Figure 1.6. Once the connections are completed, click on the noise generator box for `h` and initialize the `Noise Power` to `0.1*h` and set the `Seed` to `0`. To initialize the noise-generator box for `q`, set the `Noise Power` to `0.1*q` and the `Seed` to `1`. Using different seeds for the generators implies that they will generate different sequences of random numbers. Click on the `Integrator` box and initialize it to the equilibrium body temperature by typing `(q/h) + a`, which MATLAB evaluates to a number because `q`, `h`, and `a` have been defined as global variables. Next, bring up the `Parameters` submenu of the `Simulation` menu and set the start time to `0` and the stop time to `6`, and the maximum step size to `0.05`. The simulation will thus represent 6 hours of time. Similarly, click on the `Graph` box and set the `Time range` to `6`, `y-min` to `30`, and `y-max` to `40`. At this point the model should be saved; I've called it `heatran.m`.

Now we can run the model. We assign the global constants

```
>> global a h q w
>> a = 15
>> h = 50
>> q = 1000
>> w = 10
```

Here `a` has been set to `15` so that the equilibrium body temperature is 35°C. Bring up the `Simulation` menu again and start the simulation. The result appears as the middle panel

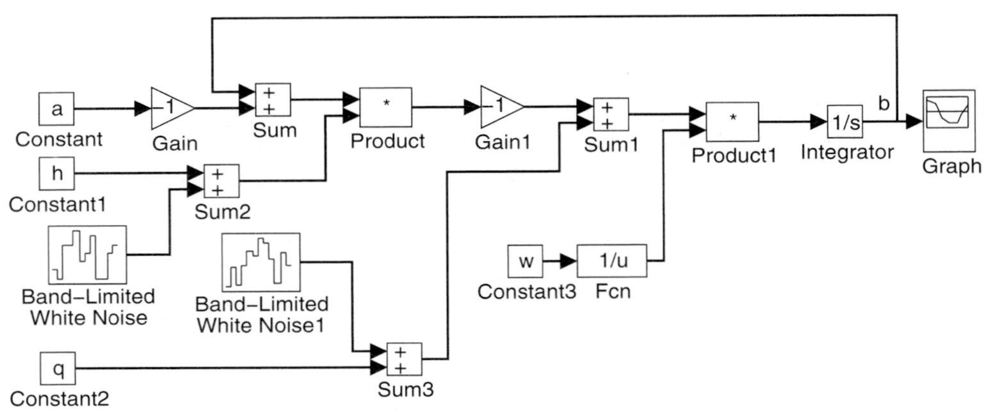

Figure 1.6: System diagram that includes fluctuations in wind and sunlight. It can be used for simulating the transient response of body temperature to heating.

of Figure 1.7 (except for the label on the x-axis). The result shows that the 1-degree margin of safety is not perfectly safe. The organism must move out of the spot into somewhere cooler about 10 different times during the 6 hours of the simulation.

We don't have to leave the matter at this point, though; there's more to discover. If `w` is set to 5 g, to simulate a small organism, the top panel in Figure 1.7 results, and with `w` set to 20 g, to simulate a large organism, the bottom panel in Figure 1.7 is produced. Taken together, the panels in Figure 1.7 show that a small animal must move around a lot to stay comfortable, whereas a large animal can stay at at the spot where the equilibrium body temperature is 35°C and rarely be forced to leave because of overheating by chance fluctuations in wind and sunlight. This trend reveals that size confers "thermal inertia" to an organism, everything else being equal. A more accurate simulation would have to allow for both q and h varying with body size too.

This example barely hints at the potential of a systems or synthetic approach. Additional commands in MATLAB bundle together the components in a system diagram to make it into a "subsystem." Multiple subsystems can then be hooked up to one another, into what is called a hierarchical systems model. Different groups of workers could be responsible for each subsystem. For example, to predict how much crop damage caterpillars cause in a field of lettuce plants, one subsystem model could predict leaf temperature: it could take into account the reflectivity of the leaf's surface, the wind, the sun, and cooling from evapotranspiration through the stomates. Another subsystem model could predict the body temperature of the caterpillars: it could take into account their reflectivity, wind speed, sun, and contact with the leaf's surface. Then the mobility and digestion rates of a caterpillar as a function of temperature could be measured. A combined systems model for both the

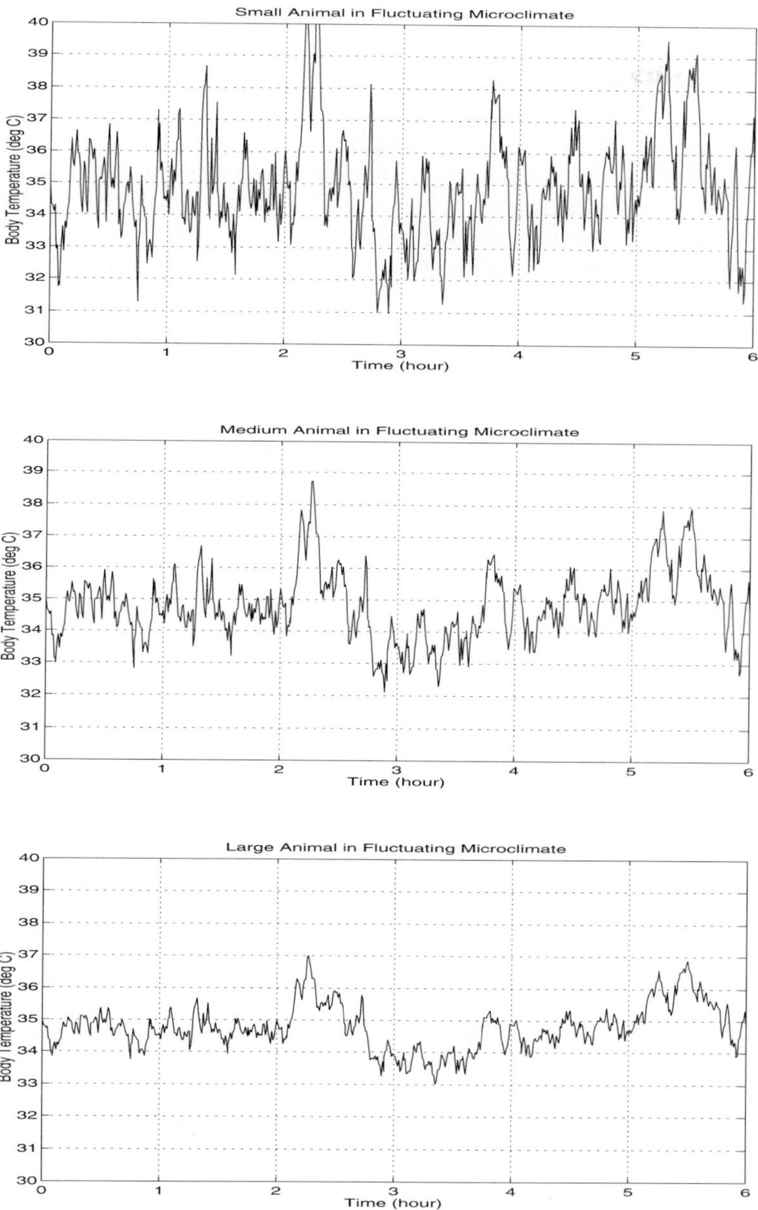

Figure 1.7: Transient body temperature during 6 hours at a spot where the equilibrium body temperature is 35°C. Top panel is a 5g animal, middle panel is a 10g animal, and bottom panel is a 15g animal.

leaves and the caterpillars together with the data on eating rates and a census of the number of leaves and caterpillars in the field could then predict the weight of the crop being lost daily to caterpillar herbivory.

The work of two groups of investigators, botanists working on plants and entomologists working on the caterpillars, would be synthesized in the systems model, and each group would have the task of delivering a subsystem for the full systems model. The only downside to this approach is cost. Scientists and programmers need a lot of time to develop a large and correct systems model, and the answer provided by the model must be valuable enough to justify a large effort. Anyone who insists on a lot of reality in a model has to have deep pockets.

Plants and animals exchange several flows, not only heat, with the environment. They exchange carbon and water with the environment as well. While the exchange of heat can be modeled as a passive diffusion process, the exchange of carbon and water takes place through eating and drinking, which are behaviors that involve decision making by the organism. Decision making takes us to the next chapter. There we will focus on food and what an animal should decide to eat and not to eat. From this we will be able to calculate the animal's eating rate. The same considerations could also be applied to the drinking rate. After the next chapter has been completed, you could then revisit this chapter and reconsider how to formulate a model for all the coupled flows of heat, food, and water, and I hope some of you do this.

1.4 Further Readings

Campbell, G. S. (1977). *An Introduction to Environmental Biophysics.* Springer-Verlag.

Gates, D. M. (1965). Energy, plants, and ecology. *Ecology,* 46:1–14.

Gates, D. M. (1980). *Biophysical Ecology.* Springer-Verlag.

Gates, D. M. (1993). *Climate Change and its Biological Consequences.* Sinauer Associates, Inc.

Odum, H. T. (1983). *Ecological and General Systems.* University Press of Colorado.

Porter, W. P. and Gates, D. M. (1969). Thermodynamic equilibria of animals with the environment. *Ecological Monographs,* 39:227–244.

Tracy, C. R. (1982). Biophysical modeling in reptilian physiology and ecology. In Gans, C. and Pough, F. H., editors, *Biology of the Reptilia, Vol. 12,* pages 275–321. Academic Press.

Tracy, C. R. (1992). Ecological responses of animals to climate. In Peters, R. L. and Lovejoy, T. E., editors, *Global Warming and Biological Diversity,* pages 171–179. Yale University Press.

1.5 Application: Tomorrow's Extinctions of Mammals

Rarely do environmental crises yield unambiguous predictions about the risks and threats we face[6]. Instead, an oil spill, the escape of an invasive plant, or logging of old-growth forests are each expected to have some negative impact—with the magnitude of the impact being uncertain because it depends on critical assumptions or estimates of parameters. One of the best uses of models is to explore the consequences of different assumptions or different estimates of environmental variables when environmental risks are being examined. A particularly clear example considers the implications of different scenarios of global warming for the extinction of mammals who live on mountaintops. The idea is simple. We are not sure by how much temperatures will rise, but we have well-accepted predictive formulas that relate temperature to elevation, and we also have field data on the distribution of several mountaintop mammal species.

If we assume the lower elevational limit of mammals on mountaintops is determined by temperature, then we can ask whether a species will be driven to extinction because climate change warms up mountaintops and species living on those mountains cannot disperse across the lowlands (a reasonable assumption for many mountain mammals). The ecologists Kelly McDonald and James Brown applied this approach to 14 mammal species living in the western United States.

Although they considered only a 3-degree warming, we can take advantage of MATLAB to quickly consider three possible warming scenarios, which effectively span the best-case and worst-case scenarios for global climate change. Specifically, we construct a vector of 1.5, 3, and 4.5 degrees of possible warming:

```
>> temp = [1.5 3 4.5];
```

Temperatures on mountains decrease with elevation in a predictable way that is quantified by what is called Hopkin's rule: We can expect, in humid conditions, a 6-degree C decrease in temperature for every 1-km increase in elevation; we can expect in arid conditions, a 10-degree C drop with every 1-km increase in elevation. Thus, any temperature increase can be translated to a predicted elevational shift. To combine the above three warming scenarios with the two versions of Hopkins rule, we define an elevation-per-degrees vector and multiply it by the warming scenarios:

```
>> H = [1/6;1/10];
```

and then type in

```
>> H*temp
```

to get the

```
ans = 0.2500    0.5000    0.7500
      0.1500    0.3000    0.4500
```

[6]This section is contributed by Peter Kareiva.

Table 1.1: Distribution of 12 small boreal mammal species on 6 isolated mountain ranges in the Great Basin. The predicted effect of a 3-degree C increase in global temperatures is the raising of the lower border of suitable habitat approximately 500 m, from 2280m to 2745m. For each species on each mountain, a P indicates that it is expected to persist if the species border raises 500 m, whereas an E indicates that it is expected to become extinct.

Species	Ruby	Toquima	Snake	Spring	Grant	Pilot
Eutamius umbrinus	P	P	P	P	P	
Neotoma cinerea	P	P	P	P		P
Eutamius dorsalis		P	P	P	P	P
Spermophilus lateralis	P	P	P	P	P	E
Microtus longicuadus	P	P	P		E	
Sylvilagus nuttalii	P	P	P	P	E	
Marmot flaviventris	P	P	P			
Sorex vagrans	P	P	P	E		
Sorex palustris	P	P	E			
Mustella erminea			E			
Ochotona princeps	E	E				
Zapus princeps	E	E				
Spermophilus bendengi	E					
Lepus townsendii	E					
Area(km^2) $>$ 2280m	141	455	161	48	58	5
Area(km^2) $>$ 2745m	42	230	65	14	16	1.2
Present species	12	11	10	6	5	3
Predicted species	8	9	8	5	3	2

where the rows correspond to the two versions of Hopkins rule, each column is a different warming scenario, and the actual numbers printed out are the expected upward shift in elevation (in kilometers) for each climate scenario. Note that we are simply pointing out that if a 6-degree cooling is associated with an elevational rise of 1 km, then, when there is a 3-degree warming, maintenance of a stable temperature will require a 3/6 (or, a half-km) rise in elevation. The key idea here is that a temperature change can be translated into an elevational shift via a simple rule.

The accompanying table shows how predictions for mammals in selected mountain ranges were produced with this simple rule, together with information on the elevation of each mountain and assumptions about mountain shape (so that the area above each elevation could be calculated). This table uses the 0.5-km prediction from the above answer matrix. After going back to the data collated by McDonald and Brown, one could use each of the other possible scenarios to generate a different table.

McDonald and Brown examined only the 3-degree warming scenario, but applied it to 19 different mountain ranges in the Great Basin of the U.S. They predicted that, depending on the mountain range being considered, as many as 60% or as few as 10% of the mammal species are expected to disappear because warming pushes their lower elevational limit above the highest mountain top.

With MATLAB one may revisit McDonald and Brown's data set and quickly predict the extinctions for a wide variety of warming possibilities (as opposed to only a 3-degree change). Armed with different predictions of risk for species due to global warming, we can better plan for the future when we design our nature reserves or manage populations for some minimum viable size.

The key reference for this application is

McDonald, K and J. Brown. 1992. Using montane mammals to model extinctions due to global change. *Conservation Biology* 6:409–415.

Chapter 2

Time For a Byte of Food?

Organisms differ from the objects modeled in other sciences because organisms take actions, and even make decisions, that promote their own welfare. An organism is not simply a gargantuan macromolecule. Instead, it is a living creature with desires and wants of its own, and it takes steps to further its own agenda. To understand an organism you should learn to put yourself in its place, and in that context what it does usually makes perfect sense. Here we will see the most elementary of the models that looks at what an animal does from its point of view. We will see that what an animal eats, and how fast it harvests food, can be predicted by our considering what it should do if it knows what's best for itself.

Looking for food, and eating when it is found, is called foraging. Here we focus on a predator, such as a bird or lizard eating insects, or fish eating small crustaceans. It is customary to recognize two styles of foraging. A "searcher" is continually on the move and picks up prey items that it wants when it encounters them. Birds called foliage gleaners are searchers because they hop from branch to branch, peering under leaves and probing the bark, all in search of bite to eat. A "sit-and-wait" predator positions itself on a high vantage point and waits for an unsuspecting prey item to wander or to fly nearby. It then jumps, swims or flies out to pounce on the item, eats it, and then returns to its vantage point (which is also called a "perch"). We'll look into both these styles of predators separately.

2.1 Searching Predator

We begin by postulating that an organism has an overall objective while foraging. Two possible objectives have been widely considered. The first is to find and to catch whatever prey make the average time spent on each one of them as low as possible. The idea is that catching food is dangerous, and the quality of the food is secondary to how long it takes to get the food. Indeed, in this situation the quality of the food *is* the time taken to get it, not its caloric or nutritional content. This objective, catching prey that minimize time per item, is the first of the possible objectives to consider. The second possible objective is to maximize the net energy harvested per unit time. If foraging is not particularly dangerous as such, then a reasonable goal is to locate and to eat those things that return the highest calories, or other nutritional value. Foraging that maximizes energy per time is probably more common than foraging to minimize time per item, but the calculations for a time-minimizer are also used for an energy-maximizer, so we'll begin with the time-minimizer.

2.1.1 What Should Be Done

Foraging models now come in two varieties. The first is called an "optimal foraging model," and with it we will predict what an animal *should* do. This can then be compared with what it does do. The second, and more recent style, is called a "rule of thumb," which predicts *how* an organism can go about doing what it should do. If the animal follows the simple rule then its behavior converges quickly to being what it should be. Therefore, the animal doesn't have to know optimal foraging theory itself to know how best to live, all it has to know is a comparatively simple rule to follow. If it does follow the rule, the animal in effect learns how to forage optimally as it goes along.

Here we start with the optimal foraging models for both the time-minimizing and energy-maximizing objectives, and then move to the rules of thumb.

Minimizing Time per Item

The basic question faced by an animal that is continually moving through the habitat and searching for prey, is whether to stop and eat an item that it has encountered. The human equivalent to this situation is a sushi bar. Little goodies come parading past on a conveyer belt, and you have to decide whether to pick one up and eat it. The items are assumed to be equally nutritious and tasty, but to differ in how hard they are to eat. With two types of items, h_1 is called the handling time for type-1 and h_2 the handling time for type-2. For animals, it might take a second for a gobble a small bug, but upwards of 30 minutes to consume a big caterpillar. So, here the better prey is the one with the smaller handling time, and we label type-1 as the prey with the lower handling time, so $h_1 < h_2$ by the labeling convention. The two prey types differ in abundance. a_1 is the number of times per second that type-1 appears, and a_2 is the rate for type-2.

Next we enumerate all the possible strategies that the animal might take. First, it could take only type-1 and ignore type-2. This is a good idea if type-1 is really abundant, because whenever it takes a type-2, time is wasted while the item is consumed, time that could have been spent taking more of type-1. Second, it could take both type-1 and type-2 whenever they turn up. After all, if type-1 is sufficiently rare, then the animal has no choice but to take type-2 if it wants to eat at all, rather than hold out for a very rare type-1. Third, the animal could take only type-2. Taking only the worst type of prey is always a bad idea, but is mentioned for completeness. Thus there are three strategies, and what we need to do is to rank them in order of the average time per item attained by following each. Then we'll choose the strategy with the lowest average time per item. This will be called the optimal foraging strategy.

The average time per item for the strategy of taking only type-1 is the sum of the time spent waiting for the item plus the time spent eating the item once it has turned up. If the abundance of type-1, a_1, is in units of items per second, then $1/a_1$ is in units of seconds per item and is the waiting time. So, the average time per item in the strategy of taking only type-1 is ti_1, the sum of $1/a_1$ plus h_1. In MATLAB, this would be entered as

```
>>  ti1 = '1/a1 + h1'
```

The average time per item for the strategy of taking both prey types is a bit more complicated. The total abundance of both prey types is $a_1 + a_2$, which is again in units of items per time, so the waiting time for some item, either type-1 or type-2, is $1/(a_1 + a_2)$. The handling time depends on which type the item is. With probability $a_1/(a_1 + a_2)$ the item is a type-1 and its handling time is h_1, and with probability $a_2/(a_1 + a_2)$ the item is a type-2 with handling time h_2. So the average handling time[1] is $h_1a_1/(a_1 + a_2) + h_2a_2/(a_1 + a_2)$. Therefore, the average time per item in the strategy wherein both prey are taken is the sum of the waiting time and the handling time—in MATLAB's notation this is

```
>>  ti12 = '1/(a1 + a2) + h1*a1/(a1 + a2) + h2*a2/(a1 + a2)'
```

And for completeness, the average time per item in the strategy of taking only type-2 is

```
>>  ti2 = '1/a2 + h2'
```

So, at this point we have the time per item for all three strategies, ti_1, ti_{12}, and ti_2. So, now let's see which of these is best. Let's substitute some values for prey abundance and handling time, make a graph of ti_1, ti_{12}, and ti_2, and see which is the smallest.

We now want to plot the formulas for ti_1 and ti_{12} to compare them with one another. To do this we need to change the formulas into versions for graphing, and what must be done is to change every occurence of `'/'` into `'./'`, of `'*'` into `'.*'` and of `'^'` into `'.^'` because we want division, multiplication, and exponentiation to take place element-wise in the array of numbers. Because we will need to do this many times in the book we might as well set up a function that will do this for us in general. To set up a user-defined function, say `fun`, make a file with the name `fun.m`, that accepts the input and returns the result. In our case, the function will accept a string formatted for symbolic manipulation, and return another string formatted for array manipulation. Let us call this function `sym2ara` (for "symbolic into array"). Then with a text editor, create the separate file in your directory (or folder) named `sym2ara.m` as follows:

```
function ara = sym2ara(sym)
 ara = strrep(sym,'/','./');
 ara = strrep(ara,'*','.*');
 ara = strrep(ara,'^','.^');
```

This user-defined function uses MATLAB's `strrep` command to substitute the division, multiplication, and exponentiation operators with their array-equivalents (i.e., the same

[1] Also called the weighted average of the handling times.

operators preceded by a period). The function receives as input a string called sym and returns another string ara as its result[2].

Once this file resides in your directory (or folder), then you can type

```
>>  ti1plot = sym2ara(ti1)
>>  ti12plot = sym2ara(ti12)
```

and obtain versions of the three formulas that are perfect for making the graphs.

Take handling times of 1 s and 60 s for type-1 and type-2 respectively,

```
>>  h1 = 1
>>  h2 = 60
```

Next, initialize a graph with

```
>>  figure
>>  hold on
```

Now let's look at ti_1 as a function of a_1. Let's let a_1 vary from 0.005 to 0.030 prey per second in steps of 0.0005

```
>>  a1 = .005:.0005:.030;
```

The graph of average time per item for the strategy of taking only type-1 is then produced with

```
>>  plot(a1,eval(ti1plot),'w')
```

The curve will be in white on the computer screen, and automatically converted to black if the graph is printed on paper.

Now, let's move to graphing the performance of the other strategy, in which both type-1 and type-2 are taken. The formula for ti_{12} requires both a_1 and a_2, so let's consider three choices for a_2—one where a_2 is half that of a_1, another where a_2 equals a_1, and another where a_2 is twice a_1. To define and graph these choices as cyan, magenta, and yellow lines, respectively, you type

```
>>  a2 = 0.5*a1;
>>  plot(a1,eval(ti12plot),'c')
>>  a2 = a1;
>>  plot(a1,eval(ti12plot),'m')
>>  a2 = 2*a1;
>>  plot(a1,eval(ti12plot),'y')
```

The result appears in Figure 2.1.

[2] It turns out that MATLAB already had this function, but nobody knew about it. The undocumented function vectorize() does what sym2ara() does. Therefore, you can use vectorize if you don't want to bother typing in the code for sym2ara.

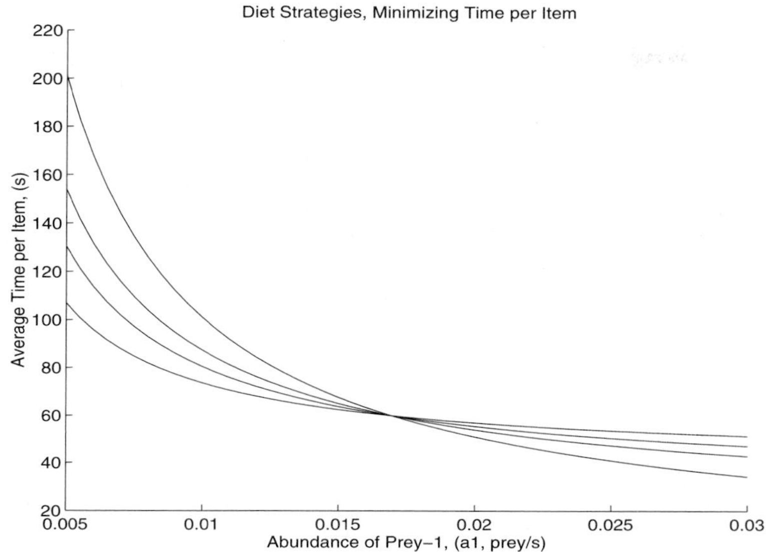

Figure 2.1: Average time per item from two foraging strategies. The curve intersecting the y-axis at 200 is for the strategy of taking only type-1. The curves intersecting the y-axis at about 155, 130, and 108 are for the strategy of taking both type-1 and type-2, where a_2 equals 0.5, 1, and 2 times a_1, respectively. In all graphs h_1 is 1 s and h_2 is one min. If the abundance of type-1 exceeds 0.017, the strategy of taking only type-1 is superior to that of taking both prey types; otherwise taking both prey types is superior. The relative superiority of taking only type-1 vs. taking both is independent of a_2.

Figure 2.1 shows that the strategy of taking both prey is better if a_1 is lower than about 0.017 prey per second because all three curves for this strategy are lower than the curve for taking type-1 only. But if a_1 is higher than about 0.017, the curve for taking only type-1 is lower than the curves for the strategy of taking both types, and therefore taking type-1 only is the better strategy when a_1 is high enough. The fact that all the curves intersect at the same point suggests that which of these strategies is best depends only on a_1 and not on a_2 at all. To see if this conjecture is true we can solve for the a_1 that makes both strategies equal and see if the answer depends on a_2. So, type

```
>>  a1hat_ti = solve(symop(ti1,'-',ti12),'a1')
```

and the answer that is returned is

```
a1hat_ti = 1/(h2-h1)
```

This is the value of a_1 at which the best strategy switches from taking both to taking type-1 only. It does not depend on a_2. The value of the switchover is found from

```
>>  eval(a1hat_ti)
```

which yields

```
ans = 0.0169
```

That the switchover between taking both and taking only type-1 depends on a_1 and not on a_2 is one of the basic predictions of optimal foraging theory. It's only the abundance of the better prey that determines the switch point—if there are enough of type-1 to make taking only type-1 the best strategy, then adding or removing any amount of type-2 doesn't matter. Type-2 can't be forced into, or removed from, the optimal diet by a change in its abundance.

Maximizing Energy per Time

Although some animals live in constant fear of predation, always looking over their shoulders for hawks, and foraging to minimize exposure to predation, others, perhaps most, forage to maximize their energetic yield. After all, it is the energetic return from foraging that is ultimately converted into eggs or other forms of reproductive activity, and that is therefore relevant to success in evolution. Now we'll extend foraging theory for animals whose goal is to maximize energy per unit time foraging. The extension is rather straightforward, given what we've already done. The idea is to look at energy per time computed as energy per item divided by time per item. The item divides out and leaves energy per time. We've already computed the time per item in the last section, so now all we need to figure out is the energy per item.

The energy per item for the strategy of taking only type-1 prey equals the energy of the item itself, e_1 in calories, minus the energy spent waiting for it to appear, minus the energy spent eating it. The energy spent waiting is e_w in cal per s, times the waiting time in seconds. The waiting time, as we saw earlier, is $1/a_1$. The energy spent eating the item once it is caught is e_h in cal per second, times the handling time in seconds, which is h_1 for type-1 prey. So the average energy per item for the strategy of taking only type-1 prey is

```
>>  ei1 = 'e1 - ew/a1 - eh*h1'
```

The average energy per item for the strategy of taking both type-1 and type-2 prey is more complicated. $a_1/(a_1 + a_2)$ is the probability that a prey is type-1 and $a_2/(a_1 + a_2)$ is the probability that a prey is type-2. Therefore, the average energy per prey item is $e_1a_1/(a_1 + a_2) + e_2a_2/(a_1 + a_2)$. The average waiting time for either prey is $1/(a_1 + a_2)$, and this is multiplied by e_w to obtain the energy spent waiting. The average handling time is $h_1a_1/(a_1 + a_2) + h_2a_2/(a_1 + a_2)$, and this is multiplied by e_h to obtain the energy spent eating the item. All this is added up in the formula for the average energy per item for the strategy of taking both prey types:

```
>>  ei12 = ['e1*a1/(a1 + a2) + e2*a2/(a1 + a2) - ew/(a1 + a2) ' ...
            '- eh*(h1*a1/(a1 + a2) + h2*a2/(a1 + a2))']
```

(Notice how the formula is spread over two lines.) Finally, for completeness we have the average energy per item for the strategy of taking only type-2 prey.

```
>>  ei2 = 'e2 - ew/a2 - eh*h2'
```

Now that we have the average time per item and the average energy per item for each of the strategies, all we have to do is divide them to obtain the average energy per time for each of the strategies. For the strategy of taking only type-1 the average energy per time is

```
>>  et1 = symop(ei1,'/',ti1)
```

To see the formula written out in full, type

```
>>  pretty(et1)
```

and MATLAB responds with

```
           ew
    e1 - ---- - eh h1
           a1
    -----------------
           1
         ---- + h1
          a1
```

Similarly, the average energy per time for the strategy of taking both prey types is

```
>>  et12 = symop(ei12,'/',ti12)
```

and the formula written out in full (by typing `pretty(et12)`) is

```
  e1 a1      e2 a2        ew       eh h1 a1    eh h2 a2
 ------- + ------- - ------- - -------- - --------
 a1 + a2   a1 + a2   a1 + a2    a1 + a2     a1 + a2
 ------------------------------------------------
              1        h1 a1      h2 a2
           ------- + ------- + -------
           a1 + a2   a1 + a2   a1 + a2
```

And finally, the average energy per time for the strategy of taking only prey-2 is

```
>>  et2 = symop(ei2,'/',ti2)
```

which pretty-prints as

```
           ew
    e2 - ---- - eh h2
           a2
    -----------------
           1
         ---- + h2
          a2
```

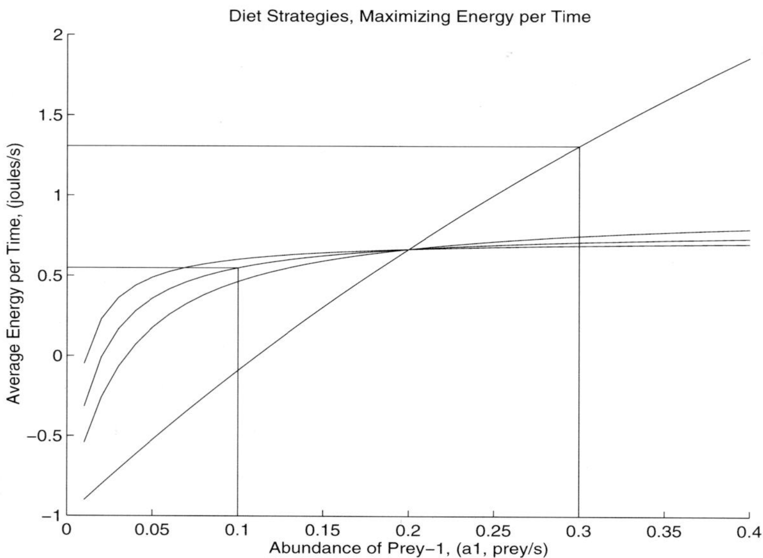

Figure 2.2: Average energy per time for two foraging strategies. The curve ending near the y-axis at about −1 is for the strategy of only taking type-1. Curves ending near the y-axis at −0.5, −0.3, and −0.1 are for the strategy of taking both type-1 and type-2, where a_2 equals 0.5, 1, and 2 times a_1 respectively. In all graphs h_1 is 1 s, h_2 is 1 min, e_1 is 10 J, e_2 is 100 J, and both e_w and e_h are 1 J/s. If the abundance of type-1 exceeds 0.2, then the strategy of taking only type-1 is superior to that of taking both prey types. Otherwise, taking both prey types is superior. The relative superiority of taking only type-1 vs. taking both is independent of a_2. In particular, if the abundance of both prey is 0.1, then both prey should be taken, and the yield from this strategy is 0.55; whereas if the abundance of both prey is 0.3, then only type-1 should be taken, and the yield is 1.3.

To find out which of the strategies is best, let's graph them and see which yields the highest average energy per time. To make plottable versions of the formulas, we'll use the `sym2ara` function introduced in the last section:

```
>>  et1plot = sym2ara(et1)
>>  et12plot = sym2ara(et12)
```

Now for some values. As before, the handling times are 1 s for type-1 and 60 s for type-2.

```
>>  h1 = 1
>>  h2 = 60
```

The energy content of a 1-mm insect is 10 J, and for an 8-mm insect is 100 J, so we let

```
>>  e1 = 10
>>  e2 = 100
```

We let the energetic cost of waiting and also of eating be 1 J/s,

```
>>  ew = 1
>>  eh = 1
```

The figure is initialized with

```
>>  figure
>>  hold on
```

Starting with the strategy of taking only type-1, a_1 will be plotted from 0.01 to 0.40 in steps of 0.01. A vector with these values is generated with

```
>>  a1 = .01:.01:.40;
```

The curve is then plotted in white with

```
>>  plot(a1,eval(et1plot),'w')
```

Moving now to the curve for the strategy of taking both type-1 and type-2, we'll again plot three curves, with a_2 equal to 0.5, 1, and 2 times a_1, respectively. These are drawn in cyan, magenta, and yellow with the following:

```
>>  a2 = 0.5*a1;
>>  plot(a1,eval(et12plot),'c')
>>  a2 = a1;
>>  plot(a1,eval(et12plot),'m')
>>  a2 = 2*a1;
>>  plot(a1,eval(et12plot),'y')
```

The result appears in Figure 2.2.

Let's consider two particular cases, where the abundance of both prey is 0.1 and is 0.3, respectively, because we'll illustrate these examples in more detail in the next section. With the relatively low prey-abundance of 0.1 for both prey types, the optimal strategy is to take both prey types, and to find the net foraging yield from doing this, we evaluate et_{12} with this a_1 and a_2 as follows:

```
>>  a1 = .1; a2 = .1;
>>  et_expect = eval(et12)
```

resulting in

```
et_expect = 0.5493
```

This case can be added to the graph in red, by our typing

```
>>  plot([a1,a1],[-1,et_expect],'r')
>>  plot([0,a1,],[et_expect,et_expect],'r')
```

In contrast, when there is the relatively high prey-abundance of 0.3, the optimal strategy is to take only type-1, and we find the foraging yield by typing

```
>>  a1 = .3; a2 = .3;
>>  et_expect = eval(et1)
```

resulting in

```
et_expect = 1.3077
```

This case is added to the graph in green, by our typing

```
>>  plot([a1,a1],[-1,et_expect],'g')
>>  plot([0,a1,],[et_expect,et_expect],'g')
```

These two cases provide a benchmark that can be compared with the simulations of behavior in the next section.

Although there is much more detail in the model now that energy is being considered, the overall result is the same as the earlier formulation that focused only on time. Here, the strategy of taking only type-1 prey is the best if the abundance of type-1, a_1, is higher than about 0.2, because the curve for this strategy is above that of the other strategies. If a_1 is lower than 0.2, then the strategies of taking both type-1 and type-2 are above that for taking time-1 only, and are therefore better. The curves intersect at the same point, again suggesting that the best strategy switches from taking both prey types to taking only type-1 at a value of a_1 that is independent of a_2. To confirm this we can solve for the switch point and see if it depends on a_2. So, typing

```
>>  a1hat_et = solve(symop(et1,'-',et12),'a1')
```

yields

```
                    e2 + (ew - eh) h2
                    -----------------
                     e1 h2 - e2 h1
```

Using the numbers we have assumed, the switch point is evaluated with

```
>>  eval(a1hat_et)
```

yielding

```
ans = 0.2000
```

To find the overall rate at which the animal is harvesting energy from the environment, we evaluate *et* with the strategy the animal is actually using. In this way, predictions from foraging theory can be incorporated into other models, such as those concerned with predicting how many animals the environment can support, and models concerned with the total energy flow through the ecosystem.

If you're new to modeling, this extension from a time-minimizing to an energy-maximizing formulation illustrates a very common outcome—that adding realism doesn't change the qualitative answer we obtained from a more simple formulation, but it does make the answer more applicable. For this reason, it's good craftmanship when you are doing models to start with the simplest formulation that can reasonably be imagined, work it through, and then add the realism as a second step.

2.1.2 How To Go About It

How can an animal know how to forage optimally? Of course, a very intelligent animal could read this book and compute for itself what the optimal strategy is. It's more likely though, that an animal finds out what's best for itself through progressive discovery, and if so, we need to investigate how an animal can learn its optimal foraging strategy as it goes along.

A simple rule that leads to optimal foraging behavior may be called a "rule of thumb" and here is an example of a rule that seems to work. The animal must somehow keeps track of two quantities since the beginning of the day: the total time that has elapsed since the beginning of the day, T, and the total energy captured during that time, E. These could be remembered in neural circuitry, or in the level of blood sugar—it doesn't matter exactly how these are remembered, just that they are. Whenever a prey item is encountered, the animal is assumed to consider how well off it will be if it takes the item, and how well off it will be if it ignores the item. Then it takes or ignores the item depending on which it thinks will make it better off. To be better off means having a higher E/T, which represents the goal of maximizing yield during the time spent foraging.

To project how well off it will be if it accepts the item, the animal guesses what its new T will be, and what its new E will be, and considers the ratio of the new E over the new T. The new T will be the present T plus the time spend waiting for this item plus the time that will be spent handling the item. The new E will be the present E plus the energy gained from eating the item, minus the energy lost while waiting for this item, and also minus the energy lost while handling the item. In symbols,

$$\left(\frac{E}{T}\right)_a = \frac{E + e_i - e_w t_w - e_h h_i}{T + t_w + h_i}$$

where e_i is the energy of the present item, e_w is the energy expended per second of waiting, t_w is the time spent waiting for this item, e_h is the energy spent per second handling the item, and h_i is the time spent handling the item.

Similarly, if the animal ignores the item, its new T will be just its present T plus the time it spent waiting for this item, and its new E will be its present E minus the energy spent waiting for this item. In symbols,

$$\left(\frac{E}{T}\right)_i = \frac{E - e_w t_w}{T + t_w}$$

The rule of thumb is, for each item, one after the other,

$$\begin{aligned} \text{Accept if}: &\quad (E/T)_a > (E/T)_i \\ \text{Ignore if}: &\quad (E/T)_a < (E/T)_i \end{aligned}$$

If an animal follows this rule all day long, then, as we will see, its foraging behavior quickly becomes optimal.

To see the rule of thumb in action we'll do a simulation. We'll write a little computer program in MATLAB that will mimic a day in the life of an animal who is searching for food, and who encounters two types of food on a regular basis, and has to decide at each encounter item whether to eat or to ignore the item. We'll write the program in a separate file that will be called `thumdiet.m`. That is, the following MATLAB commands, all of which are not preceded with `>>`, should be entered with text editor and saved as a separate file called `thumdiet.m`.

The program begins with the definition of its name and with the specs on what is given as input and what must be returned as output.

```
function [a_his,d_his,t_his,e_his] = thumdiet(tot,a1,a2,h1,h2,e1,e2,ew,eh)
```

This line means that the program will be invoked as a user-defined function called `thumdiet`, and it will be given as arguments `tot`, which is the number of prey items to consider; `a1`, `a2`, `h1`, `h2`, `e1` and `e2` which are the abundances, handling times, and energy contents of type-1 and type-2 prey respectively; and `ew` and `eh`, which are the energy expenditures per unit waiting time and handling time. The function will return as its output four vectors, `a_his` which is the history of the prey types that appeared (type-1 or type-2); `d_his`, which is the history of the decisions (accept or ignore) made for each item; `t_his`, which is the history of the total elapsed time after each item has been acted upon; and `e_his`, which is the history of the cumulative energy harvested after each item has been acted upon. These vectors contain the history of everything that happened during the foraging, and we will then prepare graphs to illustrate what occurred. We'll now go through the program, line by line. The following initializes the output vectors as empty.

```
a_his = []; d_his = []; t_his = []; e_his = [];
```

We assume that the animal begins the day by waiting for one second while it's still waking up and rubbing its eyes. These lines initialize the elapsed time counter `elap` to 1 s, and the total energy counter `energy` to the energy spent while waiting this 1 s:

```
elap = 1;
energy = -ew;
```

The next line starts up a loop to count from `1` to `tot` prey items:

```
for appearance = 1:tot
```

Now we'll wait for a bug. `wait` is the waiting time counter, which starts off at `1` s. `rand` is MATLAB's built in random number generator—each call to `rand` returns a random number between 0 and 1, as though a many-sided coin were being tossed. The abundance is in units of prey per second, so imagine we're tossing a coin each second, and asking whether an insect pops up. If the abundance is, say 0.25 insects per second, then when the random number is less than 0.25 we say an insect did appear, otherwise we say no insect appeared, and the wait continues. The total abundance is `(a1 + a2)` so we call `rand` over and over again until the random number is less than `(a1 + a2)` and keep track of how long this takes, which is the waiting time. The code to do this in MATLAB is

```
wait = 1;
while rand > (a1 + a2)
  wait = wait + 1;
end;
```

Now that a bug has appeared, we must determine its identity. The probability that the bug is type-1 is $a_1/(a_1 + a_2)$, so if another call to `rand` returns a number between 0 and $a_1/(a_1 + a_2)$ we can say the bug is type-1, otherwise it's type-2. The code to determine the bug's identity is

```
if rand < a1/(a1 + a2)
  bug = 1;
else
  bug = 2;
end;
```

Next, let's compute the projected E/T. Depending on the bug's type, we want to use either `e1` and `h1`, or `e2` and `h2` in the formula for E/T corresponding to accepting the item. The E/T corresponding to ignoring the item is the same regardless of the bug's type. So, the code to calculate the projected E/T for accepting and for ignoring the item is

```
if bug == 1
  et_accept = (energy + e1 - ew*wait - eh*h1)/(elap + wait + h1);
else
  et_accept = (energy + e2 - ew*wait - eh*h2)/(elap + wait + h2);
end;
et_ignore = (energy - ew*wait)/(elap + wait);
```

Now we see whether it's better to accept the item or to ignore the item. If E/T for accepting is higher than that for ignoring, we set the `accept` flag equal to 1, and update the total elapsed time and total accumulated energy according to the type of bug that has been taken

```
if et_accept > et_ignore
  accept = 1;
  if bug == 1
    elap = elap + wait + h1;
    energy = energy + e1 - ew*wait - eh*h1;
  else
```

```
    elap = elap + wait + h2;
    energy = energy + e2 - ew*wait - eh*h2;
  end;
```

If E/T for accepting is lower than for ignoring, we set the `accept` flag to 0, and update the totals accordingly:

```
 else
   accept = 0;
   elap = elap + wait;
   energy = energy - ew*wait;
 end;
```

Finally, we append the information from this encounter with a prey item to the history vectors. In MATLAB the construction `x = [x a]` appends the element `a` to the vector `x`. So to do this for all the history vectors the code is

```
 a_his = [a_his bug]; d_his = [d_his accept];
 t_his = [t_his elap]; e_his = [e_his energy];
```

And the conclusion to the loop is

```
end;
```

This is the end of the program. Once it has been saved in your directory as `thumdiet.m`, it is ready to use. To use it, fire up MATLAB and call `thumdiet` from the command line. To emphasize that we're now back to typing in response to MATLAB's prompts, the commands are again preceded with `>>`.

Recall that we have already defined the coefficients for the handling times, energy contents, and metabolic rates as follows:

```
>> h1 = 1; h2 = 60; e1 = 10; e2 = 100; ew = 1; eh = 1;
```

With these coefficients, the foraging theory of Figure 2.2 predicts that the optimal strategy if a_1 is less than 0.2 is to take both prey types, and to take prey-1 only if a_1 is greater than 0.2. If we type

```
>> [a_his d_his t_his e_his] = thumdiet(100,.1,.1,h1,h2,e1,e2,ew,eh);
```

we will obtain the foraging history for 100 prey encounters with both a_1 and a_2 equal to 0.1. The optimal strategy is to take both prey types in this case, and let's see if our forager does in fact do this. To inspect the numerical output from `thumdiet`, just type the name of each history vector. Type `a_his` to see the sequence of prey id's that appeared, type `d_his` to see the sequence of accept/ignore decisions, and so forth. To make a picture of the output, we can write a small loop to go through the history vectors and graph what happened. The following instructions produce the graph of Figure 2.3:

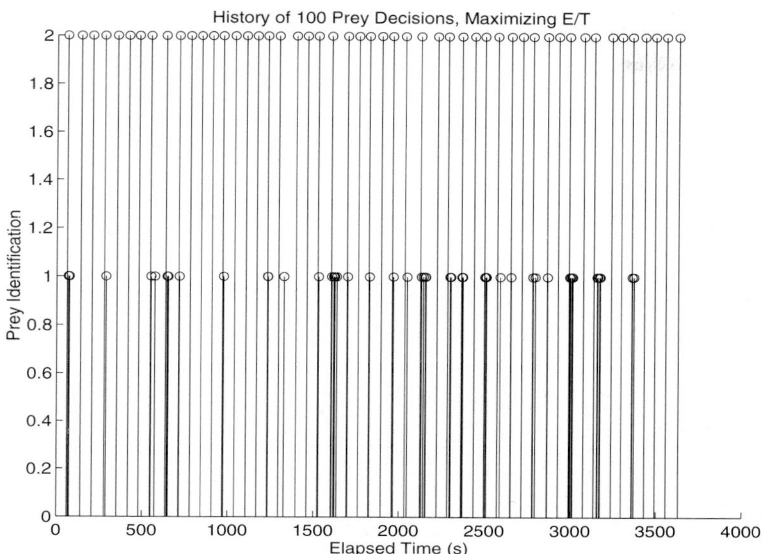

Figure 2.3: History of prey encounters with `a1` $= 0.1$ prey/s. The vertical axis is the prey identification number, and the horizontal axis is the total elapsed foraging time. Circles indicate the prey that have appeared since the day began, and a vertical line is drawn to those prey that are taken. Other parameters are `a2` $= 0.1$, `h1` $= 1$, `h2` $= 60$, `e1` $= 10$, `e2` $= 100$, `ew` $= 1$, and `eh` $= 1$. For these parameters the optimal strategy is to take both types of prey.

```
>> figure;
>> hold on
>> for i=1:length(a_his)
>>     if a_his(i) == 1
>>       plot(t_his(i),a_his(i),'ob');
>>     else
>>       plot(t_his(i),a_his(i),'oc');
>>     end;
>>     if d_his(i) == 1
>>        plot([t_his(i), t_his(i)], [0 a_his(i)],'y');
>>     end;
>> end;
```

This will produce blue circles for prey of type-1 and cyan circles for prey of type-2, with yellow vertical lines drawn to any prey item that is accepted. The vertical axis is the prey identification number. The vertical line is drawn at the time after the prey item has been acted upon; i.e., either after the forager has decided to ignore the item, or after the forager

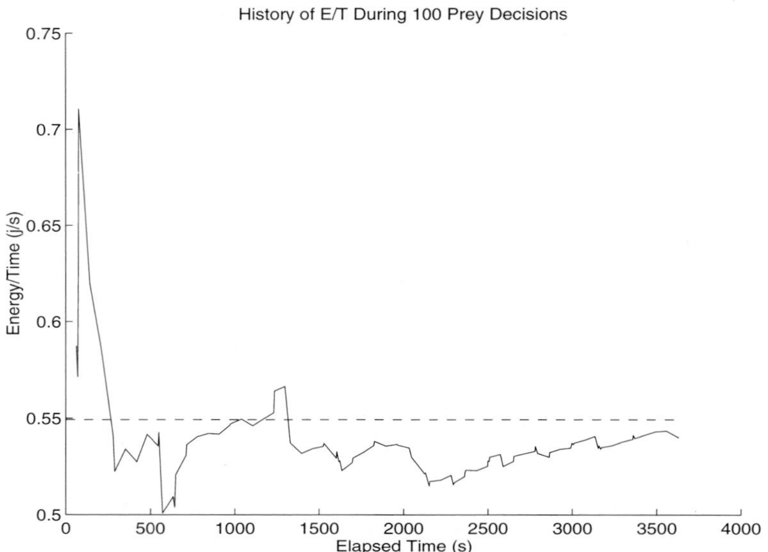

Figure 2.4: Cumulative E/T for the simulation shown in the preceding figure. The vertical axis is E/T and the horizontal axis is total elapsed foraging time. The dashed horizontal line is the expected foraging yield produced by the optimal foraging strategy.

has captured and fully handled the item. The forager definitely takes both prey types, as predicted of an optimal forager when a_1 equals 0.1.

Although the forager does take both types of prey as expected of an optimal forager, does its foraging yield also equal that expected of an optimal forager? Recall that the expected foraging yield is found as

```
>> a1 = .1; a2 = .1;
>> et_expected = eval(et12)
et_expected = 0.5493
```

A graph of the cumulative energy per time during the 100 prey encounters is produced by typing

```
>> figure;
>> hold on
>> plot(t_his, (e_his ./ t_his), 'y')
>> plot([0 t_his(length(t_his))],[et_expected et_expected], '--r');
```

Figure 2.4 shows that the rate of energy capture after several minutes approximately reaches the yield of 0.5493 J/s expected in these conditions. (The agreement would be even closer if the waiting time were computed in tenths of a second instead of whole seconds.)

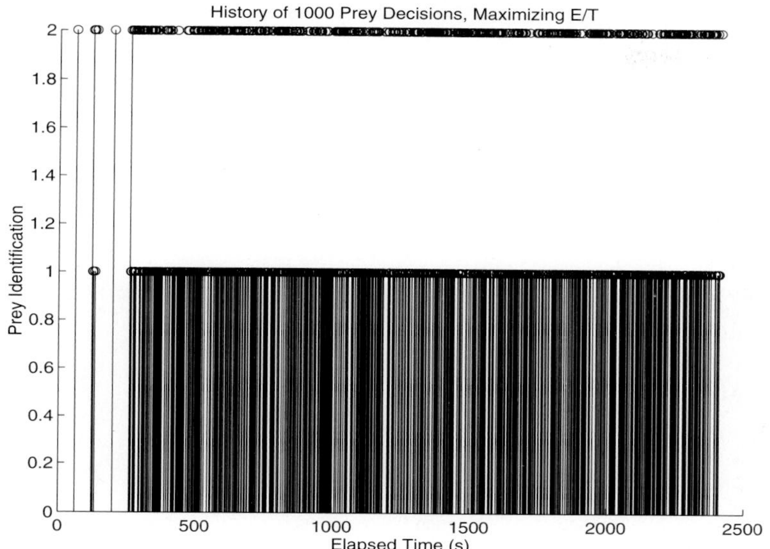

Figure 2.5: History of prey encounters with `a1` $= 0.3$ prey per second. The vertical axis is the prey identification number, and the horizontal axis is total elapsed foraging time. Circles indicate the prey that have appeared since the day began, and a vertical line is drawn to those prey that are taken. Other parameters are `a2` $= 0.3$, `h1` $= 1$, `h2` $= 60$, `e1` $= 10$, `e2` $= 100$, `ew` $= 1$, and `eh` $= 1$. For these parameters the optimal strategy is to take only type-1 prey.

But this is only half the story. We should also check out $a_1 = 0.3$ because here the optimal forager should take only type-1 and ignore type-2. By typing

```
>> [a_his d_his t_his e_his] = thumdiet(1000,.3,.3,h1,h2,e1,e2,ew,eh);
```

we obtain the foraging history of 1000 prey encounters with a prey abundance of $a_1 = a_2 = 0.3$. Figures 2.5 and 2.6 are prepared as before, and illustrate the sequence of decisions and the cumulative rate of energy yield. After some initial errors, the animal does settle down into taking only type-1 prey, and ignoring type-2 prey. Again, following the rule of thumb leads to optimal diet. The expected yield in this case is found as

```
>> a1 = .3;
>> et_expected = eval(et1)
et_expected = 1.3077
```

Figure 2.6 shows that the expected yield of the optimal forager is attained after about a half an hour.

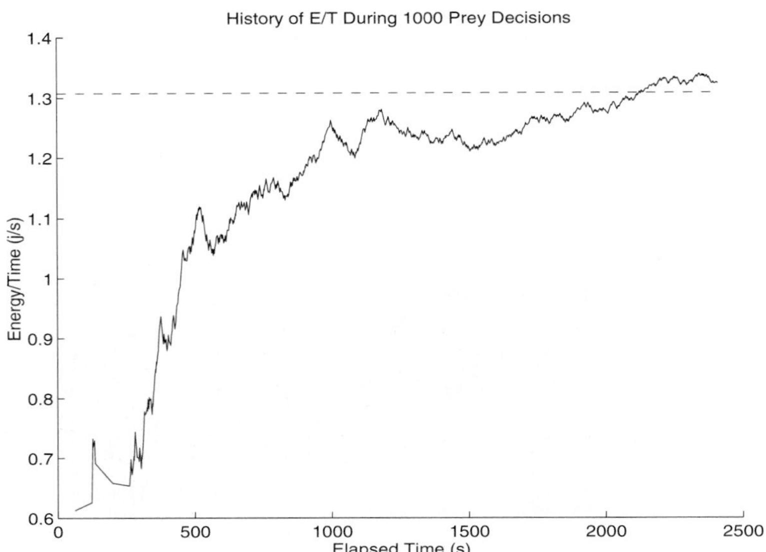

Figure 2.6: Cumulative E/T for the simulation shown in the preceding figure. The vertical axis is E/T and the horizontal axis is the total elapsed foraging time. The dashed horizontal line is the foraging yield expected when the optimal foraging strategy is used.

2.2 Sit-and-Wait Predator

2.2.1 What to Do

While some animals are continually poking around for prey, others occupy the sitting Buddha position and gaze alertly until a tasty morsel alights within striking distance. This is the sit-and-wait forager, and it too is continually making decisions. The issue is familiar—to accept or to ignore prey items.

For simplicity, we'll assume that all the prey are the same kind, but appear at different distances away. The forager must run to the item and then back to its perch. For this situation, the counterpart of the handling time is the pursuit time. There is also a waiting time to be taken into account. Here then is the specific problem to be solved: a strategy is to accept all prey out to a "cutoff distance" and to ignore all prey beyond the cutoff distance. Therefore, the optimal strategy is the strategy having the optimal cutoff distance, and we need to determine what the optimal cutoff distance is relative to some goal such as maximizing the harvest rate, E/T.

Finding the optimal cutoff distance involves trading off two conflicting effects. Increasing the cutoff distance decreases the average time spent waiting for prey to appear within this distance, but increases the average time spent running for prey. Decreasing the cutoff distance increases the waiting time, but decreases the running time. The optimal cutoff distance strikes the best compromise between minimizing waiting time and minimizing pursuit time. As before, we can use time per item as the numerical goal, or energy per time. Because energy per time is probably the most relevant, we'll go for it.

Picture an animal sitting on a perch who can look around and scan a semicircle in front of it. The radius of the semicircle will be the cutoff radius, denoted r_c. Because the area of a circle is πr^2, where r is the radius, the area of half a circle with radius r_c is $(1/2)\pi r_c^2$. If we ask how much the area changes when we vary the cutoff radius, then the change in area per tiny change in radius is the derivative of the area with respect to r_c, which is just πr_c. We need to know how the animal's foraging area changes when its cutoff radius changes because we will need to find the best cutoff radius, and take into account how the variation in radius affects both waiting time and pursuit time.

The abundance of prey is specified with a, which is in units of prey items per time per area. To find the total abundance in units of prey per time throughout the animal's semicircle, we integrate the abundance over the semicircle. Then the waiting time is the reciprocal of the total abundance, in units of time per prey item:

$$t_w = \frac{1}{\int_0^{r_c} a(\pi r)dr}$$

This formula is entered in MATLAB as

```
>> tw = symop('1','/',int('a*pi*r','r','0','rc'))
```

MATLAB proceeds to evaluate the integral and returns

```
       2
   --------
          2
   a pi rc
```

You can check that this is obviously correct because it is simply 1 over `a` times the area of a semicircle of radius `rc`. Notice that the average waiting time gets smaller as `rc` gets bigger.

The average pursuit time is calculated in the same spirit. The animal is assumed to be able to run, fly, or swim to the prey at a velocity, v. The time to run to an item at distance r away is r/v, so the time to run there and back again is $2r/v$. The average pursuit time has to include all items from those immediately adjacent to the animal out to the cutoff radius.

$$t_p = \frac{\int_0^{r_c} (2r/v)a\pi r dr}{\int_0^{r_c} a\pi r dr}$$

In this formula you can think of $(a\pi r)/(\int_0^{r_c} a\pi r dr)$ as the probability that an item lands at distance r from the animal—this is the number of items at distance r divided by the total number of items in the semicircle. So integrating this probability of being at distance r, times the pursuit time to distance r, over all r from 0 to r_c, yields the average pursuit time. This formula is entered in MATLAB as

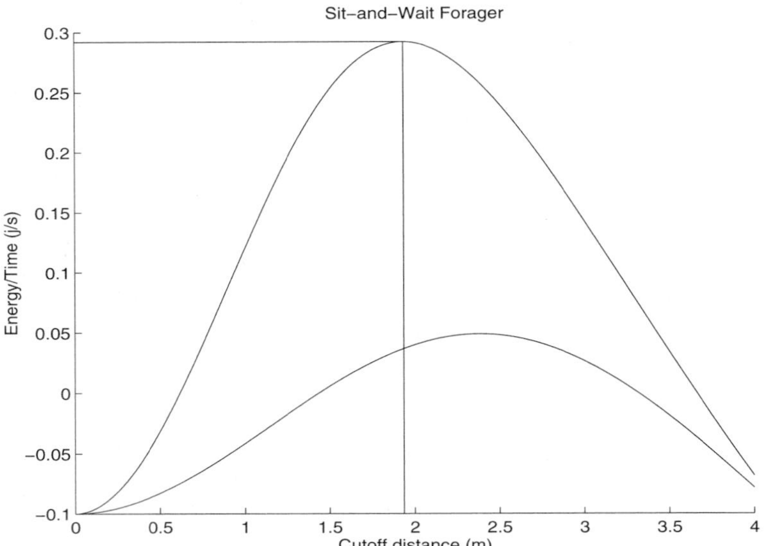

Figure 2.7: The foraging yield in J/ as a function of the cutoff radius for a sit-and-wait forager. The top curve is for a wet habitat, bottom for a dry habitat. The cutoff radius that produces the maximum foraging yield is the optimal cutoff distance. For both curves, parameters are `e` = 10, `ew` = 0.1, `ep` = 1, and `v` = 0.5. For the wet habitat `a` = 0.02 and for the dry habitat `a` = 0.005. In the wet habitat the optimal cutoff distance is 1.9349m, which produces a net foraging yield of 0.2921 J/s.

```
>> tp = symop(int('a*pi*r*(2*r/v)','r','0','rc'),'/', ...
              int('a*pi*r','r','0','rc'))
```

MATLAB returns the answer as

```
           rc
     4/3 ----
           v
```

Notice that the average pursuit time get bigger as `rc` gets bigger. Therefore the average waiting time and average pursuit time vary in opposite directions as `rc` is changed.

With formulas for the average waiting time and handling time as a function of `rc`, the rest is easy. We now form the expression for the average time per item as the sum of the average waiting and average pursuit times:

```
>> ti = symop(tw,'+',tp)
```

yielding

```
       2           rc
   -------- + 4/3 ----
          2         v
   a pi rc
```

Next, we form the expression for the average energy per item as

```
>> ei = symop('e','-','ew','*',tw,'-','ep','*',tp)
```

yielding

```
               ew            ep rc
      e - 2 -------- - 4/3 -----
                   2          v
             a pi rc
```

Here e is the energy per prey item, ew is the energy expenditure per unit waiting time, and ep is the energy expenditure per unit pursuit time. Finally, the average energy per time harvested from a semicircle of radius rc is simply the average energy per item divided by the average time per item:

```
>> et = symop(ei,'/',ti)
```

yielding

```
               ew            ep rc
      e - 2 -------- - 4/3 -----
                   2          v
             a pi rc
      ---------------------------
              2           rc
          -------- + 4/3 ----
                 2         v
          a pi rc
```

Although this formula for et looks a bit imposing, remember that it's built up from simple pieces, and as we will see, its graph is actually quite a simple shape.

To inspect the shape of et as a function of rc, let's convert the formula into a plottable version with

```
>> etplot = sym2ara(et)
```

Now let's substitute some parameters, let the energy per prey item be 10 J, the energy expended while waiting 0.1 J/s, the energy expended while running 1 J/s, and the running speed 0.1 m/s.

```
>> e = 10; ew = .1; ep = 1; v = .5;
```

Let's look at the foraging yield, et, for cutoff distances of 0.005-4.000 m, in steps of 0.005 meters. We use

```
>> rc = .005:.005:4;
```

We'll plot two curves in Figure 2.7, the first for a dry habitat with an insect arrival rate of 0.005 insects/s/m^2.

```
>> a = .005;
```

After initializing the plot with figure; and hold on, the first curve is drawn in magenta with

```
>> plot(rc,eval(etplot),'m')
```

The curve is obviously dome-shaped, and shows that there is a cutoff radius that produces a highest foraging yield—this is the "optimal cutoff distance" and is about 2.5 meters in this case. Now, for a second curve in a wetter habitat with more insects, we let

```
>> a = .02;
```

and a curve is drawn in white with

```
>> plot(rc,eval(etplot),'w')
```

The curve is again dome-shaped but its peak is higher and to the left. This means that in a wetter habitat with more prey, the optimal cutoff distance is closer, about 2 meters, and that the best yield here is better than the best yield possible in the dryer habitat.

To evaluate the optimal cutoff radius explicitly, rather than merely read its value off the graph, differentiate et with respect to rc, set the derivative equal to zero, and solve for rc, all of which is carried out in MATLAB with

```
>> rc_opt = solve(diff(et,'rc'),'rc');
```

By typing rc_opt you'll see three roots, 0, and messy formulas for two more roots. These roots are entries in a symbolic matrix; that is, a matrix of formulas. This matrix is a column vector whose element at (1,1) is the symbolic constant '0'; the element at (2,1) is the formula for one of the nonzero roots; and the element at (3,1) is the formula for the other nonzero root. To access the elements of a symbolic matrix, use the sym command. For example, to evaluate the three roots, one by one, using the existing assignment of parameters (which has a = .02, corresponding to the wet habitat), type

```
>> eval(sym(rc_opt,1,1))
>> eval(sym(rc_opt,2,1))
>> eval(sym(rc_opt,3,1))
```

and observe answers of 0, 1.9349, and -0.9675 + 3.3767i. Clearly, the second root is the relevant one, and the other two are discarded. Similarly, to obtain the answers for the dry habitat, set a = .005, and evaluate rc_opt again. Furthermore, if you set rc equal to an optimal rc and then evaluate et, you'll see explicitly what the average harvest will be for that optimal forager. When a=.02, typing

```
>> rc = eval(sym(rc_opt,2,1))
>> et_opt = eval(et)
```

yields the optimal harvest rate as 0.2921 J/s in the wet habitat. For Figure 2.7, to add green lines to indicate this optimal cutoff distance and optimal harvest rate, type

```
>> plot([rc,rc],[-.1,et_opt],'g')
>> plot([0,rc],[et_opt,et_opt],'g')
```

Let's see now if a forager can learn where the optimal cutoff distance is by following a rule of thumb; we'll use the wet-habitat parameters as an example.

2.2.2 How To Do It

Optimal foraging theory is great about telling animals what they should do. The optimal sit-and-wait forager should use the cutoff distance that maximizes the foraging yield in energy per time, as was illustrated in Figure 2.7. But can an animal do this? To see if an animal can use a simple rule to come to the optimal cutoff distance, we'll adapt the rule of thumb that worked so well in the diet-choice problem.

Picture an animal gazing out onto the semicircle in front of it; an insect alights at distance r. The animal is assumed to keep track of the total elapsed time since the beginning of the day, T, and the total energy be if it pursues the item, and what its E/T will be if it ignores the item. Then it pursues or ignores according to which is expected to bring about the higher E/T.

The projected E/T if the animal pursues the item is

$$\left(\frac{E}{T}\right)_p = \frac{E + e - e_w t_w - e_p(2r/v)}{T + t_w + (2r/v)}$$

where e is the energy of the item, e_w is the energy expended per second of waiting, t_w is the time spent waiting for this item, e_p is the energy spent per second pursuing the item and returning to the perch, and $(2r/v)$ is the time to travel to the item and back at speed v. The projected E/T if the animal ignores the item is

$$\left(\frac{E}{T}\right)_i = \frac{E - e_w t_w}{T + t_w}$$

The rule of thumb is, for each item, one after the other,

$$\begin{aligned} \text{Pursue if}: &\quad (E/T)_p > (E/T)_i \\ \text{Ignore if}: &\quad (E/T)_p < (E/T)_i \end{aligned}$$

It turns out that following this rule quickly leads to an optimal cutoff distance for a sit-and-wait predator, just as it did in the problem of optimal-diet choice for a searching predator.

The rule may again be illustrated through a simulation. The program to carry out the simulation is called `thumchas.m` and is modified from the previous program we developed for diet choice. Again, this program should be created with a text editor and saved in your directory. Once it is there, it can be invoked by name from MATLAB command line. Here is the first line of the program:

```
function [r_his,d_his,t_his,e_his] = thumchas(tot,a,e,v,ew,ep,xmax,ymax)
```

The function is called with the `tot`, the total number of sightings to consider; `a`, the number of prey per second per square meter; `e`, the energy content of a prey item; `v`, the sprint speed of the animal when chasing prey and returning to the perch; `ew`, the energy expended per unit waiting time; `ep`, the energy expended per unit sprint time; `xmax` and `ymax` the size of a rectangular region in front of the animal where the prey appears. The rectangle runs from `-xmax` on the left to `xmax` on the right, and out to `ymax` in front of the animal. The function returns four history vectors: `r_his`, the list of distances at which the prey appeared; `d_his`, the history of decisions, pursue or ignore (scored as 1 and 0) for the items in order of appearance; `t_his`, the history of total elapsed times after each prey item has been acted upon; and `e_his`, the history of total accumulated energy after each item has been acted upon. Graphs of the output can then be synthesized from these history vectors.

The program begins with initializing the history vectors to empty vectors, and with the elapsed time and energy representing one second of waiting (while the animal is still rubbing its eyes after waking up):

```
r_his = []; d_his = []; t_his = []; e_his = [];
elap = 1; energy = -ew;
```

The main loop to count the appearance of insects is started with

```
for appearance = 1:tot
```

The total area in which prey are appearing is `2*xmax*ymax`. The waiting time is counted out with

```
 wait = 1;
 while rand > (a*2*xmax*ymax)
   wait = wait + 1;
   end;
```

Now that a prey has appeared, its `x` and `y` coordinates are chosen with calls to the random number generator:

```
 x = 2*(rand - 0.5)*xmax;
 y = rand * ymax;
```

Next, the distance to the item is calculated:

```
 r = sqrt(x^2 + y^2);
```

Now, the energy per time is projected for both courses of action:

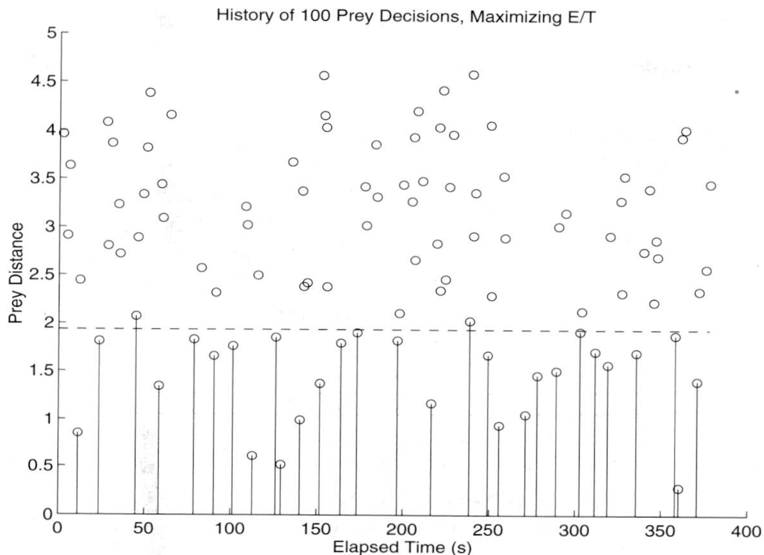

Figure 2.8: A history of prey appearances for a sit-and-wait predator. The vertical axis is distance of prey item from forager and horizontal axis is total elapsed foraging time. Circles indicate the prey that have appeared since the day began, and a vertical line is drawn to those prey that are taken. The optimal strategy is to observe a cutoff distance of 1.9349m, as indicated by the dashed horizontal line. After about a minute, foraging according to the rule of thumb extends approximately out to the optimal cutoff distance, indicating that the behavior has converged to optimal foraging. Parameters are `a` = 0.02, `e` = 10, `ew` = 0.1, `ep` = 1, and `v` = 0.5.

```
et_pursue = (energy + e - ew*wait - ep*2*r/v)/(elap + wait + 2*r/v);
et_ignore = (energy - ew*wait)/(elap + wait);
```

If pursuing is the better course of action, the elapsed-time and cumulative-energy counters are updated so that they reflect the pursuit and capture of an item:

```
if et_pursue > et_ignore
  pursue = 1;
  elap = elap + wait + 2*r/v;
  energy = energy + e - ew*wait - ep*2*r/v;
```

However, if ignoring is the better course of action, then the elapsed-time and cumulative-energy counters are updated so that they reflect this:

```
else
  pursue = 0;
```

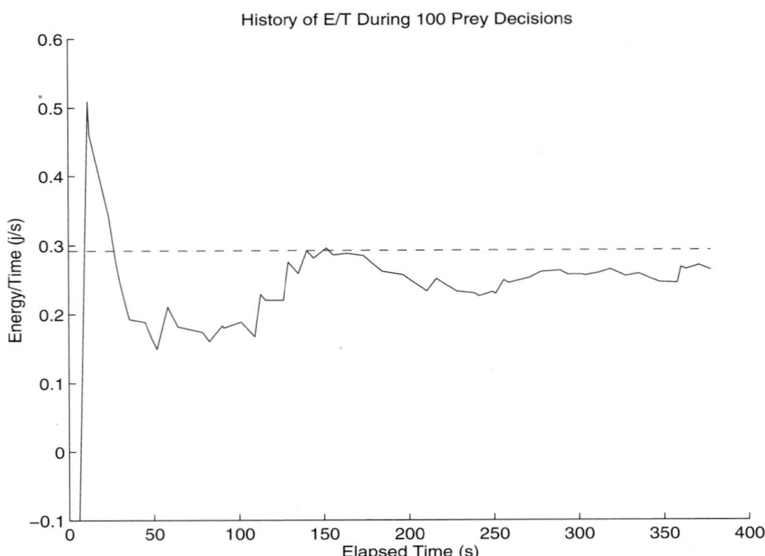

Figure 2.9: The cumulative E/T for the simulation shown in the preceding figure. The vertical axis is E/T and the horizontal axis is the total elapsed foraging time. The dashed line is the net foraging yield for an optimal forager, 0.2921 J/s. The foraging yield converges to that of an optimal forager after about 1 min.

```
    elap = elap + wait;
    energy = energy - ew * wait;
  end;
```

Finally, the history vectors are updated to record what happened at this prey sighting:

```
  r_his = [r_his r]; d_his = [d_his pursue];
  t_his = [t_his elap]; e_his = [e_his energy];
```

The loop and program conclude with

```
end;
```

Now that `thumchas.m` resides in your directory, it is invoked at the MATLAB prompt by your typing

```
>> [r_his d_his t_his e_his] = thumchas(1000,a,e,v,ew,ep,3.5,3.5);
```

When the next prompt appears, the results of the simulation are contained in the history vectors. Figure 2.8 was made by typing

```
>> figure;
>> hold on
>> for i=1:length(r_his)
>>     plot(t_his(i),r_his(i),'oc');
>>     if d_his(i) == 1
>>        plot([t_his(i), t_his(i)], [0 r_his(i)],'y');
>>     end;
>> end;
>> plot([0 t_his(length(r_his))],[rc rc],'--g');
```

The figure shows that after about 2 min, foraging according to the behavioral rule of thumb extends approximately to the optimal cutoff distance. This indicates that the actual behavior has approximately converged to optimal foraging behavior. Thus, an animal who followed a decision rule more or less like the rule of thumb being modeled in this simulation would appear to be foraging optimally, and would reap the benefit of the maximum sustainable foraging yield from its feeding territory.

To see how the foraging yield converges to that predicted for an optimal forager, we produced Figure 2.9 by typing

```
>> figure;
>> hold on
>> plot([0 t_his(length(r_his))],[et_opt et_opt],'--g')
>> plot(t_his, (e_his ./ t_his), 'y')
```

After 2 min, the foraging yield is very close to the theoretically best average yield, shown as a horizontal dashed line.

If you closely examine Figures 2.8 and 2.9, you'll notice that deviations from the optimal cutoff distances, either running too far or not far enough, are explained by whether the cumulative E/T at the time of a deviation is below or above the dashed line for the theoretically best long-term average yield. If the E/T happens to fluctuate above the dashed line in Figure 2.9, the forager is not disposed to run quite as far as it does on the average, and vice versa if E/T happens to fluctuate below the dashed line.

Foraging theory has proved to be very popular among ecologists. It's relatively easy to tailor an optimal foraging model to match the varying circumstances of different organisms—models can be devised for almost any creature faced with recurring choices for accomplishing a goal. I think the major limitation of optimal foraging models at present is that they consider the decisions of an organism all by itself. Instead, if animals interact with others when making decisions, then an interactive decision-making approach must be taken. This approach is a subject of active research. Often though, animals probably make up their minds without worrying about neighbors too much, and if so, the single-organism optimal foraging theory introduced here should work just fine.

Of course, food is not an end in itself—it's used to make offspring. The next chapter is all about how populations grow in size because of all those offspring being produced by optimal foragers.

2.3 Further Readings

Krebs, J. R. and Davies, N. B. (1984). *Behavioural Ecology. Second Edition.* Blackwell, Oxford.

MacArthur, R. H. and Pianka, E. R. (1966). On the optimal use of a patchy environment. *American Naturalist*, 100:603–609.

Pulliam, H. R. (1974). On the theory of optimal diets. *American Naturalist*, 108:59–74.

Real, L. A. (1994). Information processing and the evolutionary ecology of cognitive architecture. In Real, L. A., editor, *Behavioral Mechanisms in Evolutionary Ecology*, pages 99–132. University Of Chicago Press: Chicago, Illinois.

Roughgarden, J. (1995). Anolis *Lizards of the Caribbean: Ecology, Evolution, and Plate Tectonics.* Oxford University Press.

Schoener, T. W. (1987). A brief history of optimal foraging ecology. In Kamil, A. C., Krebs, J. R., and Pulliam, H. R., editors, *Foraging Behavior*, pages 5–67. Plenum Press, New York.

2.4 Application: The Traveling Bumblebee

One of the appeals of optimal foraging theory is that it leads to clear hypotheses that can be readily tested in the field[3]. One of the earliest tests of foraging theory examined predictions about giving up time as they applied to a common pollinator-flower interaction in the vicinity of Seattle, Washington (Best and Bierzychudek, 1982).

Bumblebees often concentrate all of their nectar-gathering on the common roadside and garden flower, foxglove. This plant produces a spike of flowers, with the flowers at the bottom of each spike secreting the largest volume of nectar, and the flowers at the top of the spike producing the least nectar. Not surprisingly, bumblebees tend to land first at the bottom flowers, and work their way up the stalk. That prediction does not require any calculations. However, these bumblebees do not visit all of the flowers on each stalk—instead they typically fly away to start foraging from a different stalk well before they have exploited all of the flowers on any single plant. Best and Bierzychudek explored the hypothesis that these bees might be foraging optimally in the sense of leaving a flower stalk when their net rate of energy intake was higher if they moved than if they kept working up a stalk.

To formalize this hypothesis, Best and Bierzychudek first used tiny capillary pipettes to sample flowers themselves and they obtained an equation that describes the expected calorie reward from a flower as a function of its position on a stalk, `i`, with the bottom position represented by `i=1`. You can explore their model by typing in the following executable MATLAB file that uses the equation for caloric reward versus flower position, together with information on the costs of flying from plant to plant, to predict the departure time from a flower stalk.

[3]This section is contributed by Peter Kareiva.

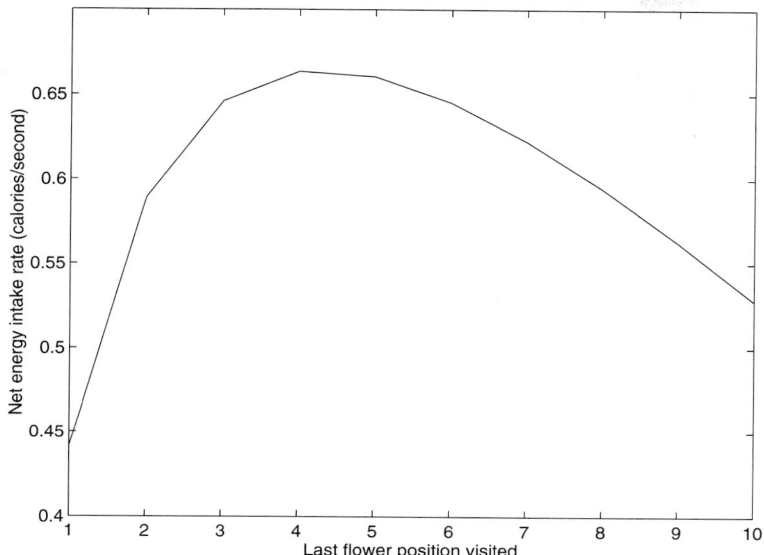

Figure 2.10: Net energy intake rate (calories/second) of a bumblebee as it forages along a single flower stalk.

The expected caloric reward for the ith flower position is calculated from the formula

```
ei=-1.7*i+20.19
```

Other parameters are `P`, the probability that plant hasn't been drained of nectar already; `eb` and `tb`, the energy and time expended flying between plants; `ew` and `tw`, the energy and time needed to get between two flowers on same plant; `ef` and `tf`, the energy and time cost of emptying "full" flowers; `ee` and `te`, the energy and time cost of sampling "empty" flowers; and `n`, which is the last flower position visited by a bee.

Now type in and save the file, which we call `two1.m`:

```
function[netE] = two1(P,eb,ew,ef,ee,tb,tw,tf,te,n)
 netE = [];
 sumi = 0;
 for i = 1:n
  sumi = sumi + (-1.7*i+20.19);
  net = (P*sumi-eb-P*ew*(i-1)-P*i*ef-ee*(1-P))/ ...
        (tb+P*tw*(i-1)+P*i*tf+te*(1-P));
  netE = [netE net];
 end;
```

The numerator in the line beginning with `net =` is the total energy gained by visits to `i` flowers, and the denominator is the total time spent foraging on that plant and moving to the next plant. When the ratio is maximized, then the bee is foraging optimally.

To run the program, we enter parameter values obtained by Best and Bierzychudek from their fieldwork:

```
>> P = .375; eb = .09; ew = .07; ef = .02; ee = .01;
>> tb = 4.4; tw = 3.3; tf = 14.7; te = 8.9; n = 10;
>> [netE] = two1(P,eb,ew,ef,ee,tb,tw,tf,te,n);
>> figure;
>> plot(1:10,netE,'r')
```

The resulting figure (Figure 2.10) shows that the optimal energy intake is obtained if the bees depart after they have visited four flowers on a stalk. In the meadows where Best and Bierzychudek estimated the parameters listed above, bumblebees left on average from flower position 4.5—which is a pretty good match to the prediction, given the possibility of their departing anywhere between flower 1 and flower 10.

Once a model like this has been coded, it is easy to explore the consequences of changing factors such as the spacing between plants. For instance, if plants are extremely widely spaced, then `eb` and `tb` would be larger, and the predicted departure time would change; to investigate these effects, simply type indifferent values for these parameters.

Best, L. and Bierzychudek, P. 1982. Pollinator foraging on foxglove (*Digitalis purpurea*): a test of a new model. *Evolution* 36: 70–79.

Chapter 3

The Sky's the Limit

First there was 1, then there were 2, then 4, then 8, 16, 32, 64 ... the sky's the limit. An *E. coli* bacterium can divide every 20 minutes. Start with 100 of these, come back in 32 hours, and see the entire earth covered one yard deep. This is called "geometric growth," provided time is marked in discrete steps, say every 20 minutes, and it's called "exponential growth" if time is recorded continuously. Geometric and exponential growth are like compound interest. A population that doubles every 20 minutes has a 100% interest rate compounded every 20 minutes. This chapter is about geometric and exponential growth in their many guises. The ecological compound interest rate is rarely as high as 100%. In fact, many populations can barely eek out interest rates much above 0. Also, many populations are spread out in space, and earn different ecological interest rates in different places and at different times. To learn how to keep a population alive, how to know when one is endangered, and how fast an invasion spreads, stay tuned.

3.1 The Population Bomb

Geometric and exponential growth are often called a "population bomb" because the formulas for population growth resemble those for an atomic bomb. In a fission atomic bomb, one particle collides with another, splitting it, and then the fragments go on to split still more particles, leading to a chain reaction that is just like population growth. The key assumption behind geometric and exponential growth is that the growth rate (or ecological interest rate) does not depend on the number of organisms there already are. This condition is called "density independence." It means that the birth and death rates do not depend on the population size, even though the birth and death rates could vary in space and time for other reasons not related to the population size. To see how geometric or exponential growth results, we'll start with discrete time steps, and then cover continuous time.

3.1.1 Discrete Time

Start by setting the time step to some convenient value, like one year, one month, one day—you get it, and census the population at the beginning of the time step. Let B stand for the number of births each animal has during a time step that live to the beginning of the next step, and D the probability that an animal dies during the time step. Because D is the probability an animal dies, $(1-D)$ is the probability that an animal survives through the time step. If there are N animals at time t, then the number at time $t+1$ is the sum of the births plus the survivors:

$$N_{t+1} = BN_t + (1-D)N_t$$

It's customary to combine the birth and death parameters into a single parameter, $R = 1 + (B - D)$ which is called the "geometric growth factor", or just "big R" for short, so that

$$N_{t+1} = RN_t$$

Now let's start with some initial population size, and call it N_0. The population size one time-step later is

$$N_1 = RN_0$$

After another time step we have

$$\begin{aligned} N_2 &= RN_1 \\ &= R^2 N_0 \end{aligned}$$

Similarly for another time step, and still another, so that in general

$$N_t = R^t N_0$$

The procedure we have just carried out, of going forward from an initial condition through successive time steps to see what happens in the end, is called "iteration."

Plotting the formula for N_t illustrates the population bomb. To do this in MATLAB, define `nt` as

```
>> nt = '(r^t)*no';
```

Then make a plottable version with

```
>> ntplot = sym2ara(nt);
```

Now assign some values to the parameters. Let the initial population size be `1`, the discrete rate of growth be `2`, and the time be a vector from `0`, with steps of `1` to `10`

```
>> no = 1;
>> r = 2;
>> t = 0:1:10;
```

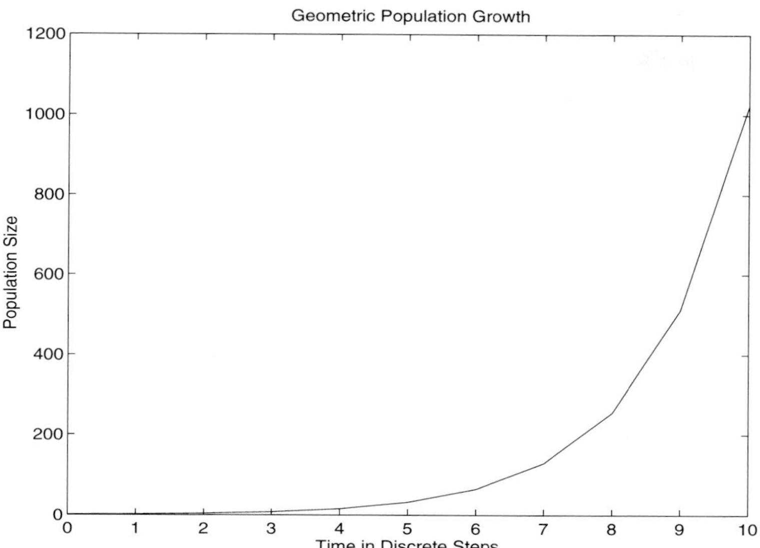

Figure 3.1: Geometric population growth, an example of the population bomb, beginning with a population size of 1. The "big R," R, is 2, indicating a population doubling each time step.

Figure 3.1 is then produced with

```
>> figure
>> plot(t,eval(ntplot));
```

Although Figure 3.1 certainly illustrates an explosion, as befits a bomb, a more useful graph is to plot the logarithm of the population size vs. time. To see why this plot is more useful, take the log of both sides of the formula for N_t and simplify, as in

$$\begin{aligned} N_t &= R^t N_0 \\ \log(N_t) &= \log(R^t N_0) \\ \log(N_t) &= \log(R)t + \log(N_0) \end{aligned}$$

That third formula above shows that geometric growth will plot as a straight line if the logarithm of the population size is used on the vertical axis and time on the horizontal axis. The slope of the line will be $\log(R)$. MATLAB has a built-in function to make a log plot, and Figure 3.2 is made by our typing

```
>> figure
>> semilogy(t,eval(ntplot));
```

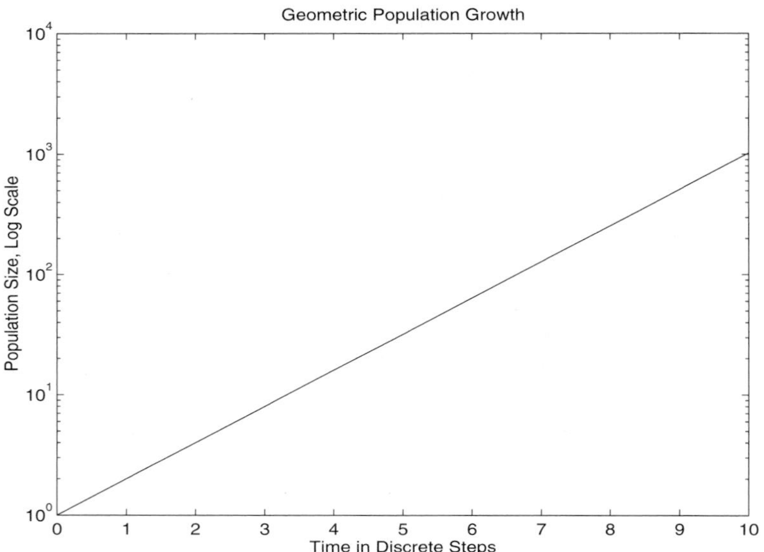

Figure 3.2: Geometric population growth on a semilog scale, using the same values as the preceding figure.

The straight-line plot on a semilog graph is *diagnostic* of geometric (and exponential) growth. If a population's size through time is a straight line on a semilog plot for some period, then it is growing exponentially during that period.

3.1.2 Continuous Time

The counterpart of geometric growth in discrete time is exponential growth in continuous time. Let the average birth *rate*, in birth per unit time per organism, be b, and the average death *rate*, in deaths per unit time per organism, be d. Then the *change* in the population's size at any time is the total birth rate minus the total death rate:

$$\frac{dN}{dt} = bN - dN$$

It's customary to combine the per capita birth and death rates into a single parameter $r = b - d$, called the "intrinsic rate of increase" or "little r" for short:

$$\frac{dN}{dt} = rN$$

You'll recognize this as a differential equation. To solve it means to find the function $N(t)$ that has the property required by the differential equation; that is, the formula for $N(t)$

must be such that its slope at each point equals r times its value at that point—this is what $dN/dt = rN$ means mathematically. As was noted in Chapter 1, MATLAB can hunt around and often come up with a formula that does satisfy the differential equation. This is as easy as they come, so type

```
>> nt = dsolve('Dn = r*n','n(0) = no','t')
```

and MATLAB replies with

```
nt = exp(r*t)*no
```

This is the formula for the population size at any time, given the initial condition of `no`.

To compare this formula in continuous time with the previous formula for discrete time, notice that "big R" is the same as e to the "little r:"

$$R = e^r$$

To make a graph of exponential growth that matches that for geometric growth, all we need to do is take "little r" as the natural logarithm of "big R":

$$r = \log(R)$$

So, let's make a plottable version of `nt` with

```
>> ntplot = sym2ara(nt);
```

assign the same initial population size; assign a "little r," `r`, that is the log of the former "big R;" and because time is now continuous, we can let time vary in tiny steps of 0.05 from 0 to 10, and obtain a smooth curve:

```
>> no = 1;
>> r = log(2);
>> t = 0:.05:10;
```

Figure 3.3 now appears when you type

```
>> figure
>> plot(t,eval(ntplot))
```

Like the discrete case, a graph of an exploding bomb, while in a sense picturesque, is not the most useful. If we take the log of both sides of the formula for $N(t)$, we obtain

$$\begin{aligned} N(t) &= e^{rt}N(0) \\ \log(N(t)) &= \log(e^{rt}N(0)) \\ \log(N(t)) &= rt + \log(N(0)) \end{aligned}$$

So again a plot of the logarithm of $N(t)$ vs. t yields a straight line. The slope is now r and the y-intercept is again $\log(N(0))$. An illustration is produced with

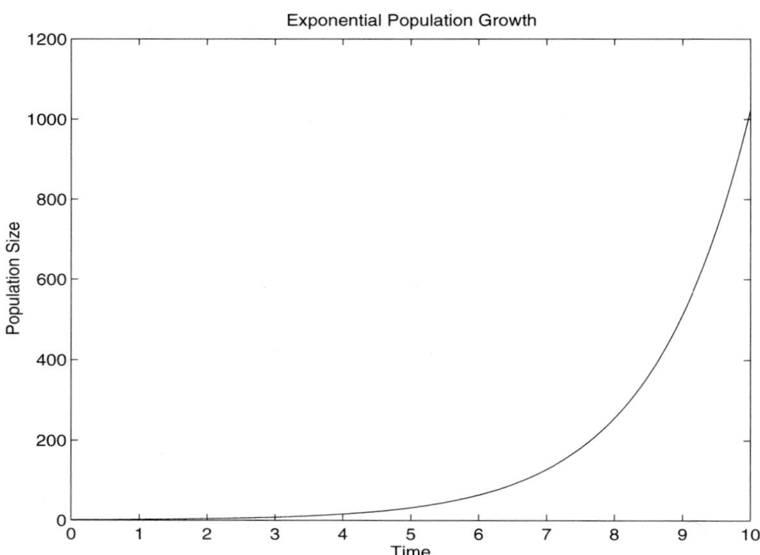

Figure 3.3: Exponential population growth, an example of the population bomb, beginning with a population size of 1. Here the "little r" or intrinsic rate of increase, r, is log(2) per time per organism, indicating a population doubling for each unit of time.

```
>> figure
>> semilogy(t,eval(ntplot))
```

and appears as Figure 3.4.

Geometric and exponential growth really amount to the same thing, and the only difference between them is the rather artificial distinction of whether we keep track of time continuously or in discrete steps. The distinction between continuous and discrete time does become biologically important in the next chapter when density dependence is considered, but here the distinction is mostly a nuisance. Geometric and exponential growth are called "linear processes" because the population size, N, is never multiplied by another N, making an N^2, or something even more complicated. For linear processes, the distinction between continuous and discrete time is important primarily to be sure one has the correct units for the population growth rate.

3.2 Keeping Track of the Years

How old are you? This is certainly an impolite question in many social contexts, but plants and animals are not offended when we ask it of them. So let us shamelessly inquire how the members of a population are apportioned among various ages, whether a population

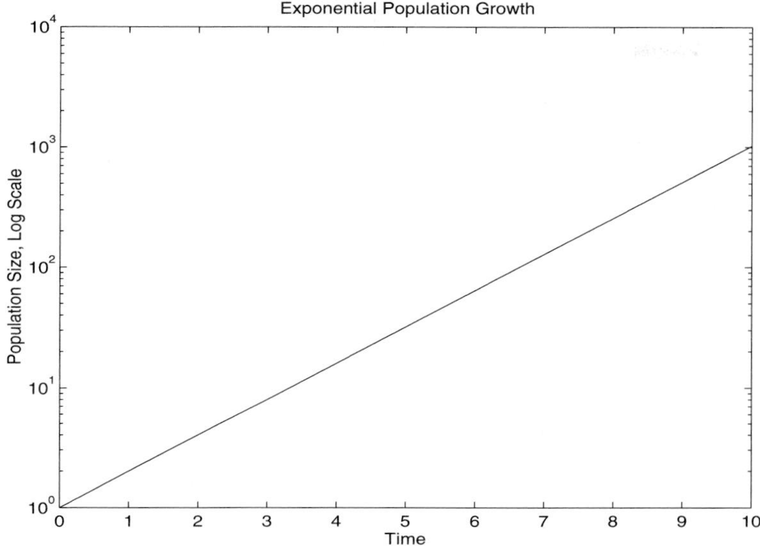

Figure 3.4: Exponential population growth on a semilog scale, using the same values as the preceding figure.

consists mostly of teenie-boppers or of wise and wizened adults.

3.2.1 The Leslie Matrix for Discrete Time

The setup for predicting population growth while keeping track of the ages of the members of the population is easier to develop in discrete time than in continuous time, and also more useful. To begin with, we must agree on the convention that the age-class width is exactly the same as the time-step length. For example, if our time step is one year, then the age classes must also be one year wide. This means that the first age class lumps together all those organisms who have just been born to those 364 days old on the census date. To keep matters from getting messy, we'll work out the formulas with three age classes—the extension to more than three age classes will then be obvious. The fertility for an organism of age class x will be labeled F_x where x is 1, 2, or 3 for the three age classes. The units of the fertility are the number of offspring born per parent of age-class x that live to be counted on the upcoming census date. This definition of fertility thus also allows for mortality that takes place between birth and the time when the population is censused. The mortality, or put more positively, the probability of survival depends on the age class. We let P_x denote the probability that an organism who starts the time step as a member of age class x survives through the entire time step. As the organism survives the time step, it also ages, so once it has survived the entire time step it is "promoted" to the next age class. Finally, let's denote the number of organisms in age class x at time t as $n_{x,t}$.

Now all that's left to do is to write down the formulas for what we've just said in words. The number in age-class 1 at time $t+1$ must be the sum of those born to parents from the three possible age classes, so

$$n_{1,t+1} = F_1 n_{1,t} + F_2 n_{2,t} + F_3 n_{3,t}$$

The number in age-class 2 at time $t+1$ must be those that survived through the time step to the next census:

$$n_{2,t+1} = P_1 n_{1,t}$$

and similarly for age-class 3:

$$n_{3,t+1} = P_2 n_{2,t}$$

Because there are only three age classes, no one survives beyond the third age class. (So if you really do want to use just three age classes, be sure they are wide enough to cover the entire life span, or otherwise use more age classes).

What we will do with these equations is to iterate them; that is, use them successively, starting with an initial condition. Given $n_{1,0}$, $n_{2,0}$, and $n_{3,0}$ at time 0, and assuming we have already measured the fertilities and probabilities of survival, F_x and P_x, we can calculate the n's at time 1, then use the formulas again to get the n's at time 2, and so forth for as long as we want. This would all be pretty boring if it weren't for the fact that some regularities emerge when we do the iteration, regularities that are not only interesting but important.

To carry out the iteration of the three equations above using MATLAB, it's helpful to take advantage of what's called "matrix arithmetic." Here's how you multiply a matrix times a vector—just memorize this. Let's start with a matrix whose elements are

$$\begin{pmatrix} a & b & c \\ d & e & f \\ g & h & i \end{pmatrix}$$

and with a column vector whose elements are

$$\begin{pmatrix} x \\ y \\ z \end{pmatrix}$$

Then *by definition* the matrix times the vector produces a new column vector with elements

$$\begin{pmatrix} ax + by + cz \\ dx + ey + fz \\ gx + hy + iz \end{pmatrix}$$

This is what you should memorize. Look at it, and see the pattern. The top element of the new vector is made from the top row of the matrix and the original column vector, the second element of the new vector is made from the second row of the matrix and the original column vector, and so forth. With a little practice, you'll find that using matrix arithmetic saves a lot of time and space, and in particular, matrix arithmetic will save time and space now.

Using matrix arithmetic, we can condense the three equations for the n's at $t+1$ into one equation. Put the F's and P's into a matrix in the following positions, with 0 everywhere else:

$$\begin{pmatrix} F_1 & F_2 & F_3 \\ P_1 & 0 & 0 \\ 0 & P_1 & 0 \end{pmatrix}$$

The F's are loaded into the top row, and the P's in a diagonal relationship called the "sub-diagonal." (The "main diagonal" runs from top left to bottom right; the first subdiagonal is immediately below the main diagonal.) Now check this out. If the n's for time t are in a column vector

$$\begin{pmatrix} n_{1,t} \\ n_{2,t} \\ n_{3,t} \end{pmatrix}$$

then multiplying the matrix with the F's and P's by this column vector yields a new column vector whose elements are exactly the n's for time $t+1$. Indeed, if we let m denote the matrix with the F's and P's, n_t denote the column vector of n's for time t, and n_{t+1} denote that for time $t+1$, we can rewrite the three equations as one equation:

$$n_{t+1} = mn_t$$

This is the way to generalize to more than three age classes, because this formula works for any number of age classes. Just use a larger matrix and column vector.

At this point you can come up for air while I mention that the matrix with F's in the top row and P's in the subdiagonal is called the "Leslie Matrix" after P. H. Leslie, who introduced this notation into the study of population growth. Also, the study of population growth with age structure is called "demography" and is widely used not only in ecology, but perhaps even more so, to project human-population growth in various countries around the world.

Now back to matrices. MATLAB does matrix multiplication for you. To confirm that it does it correctly, let's see if we can regenerate the equations that we will have to iterate. Here we'll work solely with symbols, and later move to numbers. To use the `sym` command to enter a symbolic matrix, type

```
>> m = sym('[f1 f2 f3; p1 0 0; 0 p2 0]')
```

and MATLAB replies with

```
m = [ f1,f2,f3]
    [ p1, 0, 0]
    [  0,p2, 0]
```

Similarly, to enter the column vector for n_0, type

```
>> n0 = sym('[n10; n20; n30]')
```

and MATLAB replies with

```
n0 = [ n10]
     [ n20]
     [ n30]
```

Now let's symbolically multiply the Leslie matrix and the no vector, so that we obtain the n1 vector:

```
>> n1 = symop(m,'*',n0)
```

MATLAB replies with

```
n1 = [f1*n10+f2*n20+f3*n30]
     [              p1*n10]
     [              p2*n20]
```

This is good news. MATLAB has produced for n1 a column vector whose elements are exactly those in the three equations we must iterate. So, let's move on to a numerical example.

To enter a matrix whose elements are numbers, not symbols, we simply type

```
>> m = [0.5 1 0.75; 0.6666 0 0; 0 0.3333 0]
```

and MATLAB replies with a numerical example of a Leslie matrix.

```
m = 0.5000    1.0000    0.7500
    0.6666         0         0
         0    0.3333         0
```

Here we are assuming that a member of age-class 1 has on average 0.5 offspring per time step that lives to the following census, a member of age-class 2 has on average 1 such offspring, and a member of age-class 3 has 0.75 such offspring. Also, we assume that 0.6666 of the individuals of age-class 1 survive the time step to become members of age-class 2, and 0.3333 of the individuals of age-class 2 survive to enter age-class 3. Suppose also that the population begins with 2 individuals of age-class 1, and none in the other age classes, so we type

```
>> n0 = [2;0;0]
```

and MATLAB replies with

```
n0 = 2
     0
     0
```

To separate the age distribution from the total population size, we can sum the vector n0 with

```
>> s0 = sum(n0)
```

yielding

```
s0 = 2
```

which is the total population size over all ages. The age distribution is then computed as

```
>> c0 = n0/s0
```

yielding

```
c0 = 1
     0
     0
```

The age distribution is the fraction of the population in each age class, and the sum of these fractions automatically equals one.

To do the first iteration, simply type

```
>> n1 = m*n0
```

and MATLAB replies with

```
n1 = 1.0000
     1.3332
          0
```

The total population size and age distribution now are

```
>> s1 = sum(n1)
s1 = 2.3332
>> c1 = n1/s1
c1 = 0.4286
     0.5714
          0
```

The next iteration is

```
>> n2 = m*n1
n2 = 1.8332
     0.6666
     0.4444
>> s2 = sum(n2)
s2 = 2.9442
>> c2 = n2/s2
c2 = 0.6227
     0.2264
     0.1509
```

Now do this several more times until you reach

```
>> n8 = m*n7
n8 = 4.5635
     2.5727
     0.7247
>> s8 = sum(n8)
s8 = 7.8610
>> c8 = n8/s8
c8 = 0.5805
     0.3273
     0.0922
```

You now have generated eight iterations, and have a total of nine population sizes and age distributions, including the initial condition. To see what's happened let's look at a graph of the age distributions. We can arrange them all into one big figure with nine distinct panels, one for each time. The `subplot(m,n,r)` command makes the next plot occur in the rth panel of an $m \times n$ array of graphs. The `bar` command draws a bar graph. These are used together to draw the first two of the graphs as

```
>> figure
>> subplot(3,3,1)
>> bar(c0)
>> subplot(3,3,2)
>> bar(c1)
```

Now keep it up until you reach

```
>> subplot(3,3,9)
>> bar(c8)
```

and the result appears in Figure 3.5. The figure shows that the age distribution stops changing by about the fourth time step. This ultimate age distribution that the population approaches is called the *stable age distribution.* As long as the same F's and P's are used, this particular distribution is always approached from any initial condition.

Another regularity emerges too. Let's graph the sequence of total population sizes through time on a semilog plot and see what happens. Type in the time and total size vectors as

```
>> t = 0:1:8;
>> s = [s0 s1 s2 s3 s4 s5 s6 s7 s8];
```

and make the graph with

```
>> figure
>> semilogy(t,s)
```

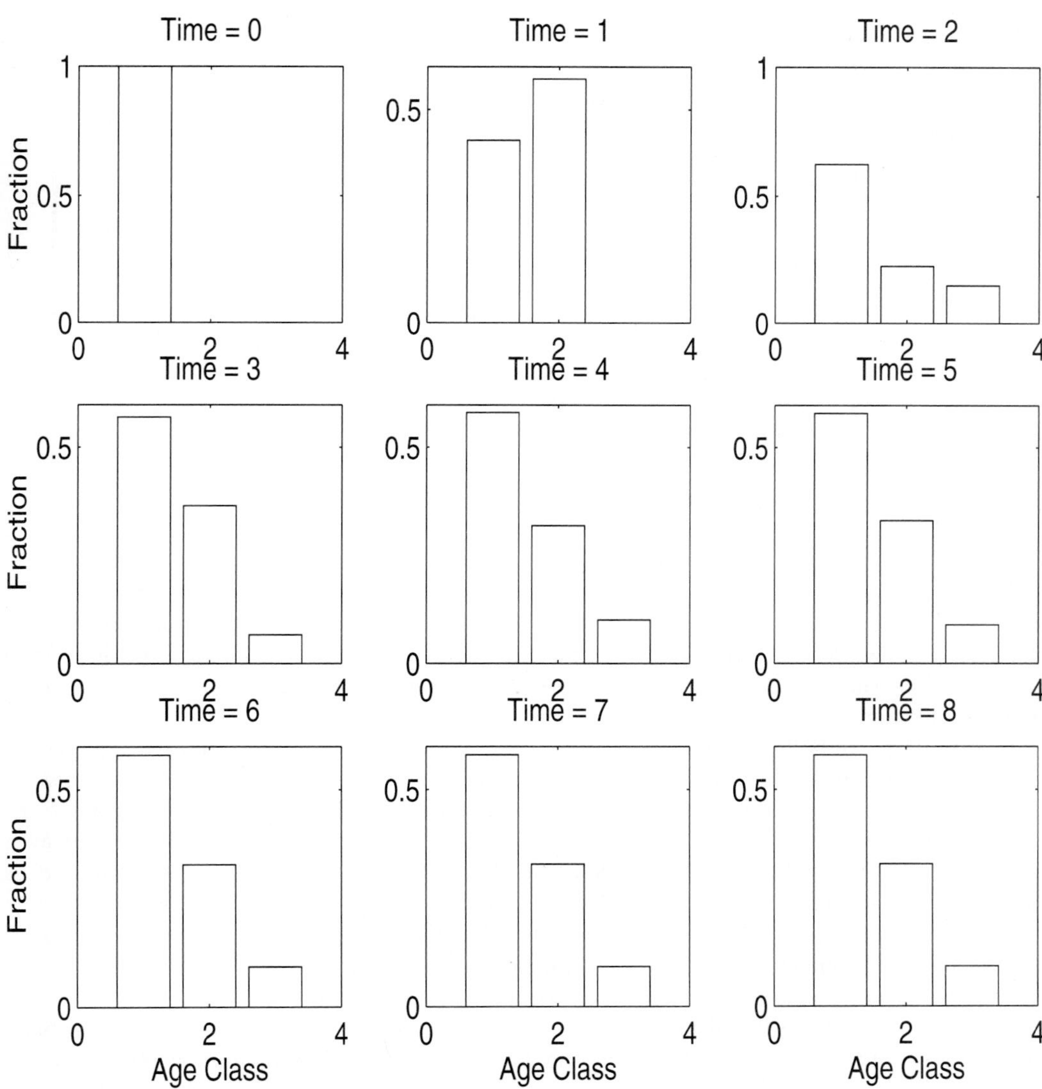

Figure 3.5: Age distributions at successive times.

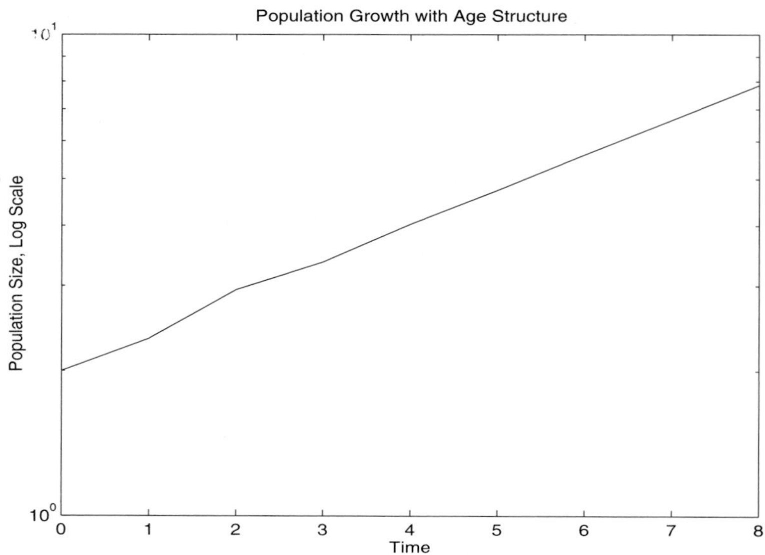

Figure 3.6: Growth of an age-structured population on a semilog scale.

The result appears in Figure 3.6. Observe that after about the fourth time step the population size increases as a straight line. What does this mean? It means that after a while the population is growing geometrically. It's no coincidence that the geometric growth happens when the population's age distribution has reached the stable age distribution—these two properties go hand in hand.

In summary, we have seen in this example that after about four time steps the age distribution has approached the stable age distribution and beginning at that time the population growth is geometric. The last age distribution that we calculated, `c8`, showed proportions of the three age classes as 0.5805, 0.3273, and 0.0922, respectively. The geometric growth factor for the population once the stable age distribution was attained can be estimated as the ratio of consecutive population sizes. With the last values we calculated, `s8/s7` is found to be 1.1828—this is the population's "big R." We will now see that the stable age distribution and the geometric growth factor can be computed directly from the Leslie matrix, without having to go though the iteration we have just conducted.

But first some more definitions. Recall that once the stable age distribution has been attained, the only effect of multiplying the Leslie matrix by the column vector, n, is to increase the population size by the geometric growth factor, R (our "big R"). Another way to say this is that once the components of n are in the stable age distribution, thereafter the operation of multiplying the Leslie matrix by n is only to lengthen it (or to shrink it) by the factor R. A vector such that the action of multiplying it by a matrix merely lengthens it or shrinks it is called an *eigenvector* of the matrix, and the factor by which it is lengthened or shrunk is called the *eigenvalue* of the matrix corresponding to that eigenvector. An $X \times X$ matrix can have up to X different eigenvectors, each with its own eigenvalue, although it doesn't have to have that many. The most important eigenvector is the one that has the biggest eigenvalue, because through time the biggest eigenvalue overshadows the others, so to speak, and so only its eigenvector matters. Using this terminology, the stable age distribution is an eigenvector of the Leslie matrix, the geometric growth factor is the eigenvalue corresponding to that eigenvector, and the geometric growth factor is the biggest of the eigenvalues.

MATLAB will find all the eigenvectors and eigenvalues of a matrix with one command, `eig`. We can use this command to calculate the stable age distribution and the population's R directly from the Leslie matrix, without having to do an iteration of the population's growth through time. Type

```
>> [v,d] = eig(m)
```

and MATLAB responds with

```
v = 0.8629          -0.2776 - 0.2712i  -0.2776 + 0.2712i
    0.4864           0.2480 + 0.6430i   0.2480 - 0.6430i
    0.1371           0.0371 - 0.6108i   0.0371 + 0.6108i

d = 1.1827                  0                  0
         0          -0.3414 + 0.1561i          0
         0                  0          -0.3414 - 0.1561i
```

The eigenvalues are on the main diagonal of the `d` matrix. The biggest one is at the top left. The other two are a conjugate pair of complex numbers, and their absolute values are smaller than the one at the top left (as can be confirmed by typing `abs(d(2,2))` and `abs(d(3,3))`). Therefore, to find the geometric growth factor, "big R,", type

```
>> r = d(1,1)
```

with MATLAB responding,

```
r = 1.1827
```

The eigenvectors are the column vectors of the `v` matrix. The first column of `v`, referred to as `v(:,1)`, corresponds to the eigenvalue at (1,1) in `d` and is the one we want. MATLAB's convention is to present eigenvectors with unit length, whereas we want the vector's components to add up to 1. To renormalize the eigenvector, type

```
>> c = v(:,1)/sum(v(:,1))
```

and MATLAB responds with the stable age distribution

```
c = 0.5806
    0.3272
    0.0922
```

If, for some reason, you wish the geometric growth factor R to be expressed in continuous time, the intrinsic rate of increase, r, is simply the natural logarithm of R, which is 0.1679 in this example.

The stable age distribution and its corresponding geometric growth factor as directly determined from the Leslie matrix, are useful because a population's actual age distribution and geometric growth factor are usually quite close to these. But if you do want to see the bumps and wiggles in the actual age distribution through time while the stable age distribution is being approached, you'll have to carry out an iteration though time. Either approach is now quite easy, with MATLAB, and earlier books in ecology often contain elaborate schemes to do the calculations that now can be accomplished in one line of typing.

If the Leslie matrix itself changes through time, then you'll have to do an iteration to predict the population's growth. Just use the correct Leslie matrix for each time step. That is, use

$$n_{t+1} = m_t n_t$$

where m_t is the Leslie matrix for time t.

Another point worth noting is that a separate Leslie matrix may be defined for males and for females if the sexes differ greatly in their survivorship. When separate matrices are used for each sex, one pretends that males produce male offspring and females produce female offspring, and the total population is obtained by summing the males and females. Alternatively, an average matrix covering both sexes can be used if it's not important to keep track of the sexes separately.

3.2.2 Have Red Deer Run Amok?

The isle of Rhum off eastern Scotland is a preserve with a population of red deer, *Cervus elaphus*. So that vegetation in a degraded habitat could be restored, the deer population was "thinned" with hunting. The question is whether the hunting is sufficient to keep the deer population in check. The deer population is considered to be in check if its geometric growth factor, R, is near one. From field data on fecundity and survivorship in the presence of hunting, a Leslie matrix was produced, which may be entered into MATLAB by our typing

```
>> red = [0   0 .26 .26 .26 .26 .35 .26 .31;
>>        1   0   0   0   0   0   0   0   0;
>>        0 .94   0   0   0   0   0   0   0;
>>        0   0 .80   0   0   0   0   0   0;
>>        0   0   0 .67   0   0   0   0   0;
>>        0   0   0   0 .61   0   0   0   0;
>>        0   0   0   0   0 .63   0   0   0;
>>        0   0   0   0   0   0 .72   0   0;
>>        0   0   0   0   0   0   0 .22   0];
```

To find the geometric growth factor and stable age distribution, we first obtain the eigenvectors and eigenvalues with

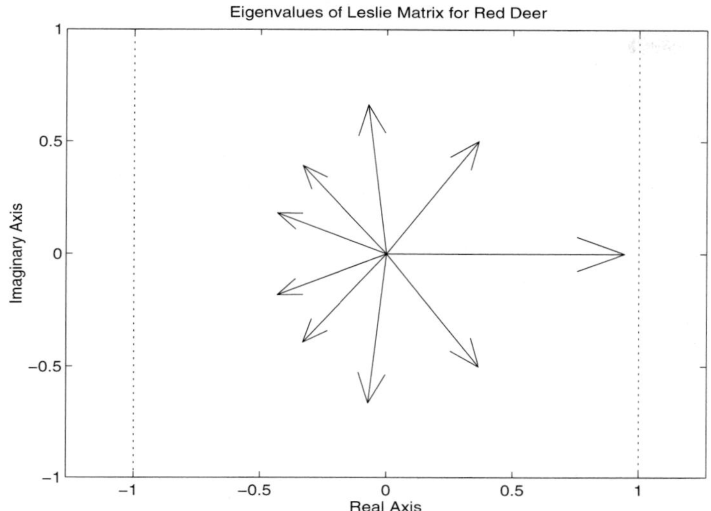

Figure 3.7: Distribution on the complex plane of eigenvalues of the Leslie matrix for red deer on the isle of Rhum in Scotland. The biggest eigenvalue, represented by the longest arrow that points to the right along the real line, is the geometric growth factor, R, for the population once the stable age distribution has been attained. The natural logarithm of this eigenvalue is the intrinsic rate of increase, r, for the population.

```
>> [v,d] = eig(red);
```

To find the geometric growth factor, R, type

```
>> r = d(1,1)
```

which leads to MATLAB's reply

```
r =       0.9439
```

This shows that the deer population is approximately holding constant, although there is a slight tendency to decline because R is a bit less than one. The intrinsic rate of increase, r, is the log of R which is -0.0578.

I should add that one may need to check whether the eigenvalue assigned to `r` is in fact the biggest. To do this check, look at the absolute value of all the other eigenvalues along the main diagonal of `d`. It seems to me that the biggest eigenvalue is always located at position (1,1). Anyway, the set of all eigenvalues can be conveniently graphed on the complex plane with

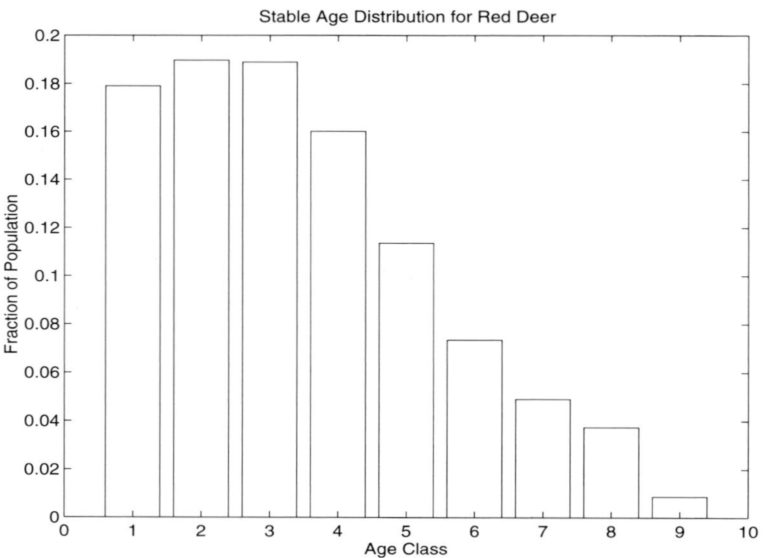

Figure 3.8: Stable age distribution for red deer on the isle of Rhum in Scotland.

```
>> figure
>> compass(d)
```

Inspecting Figure 3.7 shows that the eigenvalue equal to 0.9439 is the biggest[1].

To use this Leslie matrix to find the stable age distribution of red deer, type

```
>> c = v(:,1)/sum(v(:,1))
```

yielding

```
c = 0.1790
    0.1897
    0.1889
    0.1601
    0.1137
    0.0735
    0.0490
```

[1]The biggest eigenvalue is always positive or zero for matrices whose elements are positive or zero. If you don't want to bother graphing the eigenvalues, you can find the biggest by taking the absolute value of all the eigenvalues, and then taking the maximum of these. So instead of taking `r = d(1,1)` above, you can take `r = max(max(abs(d)))`. `max` has to be called twice because it first finds a vector which has the maximum element from each column, and then finds the maximum of these, to give the biggest element of the entire matrix, `d`.

```
    0.0374
    0.0087
```

The stable age distribution is conveniently graphed with

```
>> figure
>> bar(c)
```

as shown in Figure 3.8. The distribution has the rather unusual property of peaking at an age class beyond one. If $R \geq 1$ then the stable age distribution decreases monotonically with age, but because $R < 1$ in this example, the stable age distribution is not monotonically decreasing.

As this example shows, by using MATLAB one can now routinely determine the geometric growth factor and stable age distribution from demographic data, and avoid the tedious schemes and calculations previously used in ecology texts to accomplish this purpose.

3.2.3 Euler's Equation for Continuous Time

The theory for projecting population growth taking into account age structure was originally developed in continuous time by A. J. Lotka. The approach using the Leslie matrix that we have just covered came later, and is usually the best one to use when field data are involved. The original continuous-time version is more useful when explicit mathematical formulas are desired, and it's worthwhile looking into this approach as well. The idea here is to keep a continual census going of the number of newborn, $B(t)$, and to predict how this changes through time.

The total number newborn at any time is the sum of those produced by parents of all possible ages. In symbols,

$$B(t) = \int_0^\infty [\text{Births from parents of age } x]\, dx$$

The fertility data for this formulation is presented in a function, $m(x)$, called the "maternity function." By definition, $m(x)dx$ is the number of offspring produced by a parent of age x as it ages from x to $x + dx$. The $m(x)$ function is often a more-or-less triangle-shaped function for mammals. Therefore, the equation for $B(t)$ now becomes

$$B(t) = \int_0^\infty m(x)\, [\text{Number of parents of age } x]\, dx$$

The survivorship data for this formulation is presented as a function, $l(x)$, called the "survivorship function." By definition, $l(x)$ is the probability that an individual lives to age x or more, and $l(0)$ equals one. The $l(x)$ function is a monotonically declining curve, and often resembles (or actually is) a decreasing exponential function. Using the survivorship function, we can now say that the number of parents of age x is the number that were born x years ago times the probability of their living to age x. So now the formula for $B(t)$ becomes

$$B(t) = \int_0^\infty m(x) l(x) B(t - x) dx$$

This is the fundamental equation of Lotka's formulation for projecting population growth with age structure in continuous time. This is called an "integral equation," which is the counterpart of a differential equation but uses an integral instead of a derivative. Here, a solution is defined as a formula for $B(t)$ that satisfies the above condition on its integral.

The solution is easy to guess, with our previous experience that a population eventually approaches a stable age distribution and geometric growth. The continuous-time counterpart of geometric growth is exponential growth, so let's substitute a trial solution of the form

$$B(t) = B(0)e^{rt}$$

where r is temporarily viewed as an unknown constant. As you may suspect, it will soon emerge that r is the intrinsic rate of increase. To see if this trial solution will satisfy the integral equation, we substitute it in and see what happens:

$$B(0)e^{rt} = \int_0^\infty m(x)l(x)B(0)e^{r(t-x)}\,dx$$

Now observe that $B(0)e^{rt}$ divides out from both sides leaving

$$1 = \int_0^\infty m(x)l(x)e^{-rx}\,dx$$

This equation is called the "Euler equation." It means that the trial solution will work; i.e., it will satisfy the integral equation for $B(t)$, provided the coefficient r itself satisfies the formula above. The way to meet this requirement is then to use the Euler equation to calculate r, because if we do this, $B(t) = B(0)e^{rt}$ is automatically a valid solution.

For example, suppose the survivorship curve is an exponential decay,

$$l(x) = e^{-\mu x}$$

and suppose the maternity function is a more-or-less triangle-shaped curve called a gamma function given by

$$m(x) = a^2 x e^{-bx}$$

Then the Euler equation becomes

$$\begin{aligned} 1 &= \int_0^\infty a^2 x e^{-bx} e^{-\mu x} e^{-rx}\,dx \\ &= a^2 \int_0^\infty x e^{-(b+\mu+r)x}\,dx \end{aligned}$$

To solve this for r we need to do the integral. MATLAB has memorized a great many integrals, so let's find out what the following definite integral is:

$$\int_0^\infty x e^{-px}\,dx$$

To use MATLAB to find out, type

```
>> int('x*exp(-p*x)','x',0,Inf)
```

MATLAB's response is a rather unhelpful

```
ans = limit((-p*x*exp(-p*x)-exp(-p*x))/p^2+1/p^2,x = inf,left)
```

But type pretty and see

```
                 - p x exp(- p x) - exp(- p x)     1
        limit    ----------------------------- + ----
        x -> inf-               2                  2
                               p                  p
```

By inspection, the numerator of the first fraction is 0 in the limit as x approaches $+\infty$ from below, so the answer is $1/p^2$. Therefore the Euler equation now becomes

$$1 = \left(\frac{a}{b+\mu+r}\right)^2$$

which may be solved for r by inspection, or by typing

```
>> solve('1 = (a/(b + mu + r))^2','r')
```

Two roots are returned as answers, and the relevant one is

$$r = a - (b + \mu)$$

So, this is the intrinsic rate of increase for this example—it's an explicit formula that may be preferable to a numerical answer in some circumstances. If you can't do the integral explicitly, then you'll have to do the integral numerically, and you'd probably be better off working with the Leslie matrix.

I think the most valuable function served by the Lotka formulation is to reveal how population growth influences the shape of the stable age distribution. An age distribution may be defined as

$$c(x) = \frac{\text{Number of organisms at age } x}{\text{Total number of organisms}}$$

Recall that the number of individuals of age x is the number born x years ago times the probability of their surviving to age x, $B(t-x)l(x)$. So,

$$c(x) = \frac{B(t-x)l(x)}{\int_0^\infty B(t-x)l(x)dx}$$

Once the stable age distribution has been reached, $B(t-x) = B(0)e^{r(t-x)}$, so substituting yields

$$c(x) = \frac{B(0)e^{r(t-x)}l(x)}{\int_0^\infty B(0)e^{r(t-x)}l(x)dx}$$

The $B(0)e^{rt}$ divides out, leaving

$$c(x) = \frac{e^{-rx}l(x)}{\int_0^\infty e^{-rx}l(x)dx}$$

This is the formula for the stable age distribution, and it assumes that r has already been determined from the Euler equation. It reveals that the shape of $c(x)$ is closely related to the shape of the survivorship curve, $l(x)$. In fact, if $r = 0$, as is true in a population whose size is not changing, the shape of the stable age distribution coincides with that of the survivorship curve. If the population is growing ($r > 0$), then the age distribution is skewed to the left relative to the survivorship curve, and if the population is declining ($r < 0$), then $c(x)$ is skewed to the right relative to $l(x)$. Skewing to the left means more teenie-boppers, skewing to the right means more wise and wizened adults.

Finally, you may be wondering how to compare the fertility and survivorship data between the Leslie and Lotka formulations. An F_x of the top row of the Leslie matrix is approximately the integral of $m(x)$ from x to $x+1$, and $l(x)$ is approximately equal to the product of the P_x's from 1 up to age $x-1$, because the probability of living to age x is the product of living through the first age step times the probability of living through the second age step, and so forth on up to age class $x-1$, at the end of which the organism will have attained age x. Thus, there's nothing fundamentally different in the data that go into the Lotka and Leslie formulations; the same data are simply packaged differently.

3.3 Living Through Good Times and Bad

All this talk about exploding population bombs might lead you to think that we're about to be overcome by an avalanche of organisms any day now. Alas, this is far from the truth, because many natural populations are increasingly endangered as a result of losing their habitat to human development. But even without humans messing around, many populations are living at the edge anyway. The reason is that the best of times never sticks around for long. In the best of times everything from earthworms to elephants can have a population explosion, and most people have seen this happen where they live, say with conspicuous caterpillars, aphids, or crickets, or even with rather large mammals such as rabbits or deer. But a good moist year for earthworms might be followed by a dry year, so that what's really happening in nature is that the geometric growth factor, R, varies from year to year more or less at random because of annual variation in weather and climate. As we'll see, the net effect of fluctuations between good and bad times greatly diminishes the chance for a population explosion, and moreover, makes populations susceptible to natural extinction.

Still another source of variability in population size creeps in when the population size is low. At that time, a population may fluctuate simply because there are so few organisms that chance variation in whether and when they reproduce noticeably affects the total population size. Variation in population size caused by variation through time in the geometric growth factor, R, is called "environmental stochasticity." Variation in population size caused by random variation among individuals in their reproductive activity, even though the environment is unchanging, is called "demographic stochasticity." We'll now look into both of these sources of variability in a population's size.

3.3.1 Environmental Stochasticity

The first calculation every school child learns is how to compute an average. This is how the teacher computes grades, and everyone wants to know how the teacher comes up with their grade. But I'll bet you didn't know there are several different formulas to compute an average[2]. What you learned as a child is called the "arithmetic average:"

$$y = \frac{x_1 + x_2}{2}$$

A new kind of average that's very important in ecology is called the "geometric average:"

$$y = \sqrt{x_1 x_2}$$

In words, the geometric average of two numbers is the square root of their product. Unless x_1 and x_2 happen to be the same, the geometric average is less than the arithmetic average, and only a mean teacher would assign grades based on the geometric average. But nature does use the geometric average when grading a population's success. And here's why.

Consider two successive years, with R_1 being the geometric growth factor for the first year, and R_2 the geometric growth factor in the second year. Then the population size after the first year, when it begins at size N_0, is

$$N_1 = R_1 N_0$$

and the population size after the second year is

$$\begin{aligned} N_2 &= R_2 N_1 \\ &= R_2 R_1 N_0 \end{aligned}$$

Thus, as you already know, the R's multiply together when an iteration of geometric growth through time is carried out. Now consider a separate hypothetical population whose R is the geometric average

$$R = \sqrt{R_2 R_1}$$

If this hypothetical population goes through two years with this R, then it ends up at the same point as the original population that has different R's in each of the two years because $R^2 = R_2 R_1$. Thus, the constant-environment equivalent of the fluctuating environment is an environment with a constant R that is the geometric average of the R's in the fluctuating environment.

To illustrate the importance of the geometric average relative to the arithmetic average, let's do a simulation comparing two populations in fluctuating environments. One population sees one of two R's, depending on the flip of a coin, and the geometric average of these R's is one. The other also sees one of two R's depending on the flip of a coin, but the arithmetic average of the R's is one. Let's see which population is more successful.

[2] So that's what the teacher was doing.

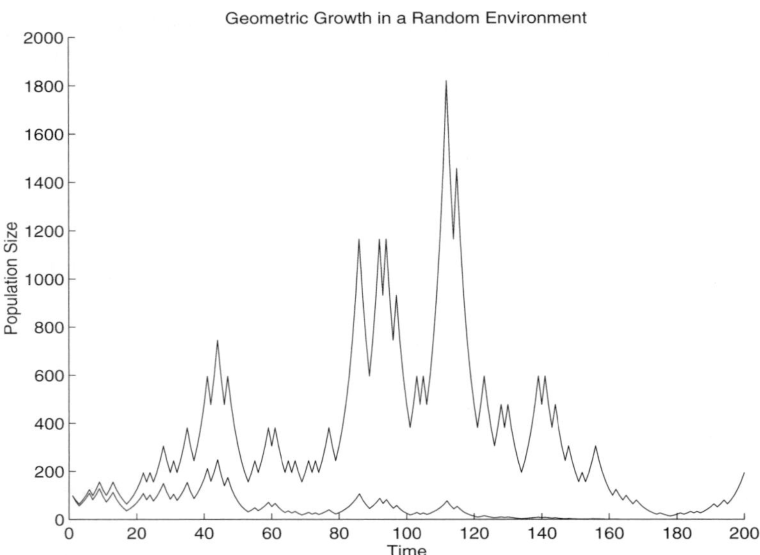

Figure 3.9: Geometric growth in a random environment. The population in the top curve has a geometric average R of one, and the population in the bottom curve has an arithmetic average R of one.

To do the simulation, we'll use the short program below. Remember to enter it with a text editor and save it as `pop1_rnd.m` in your directory. Then it can be invoked from the command line in MATLAB. The program's first line declares that it is called with the R for a good year, `rgood`; the R for a bad year, `rbad`; the initial population size, `no`; and the number of years to carry out the iteration, `runlen`. The first line also indicates that the function will return `n`, which will be a vector of successive population sizes.

```
function n = pop1_rnd(rgood,rbad,no,runlen)
```

The next line initializes the random number generator

```
rand('seed',0);
```

Next, the initial population size is placed as the first entry in the vector that will hold the successive population sizes:

```
n = [no];
```

Now the main loop begins, with time running from 1 to `runlen`

```
for t=1:runlen
```

The coin is flipped. If the random number is less than 1/2 then it's a bad year (tails) and the R to be used will be `rbad`

```
if (rand < 1/2)
   r = rbad;
```

otherwise it's a good year (heads), and the R is assigned to `rgood`

```
else
   r = rgood;
end;
```

Now the year passes. The vector `n` gets a new value appended to it, which is RN_t. This is the next value of the population size based on the R that has been selected through the coin toss.

```
n = [n (r * n(t))];
```

The loop and program concludes with

```
end;
```

To use the program, let's first investigate a population whose geometric average of the R's is one. The parameters to be used in all runs are

```
>> no = 100;
>> runlen = 199;
```

To invoke the program, type

```
>> n = pop1_rnd(5/4,4/5,no,runlen);
```

Notice that `rgood` times `rbad` is $(5/4) \times (4/5)$, which is 1. The vector `n` now holds 200 values, consisting of the initial condition together with 199 others produced during the iteration. For Figure 3.9, the population size through time is then graphed in yellow with

```
>> figure
>> hold on
>> plot(n,'y')
```

Next, let's investigate the population whose arithmetic average of the R's is one. The program is invoked with

```
>> n = pop1_rnd(5/4,3/4,no,runlen);
```

and the graph of this population's size through time is added in red to Figure 3.9, with

```
>> plot(n,'r')
```

A glance at Figure 3.9 reveals that the population whose geometric average R is one is greatly out-performing the population whose arithmetic average R is one. Indeed, the population whose arithmetic R is one is on the way to extinction because its geometric average R is less than one.

3.3.2 Spreading the Risk in a Metapopulation

Can a species escape the curse of the geometric average? Yes, by dispersal. If the entire species is spread out over a large area, then the R will vary from place to place throughout the species' range. Immigration and emigration among the various places in the species range effectively makes the entire population's R equal to an arithmetic average of the R in various local places. If the distribution of R at a single spot through time matches the distribution of R's throughout the species' range at any single point in time, then the species as a whole can enjoy the arithmetic average of the R's instead of suffering the geometric average.

A species whose members live in more-or-less isolated subpopulations is called a "metapopulation," which means a population of populations. The subpopulations exchange migrants through dispersal. When the lowered overall risk of extinction results from dispersal, the process has been called "spreading the risk."

There are actually two variants of the hypothesis that, by spreading the risk among subpopulations, dispersal buffers a species from extinction. The first version envisions that the environments of the various subpopulations are independent of one another. Imagine that there's a 50% chance of rain at each subpopulation, and that at any one time 50% of the spots are wet and the other 50% dry. If the spots are far enough away, and the rain clouds small, then the spots will be dry or wet by chance independently of one another. The second version can be combined with the first. It envisions that if one spot is good, another spot is automatically bad. Imagine hills that run east to west. In a hot year the north-facing slopes may be excellent for organisms while the south-facing slope is too dry. Conversely, in a cool year, the north-facing slopes may not warm up enough for organisms to live, but the south-facing slopes are perfect. Thus different exposures imply an inherent negative correlation between the suitability of different spots.

To see how spreading the risk works, let's illustrate with a simulation. The program to do the simulation is modified from `pop1_rnd.m` presented above and is called `pop2_rnd.m` to indicate that two populations are being iterated. The model for two subpopulations is

$$\begin{aligned} N_{1,t+1} &= R_{1,t}\left((1-m)N_{1,t} + mN_{2,t}\right) \\ N_{2,t+1} &= R_{2,t}\left(mN_{1,t} + (1-m)N_{2,t}\right) \end{aligned}$$

Here m is the probability that an organism from subpopulation-1 disperses to subpopulation-2, and vice versa—it is the probability that an organism migrates. Therefore, $(1-m)$ is the probability that an organism stays at home and doesn't migrate to the other subpopulation. $N_{x,t}$ is the number of individuals at time t in the subpopulation at location x, where x is 1 or 2. $R_{x,t}$ is the geometric growth factor in location x at time t. If m is zero, the equations describe two separate uncoupled populations, and is m is 1/2 the two populations are completely mixed and are in effect one population. Notice in the first of the equations that N at time $t+1$ at location-1 is the product of the geometric growth factor there times the population after migration has taken place. The population there after migration consists of the remaining natives, $(1-m)N_{1,t}$, plus the migrants, $mN_{2,t}$. And similarly for the other population.

The program, pop2_rnd.m starts with the line

```
function [n1,n2] = pop2_rnd(m,rgood,rbad,no,runlen,independ)
```

which indicates that it is called with the probability of migration, the geometric growth factors for good and bad times, the initial population size (which is the same for both subpopulations), and a variable that is set to 1 if the two populations have their R's chosen by independent flips of the coin, and is set to 0 if the two populations vary in opposite directions. The function returns the sequence of population sizes in each of the subpopulations. Within the function, the random number generator is initialized, the history vectors are initialized with the initial population-sizes in each subpopulation, and then the main loop is started:

```
rand('seed',0);
n1 = [no]; n2 = [no];
for t=1:runlen
```

Next, one coin is flipped to determine if subpopulation-1 is in for good times or bad.

```
  if (rand < 1/2)
     r1 = rbad;
  else
     r1 = rgood;
  end;
```

If both populations are independent, another coin is flipped to determine subpopulation-2's fate; otherwise, subpopulation-2 is assigned the opposite of subpopulation-1's fate:

```
  if independ == 1
     if (rand < 1/2)
        r2 = rbad;
     else
        r2 = rgood;
     end;
  else
     if r1 == rgood
        r2 = rbad;
     else
        r2 = rgood;
     end;
  end;
```

Now the two subpopulations are iterated. The new values for the population sizes are appended to the history vectors for each subpopulation, and then the loop terminates, and the program concludes

```
  n1 = [n1 (r1*((1-m)*n1(t) + m*n2(t)))];
  n2 = [n2 (r2*(m*n1(t) + (1-m)*n2(t)))];
end;
```

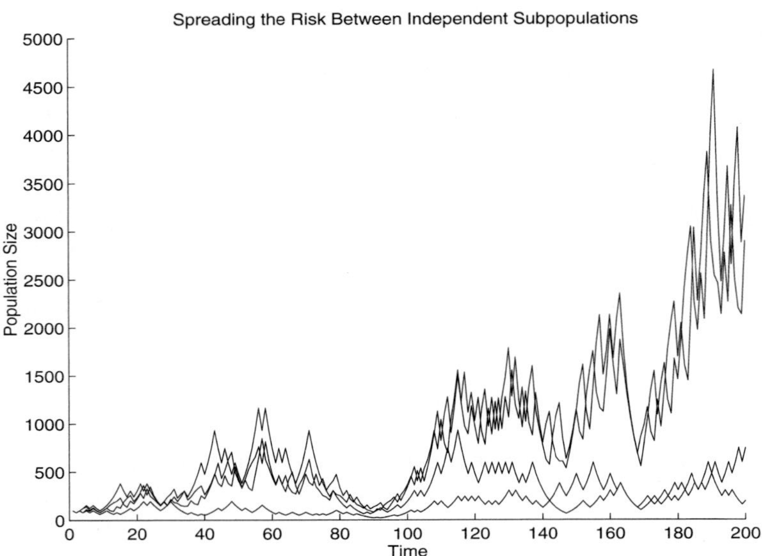

Figure 3.10: Spreading the risk between two independent subpopulations. Probability of migration between populations is 0.25 in top curves, and 0 in lower curves.

Once this program is entered with a text editor and saved in your directory as `pop2_rnd.m`, it is available for use from MATLAB's command line.

All the simulations will use an initial size in the subpopulations of 100, 199 iterations, and R's whose geometric average R is one, as assigned by

```
>> no = 100;
>> runlen = 199;
>> rgood = 5/4;
>> rbad = 4/5;
```

Now, to open the figure window that will become Figure 3.10, type

```
>> figure;
>> hold on
```

and invoke the program to generate a control run, whereby the two subpopulations are independent and uncoupled. The result is returned in the `n1` and `n2` vectors:

```
>> [n1 n2] = pop2_rnd(0,rgood,rbad,no,runlen,1);
```

The history of population sizes is graphed in red with

```
>> plot(n1,'r');
>> plot(n2,'r');
```

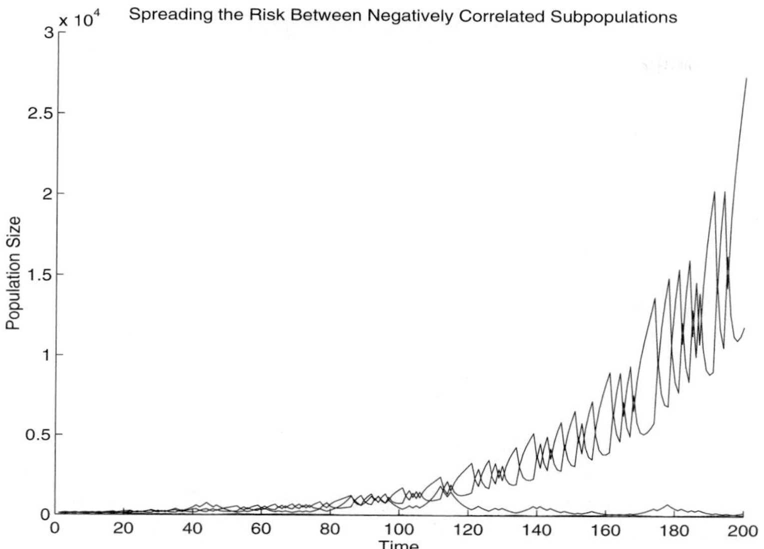

Figure 3.11: Spreading the risk between two negatively correlated subpopulations. The probability of migration between populations is 0.25 in the top curves, and 0 in the lower curves.

Coming from a control run, the results offer no surprises, and present two examples resembling what we saw in Figure 3.9 for a geometric average R of one. Now, let's couple these two populations with a migration probability of 0.25. We type

```
>> [n1 n2] = pop2_rnd(.25,rgood,rbad,no,runlen,1);
```

and add them to Figure 3.10 in yellow with

```
>> plot(n1,'y');
>> plot(n2,'y');
```

The results clearly show that both coupled subpopulations perform much better than the uncoupled controls. So spreading the risk by dispersal between subpopulations that have independent environmental variation leads to an R that is much improved over the geometric average R that the subpopulations would face in the absence of dispersal.

Now let's look at spreading the risk between subpopulations whose environmental variation is negatively correlated. Initialize what will become Figure 3.11 with

```
>> figure;
>> hold on
```

The control run, representing uncoupled subpopulations with negative environmental variation, is produced with

```
>> [n1 n2] = pop2_rnd(0,rgood,rbad,no,runlen,0);
```

and graphed in red with

```
>> plot(n1,'r');
>> plot(n2,'r');
```

Now, to make the run with the populations coupled by a migration probability of 0.25, type

```
>> [n1 n2] = pop2_rnd(.25,rgood,rbad,no,runlen,0);
```

as graphed in yellow with

```
>> plot(n1,'y');
>> plot(n2,'y');
```

Figure 3.11 shows that spreading the risk by dispersal between subpopulations whose environmental variation is negatively correlated is even better than spreading the risk between subpopulations having independent environmental variation. The reason is that with negative correlation the metapopulation is guaranteed one spot at which conditions are favorable, whereas when there is independence both spots are occasionally bad.

The conservation implications of spreading the risk are obvious. If a species is allowed only one place to live, it will labor under the yoke of the geometric average, whereas if the species is allowed two or more habitats connected with a dispersal route, then the species will enjoy a greatly enhanced chance of surviving.

3.3.3 Demographic Stochasticity

If a population drops to a very low size, say under 100, then its survival becomes precarious, because the random variation among individual circumstances may be expressed in the total population size. For example, suppose each organism produces one offspring on the average, so that the population would be self-sustaining if everyone always did what the average says. But suppose the chance is 1/4 that an individual doesn't reproduce, 1/2 that it has one offspring, and 1/4 that it has two offspring. On the average therefore, an individual has one offspring, and the population size should persist unchanged. But once the population size is small enough, the luck of the draw may intervene. Let's illustrate this simple idea with a simulation. The program `dem_rand.m` goes through all the individuals in a population, and tosses a coin for each one to see if it will have 0, 1 or 2 offspring. It adds up the total number of offspring produced and then starts over again for the next generation. The first line of the program indicates that it is called with the probabilities of having 0, or having 1 offspring, `p0` and `p1` respectively. The probability of having 2 offspring can then be calculated as `1 - (p0 + p1)`. Other calling parameters are the initial population size and the number of iterations. The function returns a vector `n` that is a history of population sizes. Here is the program.

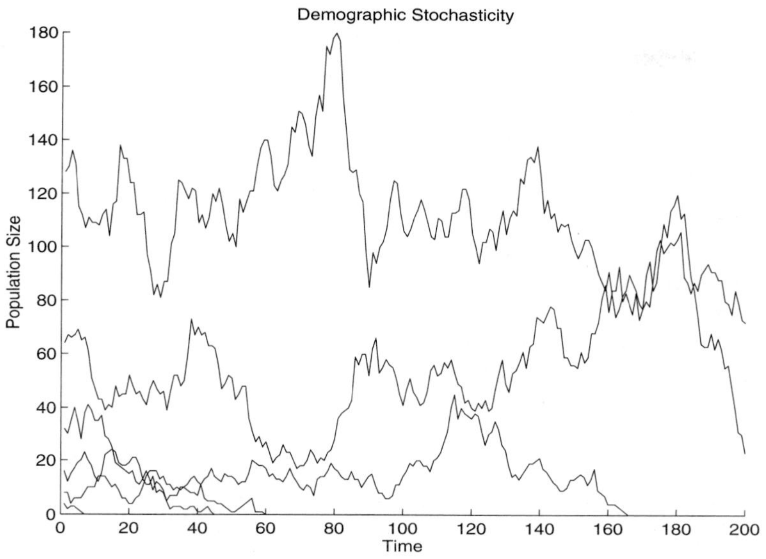

Figure 3.12: Demographic stochasticity illustrated for initial population sizes of 4, 8, 16, 32, 64 and 128. The probability of an individual having 0, 1 or 2 offspring, respectively, is 1/4, 1/2, and 1/4.

```
function n = dem_rand(p0,p1,no,runlen)
rand('seed',0); n = [no];
for t = 1:runlen
  new_n = 0;
  for i=1:n(t)
    luck = rand;
    if luck < p0
      new_n = new_n;
    elseif luck < p0 + p1
      new_n = new_n + 1;
    else
      new_n = new_n + 2;
    end
  end
  n = [n new_n];
end
```

The simulation parameters to be used in all the runs are assigned as

```
>> p0 = .25;
```

```
>> p1 = .5;
>> runlen = 199;
```

Then Figure 3.12, which illustrates the effects of demographic stochasticity for initial population sizes of 4, 8, 16, 32, 64, and 128, is prepared by

```
>> figure
>> hold on
>> plot(dem_rand(p0,p1,4,runlen),'r');
>> plot(dem_rand(p0,p1,8,runlen),'g');
>> plot(dem_rand(p0,p1,16,runlen),'b');
>> plot(dem_rand(p0,p1,32,runlen),'c');
>> plot(dem_rand(p0,p1,64,runlen),'m');
>> plot(dem_rand(p0,p1,128,runlen),'y');
```

Clearly a population under 10 is in deep trouble, whereas a population over 100 may persist for some time. You may try this out with different values. The anticipated time to extinction depends on how close everyone is to the average. If the probabilities of 0, 1 and 2 offspring are 1/8, 3/4, and 1/8 respectively, then the situation is not as dangerous for the population as it is in the illustration based on 1/4, 1/2, and 1/4. Demographic stochasticity also makes it difficult to estimate the basic birth and death rates in a tiny population, because the rates can be easily obscured by the stochasticity.

3.4 The Invading Wave

Science-fiction movies have featured invasions of aliens spreading over the world, and the introduction of an alien species from this world into a foreign habitat can have real impact almost as much as that fantasized in the movies. The spread of rabbits in Australia, of muskrats in the Balkan peninsula, of the Mediterranean fruit fly and bees, not to mention the flu and other diseases, raises legitimate alarm, and motivates the desire to predict how quickly an invading species spreads into what might be thought of as empty habitat.

The invasion of a species can be simulated with MATLAB. We want to know if an invasion speeds up after it gets started, and how fast it moves. We'll be looking at the spread of an invasion through homogeneous habitat, such as a large forest or grassland. Heterogeneous habitat can be broken into homogeneous smaller pieces and the progress of the invasion through the heterogeneous whole could then be spliced together from the invasion expected in each of the pieces.

We will again write a small program, here called `inwave.m`, to simulate the invasion. The habitat is viewed as a long strip, such as a long valley, or river bank. The invasion is seeded at one end with `no` individuals, and they can potentially spread across the entire habitat of length `hablen`. We will run the simulation for `runlen` time steps. A piece of habitat is considered to be invaded if the number of organisms there exceeds the parameter `estab`. The probability that an organism disperses to the next site is `m`, and the geometric growth factor everywhere is `r`.

The small program is called `inwave.m` and here it is. The first line of the program indicates how it is called and that it returns a vector called `edge`, which is the history of the invasion's edge.

```
function edge = inwave(hablen,runlen,estab,no,m,r)
```

The habitat is initialized, and the loop started with

```
hab = zeros(1,hablen); new_hab = hab;
hab(1) = no; edge = [1];
for t = 1:runlen
```

Now for the actions. First, there is geometric growth in each site:

```
  for h = 1:hablen
    hab(h) = r*hab(h);
  end
```

Next, the migration occurs. We have to move everyone into the vector `new_hab`, so that the contents of the vector `hab` are not destroyed before the calculation is completed. At each site in the interior of the habitat, `(1-m)` of the individuals stay at home, and `m` of them migrate. The migration is both to the right and to the left. The edges are treated separately. All this is coded as

```
  new_hab(1) = ((1-m) + m/2)*hab(1) + (m/2)*hab(2);
  for h=2:(hablen-1)
    new_hab(h) =  (m/2)*hab(h-1) + (1-m)*hab(h) + (m/2)*hab(h+1);
  end
  new_hab(hablen) = ((1-m) + m/2)*hab(hablen) + (m/2)*hab(hablen-1);
```

Now that the migration phase is completed, the vector `hab` can be reassigned to the newly calculated vector `new_hab`.

```
  hab = new_hab;
```

To find out how far the invasion has now spread, we use the `find` command to give us the indices of the sites where the abundance of invaders is still less than the establishment threshold. We then use the `min` command to find the closest such site, which is then the new edge. The new edge is appended to the `edge` vector, which records the progress of the invader's edge.

```
  edge = [edge min(find(hab<estab))];
```

The loop and program then conclude with

```
end
```

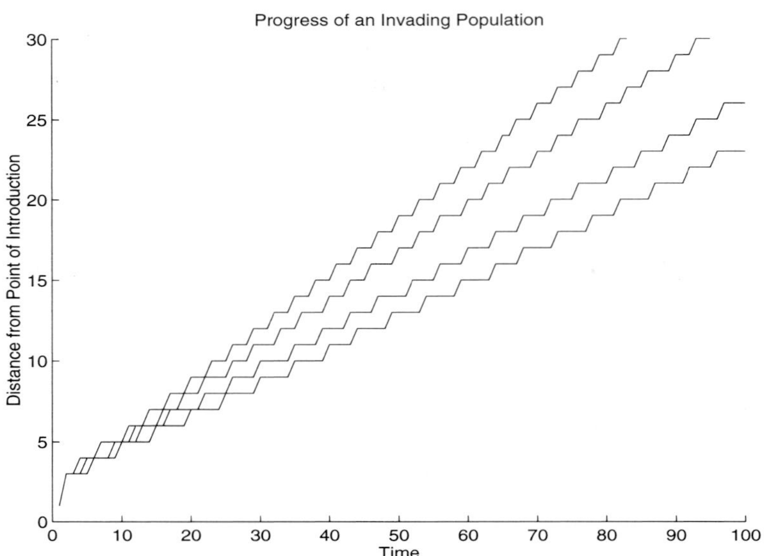

Figure 3.13: Progress of an invasion, measured by how far the invaders have spread from the place of introduction. The edge of the population, defined as the place where it exceeds 10 individuals, is plotted on the vertical axis, and time is plotted on the horizontal axis. The distance of the edge from the place of introduction increases linearly with time; this linearity indicates that the invasion spreads with a constant speed through homogeneous habitat. The speed depends on the average dispersal distance of the organisms and on the invading population's geometric growth factor. Curves from top to bottom are $m = 0.33$, $R = 1.2$; $m = 0.25$, $R = 1.2$; $m = 0.33$, $R = 1.1$; and $m = 0.25$, $R = 1.1$.

To illustrate how an invasion spreads, this program is invoked four times, with the results graphed in Figure 3.13. The parameters in common to all the runs are a habitat length of 30 sites, a run length of 99 time steps, an establishment threshold of 10 organisms, and an initial introduction of 100 individuals that are put in site-1 at the beginning of the run.

```
>> hablen = 30; runlen = 99; estab = 10; no = 100;
```

The figure is produced with

```
>> figure; hold on;
>> plot(inwave(hablen,runlen,estab,no,.25,1.1),'c')
>> plot(inwave(hablen,runlen,estab,no,.25,1.2),'g')
>> plot(inwave(hablen,runlen,estab,no,.33,1.1),'r')
>> plot(inwave(hablen,runlen,estab,no,.33,1.2),'y')
```

The simulations show the edge of the invader's range spreading linearly with time—this means the invasion moves at a constant speed, once the first few time steps have passed[3]. Thus the invasion does not speed up as it progresses, nor does it run out of steam and stop of its own accord. Instead, its progress is regular and steady. The speed of the spread depends on both the probability of migration and on the population growth rate. Figure 3.13 shows that increasing both `m` and `r` raises the speed of the invasion's spread.

But if you're worried that the invasion will never stop, then just keep reading. The next chapter is all about limits to growth.

3.5 Further Readings

Caswell, H. (1989). *Matrix Population Models.* Sinauer Associates.

Cohen, J. E. (1979). Long-run growth rates of discrete multiplicative processes in Markovian environments. *J. Math. Analysis Applic.*, 69:243–251.

den Boer, P. J. (1981). On the survival of populations in a heterogeneous and variable environment. *Oecologia*, 50:39–53.

Hanski, I. and Gilpin, M. (1991). Metapopulation dynamics: brief history and conceptual domain. *Biol. J. Linn. Soc.*, 42:3–16.

Harper, J. and White, J. (1974). The demography of plants. *Annual Review of Ecology and Systematics*, 5:419–463.

Keyfitz, N. and Flieger, W. (1971). *Population: Facts and Methods of Demography.* Freeman.

Leslie, P. H. (1945). On the use of matrices in certain population mathematics. *Biometrica*, 33:183–212.

Lotka, A. J. (1925). *Elements of Physical Biology.* Williams and Wilkins (Reprinted 1956, Elements of Mathematical Biology, Dover).

Lowe, V. P. W. (1969). Population dynamics of the red deer (*Cervus elaphus* L.) on Rhum. *J. Animal Ecology*, 38:425–457.

Roughgarden, J., Iwasa, Y., and Baxter, C. (1985). Demographic theory for an open marine population with space-limited recruitment. *Ecology*, 66:54–67.

Skellam, J. G. (1951). Random dispersal in theoretical populations. *Biometrica*, 38:196–218.

Tuljapurkar, S. (1990). *Population Dynamics in Variable Environments.* Springer Verlag (Lecture Notes in Biomathematics 85).

[3]The sawtooth appearance of the curves is an artifact of using an integer threshold and an integer number of sites.

3.6 Application: Eigenvalues Save Endangered Turtles

Loggerhead sea turtles are spectacular marine predators that reach lengths of greater than a meter and can weigh well over a hundred pounds[4]. They clamber up onto sandy beaches in the southeastern United States to bury their eggs in the sand. When the eggs hatch, they rush to the water and eventually swim far out into the ocean before returning to lay eggs on the same beaches some twenty years later.

Sadly, these marine turtles have been dwindling in number and are now listed as an endangered species. One of the first approaches to protecting these animals was massive closure of the beaches where they lay their eggs, thereby protecting vulnerable newborn turtles from the disturbance and hazards associated with humans.

In the mid-1980's, a graduate student named Deborah Crouse decided to look at the conservation of this species rigorously—using the best available estimates of growth, reproduction, and survival. She pieced together the following annual projection matrix, which depicts how a stage-structured turtle population changes from one year to the next (written in MATLAB format):

```
>> A = [0     0     0     4.665 61.896;
        0.675 0.703 0     0     0     ;
        0     0.047 0.657 0     0     ;
        0     0     0.019 0.682 0     ;
        0     0     0     0.061 0.8091]
```

This matrix is nearly the same as a Leslie matrix, except that here the turtles are classified by their size rather than by their age—this is called a "stage-structured" population as compared with an "age-structured" population. Here the elements of the matrix indicate that 67.5% of the turtles survive and grow large enough during one year to be promoted from the hatchling size class into the first juvenile size class, and the remainder die. Next, 70.3% of the turtles from the first juvenile size class remain in that size class after one year, while 4.7% grow large enough to join the second juvenile size class. The final two size classes are the mature classes. Each year, 1.9% of the turtles from the second juvenile class enter the first mature size class. Each of these produces 4.665 turtle hatchlings per year, on the average. 6.1% of these small mature turtles graduate to the largest mature size class while 68.2% remain in this class each year. Each large mature turtle produces 61.896 hatchlings per year, on the average. 80.91% of the large mature turtles survive the year and remain in this size class.

Just as in an age-classified population, the asymptotic growth rate of the population is the biggest eigenvalue (called the "dominant eigenvalue" by mathematicians), and the stable size distribution, which is the counterpart of the stable age distribution, is found from the eigenvector corresponding to the biggest eigenvalue. As you learned in section 3.2.1, with MATLAB it is a simple matter to obtain the eigenvalues and eigenvectors for this matrix, and in turn to find out the asymptotic rate of change for the population, given by the dominant eigenvalue:

[4]This section is contributed by Peter Kareiva.

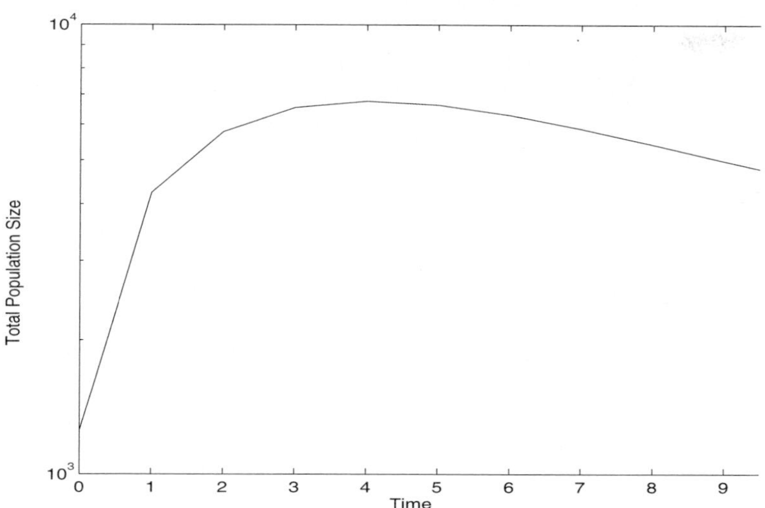

Figure 3.14: Projected decline of the turtle population based on the annual projection matrix developed by Crouse et al. (1987).

```
>> [v,d] = eig(A)
>> R = max(max(abs(d)))
```

The biggest eigenvalue is less than 1.0, indicating eventual and inexorable decline of the population towards zero. To see this, we can provide a starting population of turtles for each of the five stage-classes, and simply iteratively multiply the population by the projection matrix as follows:

```
>> n0 = [300;800;50;50;50]
>> n1 = A*n0; n2 = A*n1; n3 = A*n2; n4 = A*n3; n5 = A*n4; n6 = A*n5;
>> n7 = A*n6; n8 = A*n7; n9 = A*n8; n10=A*n9;
>> s1 = sum(n1); s2 = sum(n2); s3 = sum(n3); s4 = sum(n4); s5 = sum(n5);
>> s6 = sum(n6); s7 = sum(n7); s8 = sum(n8); s9 = sum(n9); s10 = sum(n10);
```

When we plot the summed population over all stage-classes, we see that even though there is an initial increase for this population, a linear (on a semilog plot) decline ensues within ten years:

```
>> t = 0:1:10; s = [s0 s1 s2 s3 s4 s5 s6 s7 s8 s9 s10];
>> figure; semilogy(t,s)
```

When the figure is viewed, it is clear that the above demographic rates indicate relentless decline of the turtle population. This is confirmed by the observation that, despite strict beach regulations that have been protecting newly born turtles for decades, loggerheads have still been declining.

Because loggerheads had been declining despite major efforts to protect their eggs and hatchlings, Crouse decided to explore permutations of the above projection matrix. Such an exploration could test the relative merits of different conservation strategies. First she increased juvenile survivorship in the matrix to 100% (an impossible best-case scenario), and recalculated the dominant eigenvalue:

```
>> A(2,1) = 1.0
>> R =max(max(abs(eig(A))))
```

The eigenvalue is still less than 1, meaning that if our beach protection guaranteed the survival of every single hatchling turtle, the species would still be declining towards extinction. This led Crouse to conduct several other "matrix experiments," to identify those aspects of turtle demography (entries in the matrix A) that should be targeted by conservation biologists if they wanted to have the greatest chance of turning around the species' decline.

To do similar experiments, insert changes in the matrix `A` and recalculate eigenvalues. Indeed, alterations of matrix entries corresponding to particular management efforts are remarkably practical applications of matrix theory. When Crouse attempted a wide variety of such matrix experiments, she learned that modest decreases in the mortality rates suffered by the two oldest and largest stages turtles (stages 4 and 5) could reverse the decline. Find out for yourself by how much annual mortality during stages 4 and 5 must be reduced in order to obtain a growing turtle population.

This modeling turned into some very useful recommendations for turtle protection, and ultimately changed federal regulations. In particular, analyses of the carcasses of dead turtles washing up on beaches showed that they were mostly individuals of stage 4 and 5, and the deaths were generally a result of drowning. It turns out that these large turtles were often getting caught in the nets of shrimp trawlers and were subsequently drowned. Analyses of the simple matrix model suggested that, if this drowning mortality could be reduced, the species could be saved.

Shrimp trawlers are now required to use what are called TED's (turtle exclusion devices) to reduce inadvertent turtle mortality. Because the species is so long-lived, with a 20-year wait before reproduction begins, we still do not know whether this management policy has worked as well we would expect from our calculations of dominant eigenvalues for permutations of matrix `A`. If turtles do rebound, then this will be a compelling example of how a graduate student armed with some simple analytical tools can make a major contribution to conservation.

The key reference for this application is

Crouse, D., Crowder, L. and H. Caswell, 1987. A stage-based model for loggerhead sea turtles and implications for conservation. *Ecology* 68: 1412–1423.

Chapter 4

Got'ta Stop Somewhere

Will *E. coli* cover the earth to a depth of 1 m because of geometric growth with a doubling time of 20 min? I hope you haven't lost sleep over this. Populations consume resources, and growing populations have growing needs for food and space. The gradual loss of resources to a population causes the average birth-rate per individual to decline and the average death-rate per individual to increase, both of which slow population growth. The population may then level off at a steady-state population size. If the average birth-rate per individual and/or the average death-rate per individual are functions of the population's size, the population is said to have "density-dependent population growth."

This chapter is about how the loss of resources affects population growth. The natural fluctuation of good and bad circumstances irrespective of the amount of resources available to a population may itself prevent a population bomb from exploding, as we saw in the last chapter. So resource limitation is not *necessarily* involved when a population bomb fizzles, and we will investigate how to tell if a population has density dependence in it or not. If a population's growth does have density dependence, then we can use a model for population growth that includes it, and see what happens.

Now why would you want to know what this chapter is about? Primarily, I think, because this chapter tells us how resources limit population growth. You may want to know, for example, how many bald eagles, grizzly bears, or parrots a particular habitat will contain even if conservation practices are perfect. You may want to know how many deer, or rabbits, or sheep, or cattle a forest or grassland can support. If you're harvesting fish or crabs or lobsters from the sea, you may want to know how many you could ever catch, as well as how many you should catch. This chapter is about a finite world, whereas the last chapter was about an infinite world. If you live in the real world, this chapter is relevant to you.

4.1 Measuring Abundance

Suppose you see some birds and butterflies while you're taking a walk. If you tell a friend that you saw many animals on the walk, you might be telling a half-truth. You may have seen one bird many times or many birds one time, and similarly for butterflies. This problem also interferes with taking a census of animals. Although the population size of sessile animals or plants can be measured simply by counting, more care is needed for mobile animals.

Most techniques for censusing the population size of animals involve some variant of what is called a "mark/resighting" or "mark/recapture" approach. The simplest version of this approach uses a formula known as the "Lincoln index." Suppose the unknown is N, the total population size, and that you make two visits to the population's site. On your first visit to the population, you observe and tag V_1 animals. The tag does not have to be an individual tag, although it can be if you prefer—what the tag must indicate is that an organism was observed and counted in visit-1. Thus, the tags for visit-1 can all be simply a color mark with the same color, say red for visit-1. On the second visit, the animals are resighted and remarked. The second day's mark may be simply a different color, say blue for visit-2. Therefore, at the end of the second visit, some animals will have been seen twice, and will now carry two color marks; let's call the number of doubly-marked animals V_2. Also, at the end of the second visit, some of the animals will have been seen only then, and the number of animals carrying only the second day's color we'll call V_3. The population size is then estimated by use of the following argument: the ratio of those seen on the first visit to the total number is the same as the ratio of the number of double sightings to the total seen on the second visit. In symbols

$$\frac{V_1}{N} = \frac{V_2}{V_2 + V_3}$$

To use MATLAB to solve this for the unknown, N, type

```
>> nl = simplify(solve('v1/n = v2/(v2 + v3)','n'));
```

which returns

```
nl = v1*(v2+v3)/v2
```

This formula is called the Lincoln index, and `nl` stands for the Lincoln estimate of the population size. Because this is an estimate, there is also a "standard error" associated with the estimate. The standard error means that 95% of the time the true population size is within 2× the standard error above or below the estimate. For example, if the estimate is 100 animals, and the standard error is 10, then 95% of the time the true population size is between 80 and 120 animals. (The "95%" assumes the error is approximately distributed as a bell-shaped curve, called the normal distribution, around the estimate. The standard error is a measure of the width of this bell-shaped curve.) We won't derive it here, but the formula for the standard error of the Lincoln index (from Bishop, Holland & Fienberg 1975, p. 233), is

```
>> nl_se = 'sqrt(nl*(v1-v2)*v3)/v2'
```

Here's an example. Suppose 40 butterflies are marked on the first visit, and on the second visit 25 are remarked and 25 new butterflies are marked. Then these data are entered into MATLAB as

```
>> v1 = 40;
>> v2 = 25;
>> v3 = 25;
```

The Lincoln index is found from

```
>> nl = eval(nl)
nl = 80
```

and the standard error of this estimate is

```
>> nl_se = eval(nl_se)
nl_se = 6.9282
```

So, there are estimated to be 80 butterflies, and the odds are 95% that the actual population size is between 66 and 94 individuals. In practice, the standard error is too high for the estimate to be very useful if the recapture fraction on the second visit, $V_2/(V_2+V_3)$, is less than 50%. If you're planning on doing a census, make sure you have enough people to help out, and a fast enough way of applying a tag, so that the recapture fraction on the second day exceeds about 50%.

So now that we know how to measure population size, let's see how to detect density dependence in the record of population sizes through time.

4.2 Detecting Limits to Growth

Everyone would like to think the world was infinite, or at least that we're a long way from reaching its limits. Is this wishful thinking? It's instructive to see if some species that have been on this earth long before us have reached limits of *their* own. If they've reached limits, it's especially arrogant to assume that we won't.

4.2.1 Real Birds

Figure 4.1 illustrates the abundance of birds from Buckeye Lake, Ohio, a park where the Audubon Society has been maintaining a regular count of birds. Four species are reported here, the cardinal, titmouse, nuthatch and flicker, and their abundances are entered into MATLAB with

```
>> card   = [227 310 188 186 218 239 189 383 381 309 ...
>>           142 192 207 192 402 313 224 307 351 316 ...
>>           222 593 558 354 274 355 264 238 358 348 ...
>>           314 293 267 239 351 514 426 248 283 559 ...
>>           361 236 265 373 205 251];
>> tmouse = [224 152 192 151 154 126 186 206 213 132 ...
>>           107 129  87 126 154 170 103 184 254 125 ...
>>           130 229 299 199 179 257 193 185 158  99 ...
```

```
>>             162 160 143  66 250 200  97 111 148 108 ...
>>             166  48 111 104  35  40];
>> nhatch = [ 97 198  69 104  77  89 125 199 144 123 ...
>>              36  88  52 145 141  81  97 110 168  95 ...
>>             122  75 159 124  95 203 137 119 111 145 ...
>>             123 143  89  94 181 122  44  74  98  68 ...
>>              57  81  73 109  39  60];
>> flick  = [ 45  80  15  48  14  57  30  55  40  90 ...
>>              20  35  39  61  42 101  72  68 133  78 ...
>>              68  66 185  90  70  66  96  64 142 110 ...
>>             110  80  27  86 139 106  53  96  64  37 ...
>>              56  42  47  29  47  31];
```

Figure 4.1 is then made with

```
>> t = 1935:1:1980;
>> figure
>> hold on
>> plot(t,card,'g');
>> plot(t,tmouse,'c')
>> plot(t,nhatch,'y')
>> plot(t,flick,'r')
```

How would we know density dependence if we saw it? One way is to see if the population dynamics are *not* density independent. That is, we know that if the population growth is density independent, then population's geometric growth factor, R, at any time does not depend on the population size, N, at that time. So if we find that R does depend on N, then the population growth cannot be density independent, and is density dependent instead. The other way to find density dependence is to search for the mechanisms whereby resource depletion affects birth and death rates. Which approach you take depends on the information available to you. If you have a record of population sizes through time, then checking out the relation between R and N is the way to go. If you know what the resources are that a population uses, where they're located, whether they're in fact limiting, how the loss of resources affects various aspects of reproduction (including physiology, behavior, mating success, and success in raising progeny), and also how the loss of resources affects an animal's exposure to hazard, well then, you could splice all this together to get a mechanism-based picture of density-dependent population growth. Seeing that we don't know all this about the Buckeye Lake birds, and we do know their population sizes, we'll look into how to analyze the relation between R and N through time.

The idea is to see if the low R's match up with high N's and the high R's with low N's. If the high and low R's don't match up with the N's in any way, then the population growth is density independent; whereas if low R's tend to match up with high N's and vice versa, then the population growth is density dependent. To see how the R's match up against the N's, we first calculate the average N over all the census dates. (The set of censuses on consecutive dates is called a time series.) We also calculate the geometric average of all

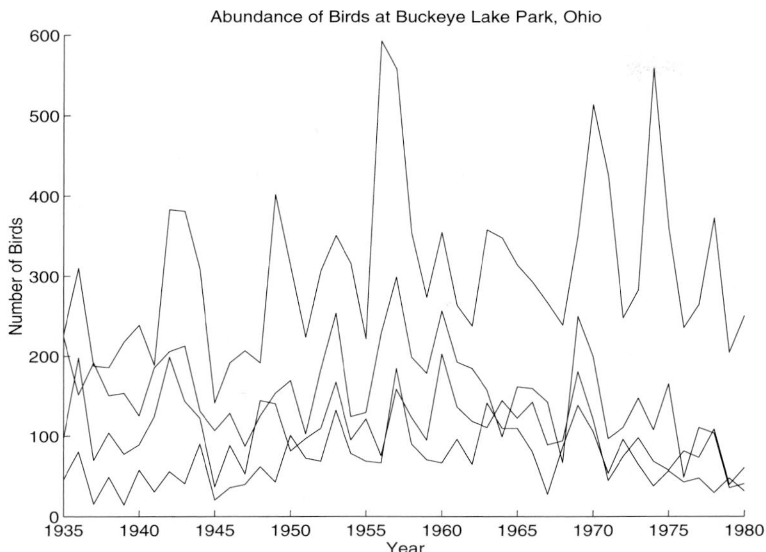

Figure 4.1: The abundance of birds from Buckeye Lake, Ohio. The curves from top to bottom are the cardinal, titmouse, nuthatch, and flicker.

the R's. To do the match up, we count the number of times that N is above average and R above average, that N is below average and R above average, that N is above average and R below average, and that N is below average and R below average—these are all the possibilities. The table of how many times each possibility occurred is called a "contingency table." Now what would be expected if R does not match up one way or the other to N? We would expect that the number of times N is above average and R is also above average to be the *product* of the probability that N is above average times the probability that R is above average. This is called "statistical independence." For example, if you toss a coin once and then again, what is the probability of getting heads both times? It's $1/2 \times 1/2$. Similarly, if the probability that N is above average is $1/2$, and that R is above its average is also $1/2$, then the probability that both are simultaneously above average is $1/4$. So we compute this probability for the other possibilities too, and compare what actually does happen with what is expected if R and N are independent. If what does happen matches with what is expected of growth with independence, then the population growth is density independent. Otherwise it's density dependent.

To do these calculations, we'll prepare a little program in MATLAB to calculate all the averages and do the counting. With a text editor, prepare a separate file called `d_detect.m` as follows:

```
function [n_ave,r_ave,observed,expected,chisq] = d_detect(n);
```

This line indicates that the function accepts as input a vector of population sizes through time, and returns the average N; the average R; the contingency table of observations; the contingency table expected if R and N are independent; and a number called `chisq` that tells us how big the discrepancy is between the observed and expected contingency tables and whether to take any difference between them seriously. If this number is big enough, it will mean that density dependence is present, because the observed contingency table does not match that expected of independent R and N. The arithmetic average N is computed with a single line:

```
n_ave = mean(n);
```

The vector of R's is computed as the ratio of successive population sizes, with

```
r = [];
runlen = length(n) - 1;
for t = 1:runlen
  r = [r n(t+1)/n(t)];
end;
```

Note that if there are, say, 10 N's, only 9 R's can be defined. The geometric average of the R's is computed with

```
r_ave = (prod(r))^(1/runlen);
```

Now the observed contingency table is found with

```
observed = [0 0; 0 0];
for t=1:runlen
  if     (n(t) < n_ave) & (r(t) < r_ave)
            observed(1,1) = observed(1,1) + 1;
  elseif (n(t) < n_ave) & (r(t) > r_ave)
            observed(1,2) = observed(1,2) + 1;
  elseif (n(t) > n_ave) & (r(t) < r_ave)
            observed(2,1) = observed(2,1) + 1;
  elseif (n(t) > n_ave) & (r(t) > r_ave)
            observed(2,2) = observed(2,2) + 1;
  end;
end;
```

The probability that N is less than average is the sum of the times that N was below average and R was also below average, plus the number of times that N was below average and R was above average, all divided by the total number of observations:

```
p_n_lo = (observed(1,1) + observed(1,2))/runlen;
```

Similarly, this produces the probability that N is above average:

```
p_n_hi = (observed(2,1) + observed(2,2))/runlen;
```

The same logic is used to compute the probabilities that R is below or above its average:

```
p_r_lo = (observed(1,1) + observed(2,1))/runlen;
p_r_hi = (observed(1,2) + observed(2,2))/runlen;
```

Now that we have these probabilities, the contingency table expected if R and N are independent is

```
expected = [(p_n_lo*p_r_lo*runlen) (p_n_lo*p_r_hi*runlen);
            (p_n_hi*p_r_lo*runlen) (p_n_hi*p_r_hi*runlen)];
```

Notice that the number of times both N and R are expected to be below average is the product of the probability that N is below average, times the probability that R is below average, times the number of observations. So, if the product of the probabilities is say, $1/2 \times 1/2 = 1/4$, and if there are 100 observations, then at 25 observations, both N and R should be below their averages, provided N and R are independent. Finally, in statistics, the difference between the contingency tables is measured with a number called the "chi-squared" which is defined by the formula

$$\chi^2 = \Sigma_i \left(\frac{(\text{observed}_i - \text{expected}_i)^2}{\text{expected}_i} \right)$$

where the summation is over all the positions in the contingency table. In MATLAB this summation is carried out with

```
chisq = sum(sum(((observed - expected).^2)./expected));
```

The program is now complete, and once it has been saved as a file in your directory, it is ready to be invoked from the MATLAB command line.

Let's see if there is density dependence in the population of cardinals from Buckeye Lake park. The following line invokes the program `d_detect.m` that we have just prepared:

```
>> [n_ave r_ave observed expected chisq] = d_detect(card)
```

MATLAB responds with

```
n_ave = 304.8913
r_ave = 1.0022
observed =  6         17
           20          2
expected = 13.2889    9.7111
           12.7111    9.2889
chisq = 19.3679
```

So the arithmetic average population size of cardinals is 304 birds, and the geometric average R is almost exactly 1. The observed contingency table shows there were 6 times when both N and R were below their averages, 2 times when both N and R were above their averages, 17 times when N was below average and R was above average, and 20 times when N was above average and R below average. OK, now compare these results with the expected contingency table. It's clear that N and R are seldom both above or below their averages relative to what is expected, and that N is high and R low or vice versa much more often than expected. Therefore, there *is* evidence of density dependence. But how important are these differences between what was observed and expected? Could differences this large occur just by chance? Here's where `chisq` is used. In statistics it has been shown that `chisq` may be as high as 3.841[1] by chance 95 out of 100 times if R and N are truly independent. But a `chisq` of 19.3679 is much too big to be consistent with chance unless you're willing to bet against deep, deep odds. In fact, the probability of a `chisq` this high by chance if N and R are independent is 0.00001, or 1 in 100,000. So, the best bet by far is that N and R are not independent, and instead that there is density-dependent population growth going on in the cardinals at Buckeye Lake.

Checking the other bird populations of Figure 4.1 also shows density dependence:

```
>> [n_ave r_ave observed expected chisq] = d_detect(tmouse)
n_ave = 153.3043
r_ave = 0.9624
observed =  6         16
           16          7
expected = 10.7556   11.2444
           11.2444   11.7556
chisq = 8.0489
>> [n_ave r_ave observed expected chisq] = d_detect(nhatch)
n_ave = 108.3261
r_ave = 0.9894
observed =  5         18
           18          4
expected = 11.7556   11.2444
           11.2444   10.7556
chisq = 16.2427
>> [n_ave r_ave observed expected chisq] = d_detect(flick)
n_ave = 68.0435
r_ave = 0.9918
observed =  9    18
           16     2
expected = 15    12
           10     8
chisq = 13.5000
```

4.2.2 Null Models

Well, it's all fine and good that a statistical test finds density dependence in birds from Buckeye Lake, but if you're like me you don't trust statistics very much, and may not be

[1] This is `chisq` associated with a probability level of $P = 0.05$ with one degree of freedom.

really convinced. So let's see what happens if we apply the test for density dependence to a population that we know doesn't have any. The way to get a population without density dependence is to generate one on a computer. Indeed we did this in the last chapter, and let's reuse the program `pop2_rnd.m` that produces two subpopulations of a metapopulation with random R's. Recall that two subpopulations may be coupled to one another by migration at each time step, and that the R's in each subpopulation may be chosen independently, or be chosen to vary in opposite directions. We can apply the density-dependence test to all of these kinds of randomly fluctuating populations, the key point being that there is no connection in any of them between the R that a subpopulation receives from the random number generator, and the subpopulation's size. A model in which some process of interest, such as density dependence in this case, is deliberately left out is called a "null model." We'll see here how a null model scores in the statistical test for density dependence.

Let's start with two subpopulations without migration between them, and whose R are chosen at each time step, independently of the other. The two subpopulations are produced with

```
>> [n1 n2] = pop2_rnd(0,5/4,4/5,100,199,1);
```

Density dependence in one of the subpopulations is checked with

```
>> [n_ave r_ave observed expected chisq] = d_detect(n1)
```

and the `chisq` is seen to be 2.8494. Density dependence is checked in the other subpopulation with

```
>> [n_ave r_ave observed expected chisq] = d_detect(n2)
```

yielding a `chisq` of 3.1056. Both these values are less than 3.841, which is our critical value: it must be exceeded for us to agree that density dependence is present.

Similarly, try two subpopulations with independent R's, where the probability of an animal's migrating is 1/4.

```
>> [n1 n2] = pop2_rnd(0.25,5/4,4/5,100,199,1);
>> [n_ave r_ave observed expected chisq] = d_detect(n1);
>> [n_ave r_ave observed expected chisq] = d_detect(n2);
```

Here the `chisq` values are 0.0391, and 2.6193, again both less than 3.841.

Finally, try two coupled subpopulations that have oppositely varying R's and a migration probability of 1/4:

```
>> [n1 n2] = pop2_rnd(0.25,5/4,4/5,100,199,0);
>> [n_ave r_ave observed expected chisq] = d_detect(n1);
>> [n_ave r_ave observed expected chisq] = d_detect(n2);
```

The `chisq` values here are 4.2278e-04 and 0.1184, both less than 3.841 again.

Thus, density dependence is not detected in populations that don't have any, which is good. Therefore, we should take it very seriously when density dependence is detected, as in the Buckeye Lake birds. The chance that these populations do not have density dependence is so low as to be negligible, and we will now proceed on the basis that density dependence is a reality. So, if it's a reality, what do we do about it? How do we predict the effect of density dependence on population growth?

4.3 The Logistic Model of Limits to Growth

The most elementary model for density dependent population growth is called the "logistic model," named for the S-shaped curve that it predicts, which in statistics is called the "logistic distribution." Different versions exist for discrete and continuous time, and we'll start with the discrete-time version.

4.3.1 Discrete Time

As in the last chapter, we let B stand for the number of births each animal has, during a time step, that live to the beginning of the next step; and D is the probability that an animal dies during the time step. The new twist is to assume that both birth and death depend on population size. Therefore, we use the parameters B_o and D_o, the density-independent components of birth and death, and the parameters b and d, the coefficients relating individual birth and death to population size. With them, we write

$$B(N) = B_o - bN$$

and

$$D(N) = D_o + dN$$

to indicate that the average number of births per time step per individual decreases with population size, N, and that the probability of an individual's dying per time step increases with population size.

As before, the number at time $t+1$ is the sum of the births plus the survivors

$$N_{t+1} = B(N_t)N_t + [1 - D(N_t)]N_t$$

which is, in full,

$$\begin{aligned} N_{t+1} &= (B_o - bN_t)N_t + [1 - (D_o + dN_t)]N_t \\ &= [(1 + B_o - D_o) - (b + d)N_t]N_t \end{aligned}$$

This model is called the "logistic model." When the population size, N_t, is low, then N_{t+1} approximately equals $(1+B_o-D_o)\times N_t$. This represents geometric growth, and $(1+B_o-D_o)$ is the geometric growth factor. On the other hand, when N_t gets big, then the population growth rate slows down. When the expression in brackets, $(1+B_o-D_o)-(b+d)N_t$, shrinks to the point where it equals 1 because N_t is increasing over time, then the population actually stops growing, and N_{t+1} equals N_t.

Now please be patient while we do some rearranging. The logistic model is usually presented in a standard form as follows. We define the "carrying capacity" as the population size at which direct replacement occurs; that is, the size at which $N_{t+1} = N_t$. This happens when the expression in brackets in the equation above equals 1:

$$(1 + B_o - D_o) - (b + d)N = 1$$

If you solve this equation[2] for N, you get the population size at which direct replacement occurs, which is the carrying capacity, and is labeled as K

$$K = \frac{B_o - D_o}{b + d}$$

Next, it is customary to rearrange the logistic model so that it focuses on how the population size *changes* at each time step. The symbol Δ is used in mathematics to indicate a difference, and the difference in population size after one time step is denoted as

$$\Delta N = N_{t+1} - N_t$$

So, if we subtract N_t from both sides of the logistic model, and then drop the t subscript because it is no longer needed, we get

$$\Delta N = [(B_o - D_o) - (b + d)N]N$$

Finally, we define a discrete-time version of the intrinsic rate of increase, r, as

$$r = B_o - D_o$$

Using just the two parameters, r and K, rather than the four parameters, B_o, D_o, b, and d, the logistic equation is usually presented in journals and textbooks as

$$\Delta N = r\left(1 - \frac{N}{K}\right)N$$

or equivalently as

$$\Delta N = r\left(\frac{K - N}{K}\right)N$$

In the literature, you'll often see reference to "r" and "K," and these symbols refer to the logistic model for population growth, as written in this format. As we have seen, r and K have been defined in terms of the underlying birth and death processes as $B_o - D_o$ and $(B_o - D_o)/(b + d)$ respectively. Usually though, r and K are simply used by themselves as summary parameters that lump together the details of what's happening at the level of individual births and deaths.

[2]The subscript t is now omitted because we're solving for an equilibrium, which by definition doesn't change in time.

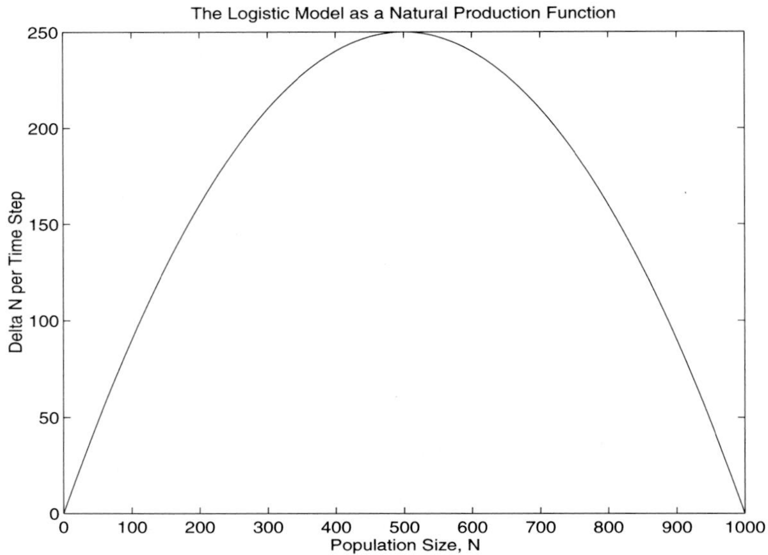

Figure 4.2: The logistic model as a natural production function. The curve is a downward-facing parabola with roots at $N = 0$ and $N = K$. The peak occurs at a population size of $K/2$ and the maximum sustainable yield that is produced at this population size is $rK/4$.

Now that we have defined the logistic model, let's see what it tells us. There are two ways to look at the logistic model, one focusing on how fast the growth occurs, the other on the cumulative effects of growth through time. The equation for ΔN tells us how many new animals will be produced by the animals that are there. In economic terms you can think of the population size as a stock or asset—the phrase "natural capital" is often used too. The growth of the stock, ΔN, is then the interest or earnings on that stock in units of organisms. In this context, the equation for ΔN is called a "natural production function."

To use MATLAB to illustrate the logistic model for ΔN as a natural production function, define the formula for `delta_n` as

```
>> delta_n = 'r*(1 - n/k)*n';
```

Then make a plottable version with

```
>> delta_n_plot = sym2ara(delta_n);
```

Some representative parameter values are

```
>> r = 1;
>> k = 1000;
```

Then Figure 4.2 is generated with

```
>> n = 0:10:1000;
>> figure
>> plot(n,eval(delta_n_plot))
```

The logistic model of a natural production function is a downward-opening parabola with roots at $N = 0$ and $N = K$. The peak of the production function is called the "maximum sustainable yield." To find the population size that produces the maximum yield, differentiate the formula for ΔN with respect to N and solve for N. This is done in MATLAB as

```
>> n_msy = solve(diff(delta_n,'n'),'n')
```

which returns the result that

```
n_msy = 1/2*k
```

The maximum sustainable yield that is attained when the population has this size is found from MATLAB as

```
>> msy = subs(delta_n,n_msy,'n')
```

resulting in

```
msy = 1/4*r*k
```

The stock size producing the maximum sustainable yield is generally not the economically optimal stock size, as will be discussed later. Still, the production function is needed when calculations of the economically optimal stock size are being made.

The logistic model may also be used to forecast the growth of a population through time: simply iterate the equation for ΔN. That is, N_{t+1} equals N_t plus ΔN. Repeating this calculation through successive time steps generates a sequence of population sizes through time. A simple program to carry out the iteration appears below. It is called `logist_d.m` and should be entered with a text editor and saved in your directory.

```
function n = logist_d(r,k,no,runlen)
 n = [no];
 for t=1:runlen
  nprime = n(t) + r*n(t)*(k - n(t))/k;
  if nprime < 0
   nprime = 0;
  end
  n = [n nprime];
 end
```

The program is like many others we have developed in earlier chapters. It takes as input the values of the intrinsic rate of increase, `r`, the carrying capacity, `k`, the initial population size, `no`, and the number of time steps to conduct the calculations, `runlen`. The program returns

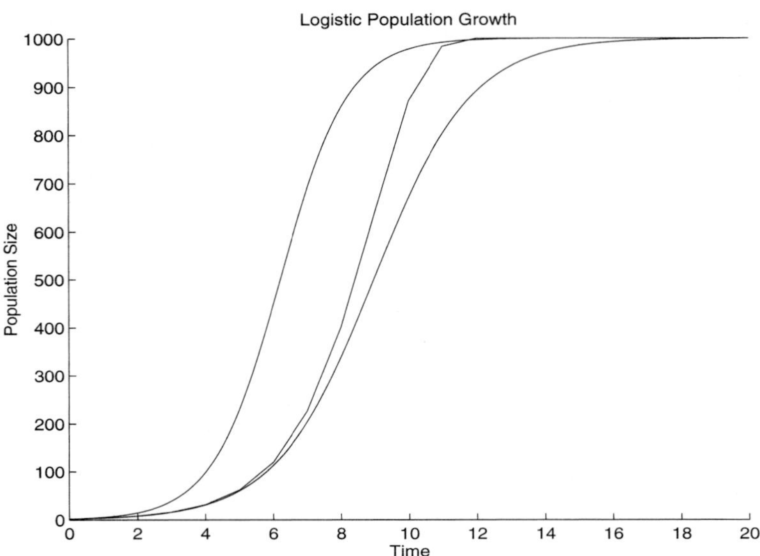

Figure 4.3: A logistic population growth through time. The population size is started with two individuals and the carrying capacity, K, is 1000. The curve in the middle is produced by the logistic model in discrete time with a discrete-time r of 1. The curve on the left is produced by the logistic model in continuous time with a continuous-time r of 1, and that on the right by the continuous-time logistic model with a continuous-time r of 0.6931.

a vector of population sizes, beginning with the initial population size. The first line below the function's declaration puts the initial population size as the first element of the vector of population sizes through time. The next line sets up a loop for time from 1 to `runlen`. Then the new population size, called `nprime`, is computed from the present population size plus the formula for ΔN. The new population size is then checked in case it is negative, and if it is, then it is set to zero. Finally, the new population size is appended to the vector of population sizes, and the loop repeats.

Now let's use `logist_d` to predict the population size through 20 time steps, starting with a population size of 2, and using the existing values of r and K already defined when Figure 4.2 was made. The vector of population sizes is produced with

```
>> no = 2;
>> n = logist_d(r,k,no,20);
```

After `figure` and `hold on` are typed, these are plotted in yellow with

```
>> plot(0:20,n,'y')
```

and the middle curve in Figure 4.3 results. The S-shaped curve shows that the population's growth initially resembles geometric growth, but density dependence applies the brakes, and the population abundance eventually levels off at the carrying capacity.

4.3.2 Continuous Time

The continuous-time version of the logistic equation is the same as the discrete-time version except that a derivative, dN/dt, is in place of the ΔN.

$$\frac{dN}{dt} = r\left(1 - \frac{N}{K}\right)N$$

where r is the continuous-time intrinsic rate of increase and K the carrying capacity. This formula can be derived by our building up from birth and death processes, just as we did for the discrete-time logistic model. If the instantaneous birth rate is $B(N) = B_o - bN$, and the instantaneous death rate is $D(N) = D_o + dN$, then the continuous-time logistic model is $dN/dt = B(N) - D(N) = (B_o - D_o) - (b + d)N$. Exactly as before, we can then rearrange the logistic model into its conventional form, when we define $r = B_o - D_o$, and $K = (B_o - D_o)/(b + d)$. The key point is that here births refer to the continuous production of offspring, and death to the continuous loss of individuals, rather than to births and deaths counted only at discrete time steps.

The continuous-time logistic model is a differential equation, and to solve it means to find a formula for $N(t)$ whose slope at every point, N, satisfies this equation. MATLAB will find the formula for us, if we type

```
>> n = simplify(dsolve('Dn = r*n*(1 - n/k)','n(0) = no','t'))
```

where `no` is the initial condition. Typing `pretty(n)` yields the formula for $N(t)$

```
                    k no
   ----------------------------------
   no + exp(- r t) k - exp(- r t) no
```

Upon inspection, it's even prettier to write

$$N(t) = \frac{K}{1 + [K/N(0) - 1]e^{-rt}}$$

To compare the predictions of the continuous-time logistic model with those of the discrete-time version , let's place some new curves into Figure 4.3. First make a plottable version of the formula with

```
>> n_plot = sym2ara(n);
```

Then

```
>> t=0:.1:20;
>> plot(t,eval(n_plot),'c')
```

puts the curve in cyan that appears toward the left in Figure 4.3. This curve accelerates more quickly than the discrete-time curve, even though both have an r of 1. The reason is that the discrete-time r and continuous-time r are not in quite the same units. In continuous time, the population growth is compounded continuously, whereas in discrete time the population growth is compounded only at the end of each time step. To adjust for this, we note that after one full time interval of exponential growth within which growth compounds continuously at rate r, the population size increases from N to $e^r N$ (This is $e^{rt}N$ with $t = 1$). So, the ΔN over this time is $e^r N - N$, which equals

$$\Delta N = (e^r - 1)N$$

To achieve this same growth by compounding only at the end of the period, we would have to use the expression in parentheses as our discrete-time r. Thus,

$$r_{\text{discrete}} = e^{r_{\text{continuous}}} - 1$$

Conversely, rearranging the formula shows that the continuous-time r that matches compounding only at the end of time steps is

$$r_{\text{continuous}} = \log(r_{\text{discrete}} + 1)$$

If a continuous-time r is desired and the data give only a discrete-time r, just convert it with the formula above, and similarly to convert in the other direction. To illustrate, let's reset the r we have used to plot a discrete-time curve to the value we should use in a continuous-time curve. Type

```
>> r = log(r+1)
```

Because r was previously equal to 1, MATLAB responds that its new value is 0.6931. Now, plot the continuous-time logistic curve with this new value of r,

```
>> plot(t,eval(n_plot),'m')
```

and obtain the curve toward the right in Figure 4.3. Its initial acceleration does match the discrete-time logistic, as desired.

Figure 4.3 shows something else too. The discrete-time logistic rises to its ultimate value more quickly than the new continuous-time logistic does. Although scaling the discrete-time and continuous-time r's appropriately makes the curves agree at the bottom, they disagree at the top. As we'll see later, the discrete-time logistic can actually overshoot its mark, whereas the continuous-time logistic always approaches K with grace and elegance.

4.4 Using the Logistic Model with Data

Now that we have a model for density-dependent population growth, let's see how the model works with some data. The first question is, should you do discrete or continuous? The continuous model is right if the organisms averaged over the whole population are having

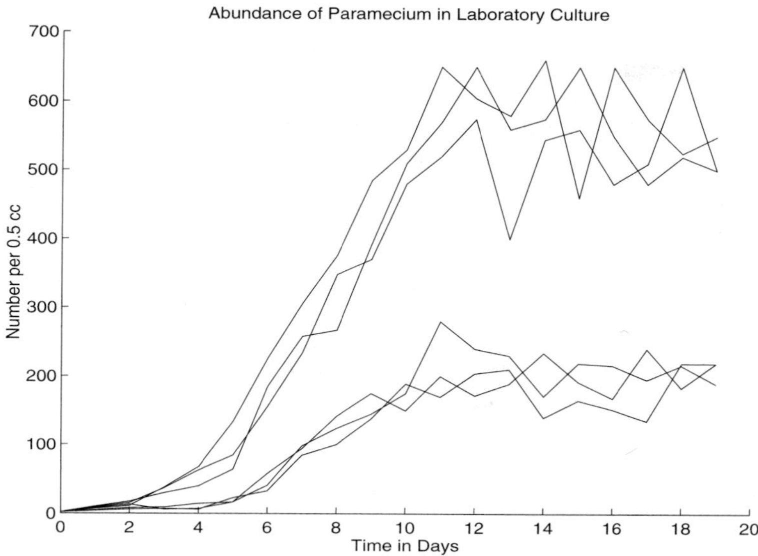

Figure 4.4: The abundance of three replicates of *Paramecium aurelia* at top, and of *P. caudatum* at bottom, through time in days. The data is taken from the laboratory cultures of G. F. Gause in 1934.

births and deaths all the time. In people, for example, births and deaths occurs 24 hours a day, and population growth really is continuous. In the temperate zones, and often in the tropics too, reproduction by plants and animals is seasonal, and it's natural to view the time step as one year, encompassing a full season. Here a discrete-time model is right. A discrete-time model can also be used to suit *our* convenience if we want to census what is truly a continuous-time process at discrete intervals. We must take care, though, to make the census interval short enough that the discrete-time model closely approximates the continuous-time model; or equivalently, we can measure the r and K with discrete time intervals and convert them to their continuous-time counterparts with the formulas of the previous section. So, to illustrate all this, let's turn to a famous set of experiments with continuously growing populations.

4.4.1 The Growth of Laboratory *Paramecium*

A classic data set on population growth was assembled by G. F. Gause in 1934 with laboratory populations of *Paramecium*, a single-celled protozoan. There are two species, *Paramecium aurelia* and *P. caudatum*,[3] with three replicates of each. The data are entered into

[3] Incidentally, the convention for referring to a species in biology is to use two names, called the Latin binomial. The first name is the genus, which is capitalized and the second is the species, which is not

MATLAB with

```
>> pa1 = [  2  10  17  29  39  63 185 258 267 392 ...
>>        510 570 650 560 575 650 550 480 520 500];
>> pa2 = [  2   9  15  36  62  84 156 234 348 370 ...
>>        480 520 575 400 545 560 480 510 650 500];
>> pa3 = [  2   7  11  37  67 134 226 306 376 485 ...
>>        530 650 605 580 660 460 650 575 525 550];
>> pc1 = [  2   5   8   9  14  16  57  94 142 175 ...
>>        150 200 172 189 234 192 168 240 183 219];
>> pc2 = [  2   8  13   6   7  16  40  98 124 145 ...
>>        175 280 240 230 171 219 216 195 216 189];
>> pc3 = [  2   4   6   7   6  22  32  84 100 138 ...
>>        189 170 204 210 140 165 152 135 219 219];
```

and plotted with

```
>> figure
>> hold on
>> plot(0:length(pa1)-1,pa1,'r')
>> plot(0:length(pa2)-1,pa2,'g')
>> plot(0:length(pa3)-1,pa3,'b')
>> plot(0:length(pc1)-1,pc1,'c')
>> plot(0:length(pc2)-1,pc2,'m')
>> plot(0:length(pc3)-1,pc3,'y')
```

yielding Figure 4.4. *P. aurelia* consistently attains a larger population size than *P. caudatum* under the conditions of the experiment.

Before you determine the r and K, it's a good idea to check for density dependence in the record of population sizes, because if there isn't any, then it would be misleading to use the logistic model with the data. So, calling again on our trusty `d_detect` function, we type

```
>> [n_ave r_ave observed expected chisq] = d_detect(pa1);
```

Then typing `chisq` yields

```
chisq = 15.3535
```

which is greater than our cutoff of 3.841, thereby confirming the presence of density dependence in the first replicate of population growth for *P. aurelia.* Similarly, running `d_detect` on `pa2`, `pa3`, `pc1`, `pc2`, and `pc3` yields `chisq` values of 15.3535, 11.6812, 6.1338, 2.5735, 4.2318, respectively. The second replicate of *P. caudatum* actually doesn't show significant

capitalized. Both are in italics because the names are considered to be Latin words, and words in any foreign language are always written in italics. Once the genus name has been introduced, in a binomial it can be abbreviated with the first initial. So once *Paramecium* has been introduced, a binomial can be abbreviated as *P.* followed by the species name written out in full, as in *P. caudatum* above.

density dependence by itself, but in context with the other experiments we will assume it does have density dependence anyway, even though this assumption is a bit risky.

Three approaches to determining r and K are readily taken, and deciding which to use depends on the kind of data available and whether we believe the population growth is really continuous or discrete.

The first approach is to calculate all the ΔN's and plot them against N. We then find the best fit of a quadratic through these points, and use r and K to describe the shape of this quadratic. We refer to this procedure as fitting natural production to stock size—it's best if we have many replicates of population growth, each involving only a short time. Imagine, for example, many populations each starting at different initial abundances, and watching them through only one or two time steps. These data provide a picture of how ΔN relates to N, which can then be fitted to the logistic natural-production function.

The other two approaches assume that a long-term census record is available. The best fit of either a discrete-time or continuous-time logistic trajectory through the entire time series is computed as a function of r and K.

The "best fit" is defined as the curve that minimizes the sum of the squared deviations between the data and the curve. For example, suppose we're trying to fit a curve, $y(x)$ to a bunch of points, (x_1, y_1), (x_2, y_2), ... We start with x_1 and evaluate the function at this point, $y(x_1)$. This is the *predicted* y at this point, whereas y_1 is the actual value of y at that point. The squared deviation between the predicted and the actual is $(y(x_1) - y_1)^2$. We want to take the square because we don't care if the prediction is above or below the actual. The sum of the squared deviations between actual and predicted over all the data points tells us how good the curve is at fitting the entire data set. Now our curve, $y(x)$, can be moved around. So we want to move it around, to minimize the sum of the squared deviations between what it predicts and what actually happens. Once we've done this, we've got the best fit to the entire data set.

To do the curve-fitting with all three approaches, we'll employ the function `fmins` from MATLAB. This function finds the parameters that minimize a function of several variables. For our application, we want to find the r and K that minimize the sum of squares between what the logistic model predicts and the data. If you pull out your dog-eared copy of the *MATLAB User's Guide*, you'll find that MATLAB is fussy about how `fmins` is invoked. First, we must write a separate function to calculate the sum of squares—this is easy, it's only four lines long. Second, we must supply initial guesses for r and K. Glancing at Figure 4.4 suggests that initial (and rather crummy) guesses for r and K are 1 and 500, respectively. Third, we must invoke `fmins` with the following syntax:

```
best = fmins('function',guess,[],[],data)
```

Here `'function'` refers to the name of the sum-of-squares function that we have written to compare the logistic model with the data; `guess` is a vector of our initial guess of r and K; the `[],[]` is to keep MATLAB smiling and happy; and `data` is a vector of the census data that is passed to `function`. After all this, `fmins` returns `best`, a vector containing the r and K that best fit the data. Don't worry, this is easier than it sounds.

4.4.2 Fitting Natural Production to Stock Size

Enough talking, let's do it. With a text editor, create a function called `rkprod.m`, as follows:

```
function sumdevsq = rkprod(guess,n)
 delta_n = diff(n);
 n = n(1:length(delta_n));
 sumdevsq = sum((delta_n-guess(1)*(1-n/guess(2)).*n).^2);
```

The function takes as input a vector, `guess`, whose first element is r and second element is K, and another vector, `n`, that contains the census data. The function returns the sum of deviations between the production based on a logistic model using this r and K and the production observed in the data. The first line of the function computes a vector of ΔN's. The length of the `delta_n` vector is one less than the length of the `n` vector. The second line of the function discards the last element of the `n` vector, so that it is now the same length as the `delta_n` vector. The third line of the function calculates the sum of the squared deviations. For example, the first element of `delta_n` is the change in population size from time 0 to time 1. The predicted ΔN is $r(1-N/K)N$, where N is the first element of `n`. And similarly for the rest of the predicted ΔN's. Because r is `guess(1)` and K is `guess(2)`, the predicted ΔN is written as `guess(1)*(1-n/guess(2)).*n`. The `.*` signifies element by element multiplication. The predicted value is then subtracted from the actual `delta_n`, and the result is squared; the `.^` indicates that each element is individually squared. Then the vector of all the squares is added up with `sum`, and the result is returned.

Once `rkprod.m` has been saved in your directory, you can type

```
>> best_pa1 = fmins('rkprod',[1 500],[],[],pa1)
```

which yields

```
best_pa1 = 0.6921  566.4558
```

These are r and K for the first replicate of *P. aurelia.* The r and K for the other two replicates are found with

```
>> best_pa2 = fmins('rkprod',[1 500],[],[],pa2);
>> best_pa3 = fmins('rkprod',[1 500],[],[],pa3);
```

We can package the three estimates of r for *P. aurelia* into a vector

```
>> rpa =  [best_pa1(1) best_pa2(1) best_pa3(1)]
```

yielding

```
rpa = 0.6921     0.9472     0.9350
```

and the three estimates of K into another vector

```
>> kpa =  [best_pa1(2) best_pa2(2) best_pa3(2)]
```

which yields

```
kpa = 566.4558  523.6727  576.8168
```

The best overall estimate of r from the three replicates is then the average, which is found as

```
>> rpa_ave = mean(rpa)
```

yielding

```
rpa_ave = 0.8581
```

The standard error of this estimate is the standard deviation of the r's divided by the square root of the sample size, which is three. So type

```
>> rpa_se = std(rpa)/sqrt(3)
```

yielding

```
rpa_se = 0.0831
```

Similarly,

```
>> kpa_ave = mean(kpa)
kpa_ave = 555.6484
>> kpa_se = std(kpa)/sqrt(3)
kpa_se = 16.2652
```

The bottom line is that the discrete-time intrinsic rate of increase for *P. aurelia* is 0.8581 ± 0.0831 organisms per day per 0.5 cc, and the carrying capacity is 555.6484 ± 16.2652 organisms per 0.5 cc.

Upon repeating this procedure with the three replicates for *P. caudatum*, we obtain

```
rpc =    1.0771    0.8166    0.8721
kpc = 196.8077  217.5368  185.5547
rpc_ave = 0.9219
rpc_se  = 0.0792
kpc_ave = 199.9664
kpc_se  =   9.3665
```

Let's now see how well the logistic model describes the *Paramecium* data with the r's and K's we've just estimated. Figure 4.5 presents a graph of all the $\Delta N's$ for *P. aurelia* together with the logistic production function using $r =$ `rpa_ave` and $K =$ `kpa_ave`. To produce it, we type

```
>> figure
>> hold on
>> plot(pa1(1:length(pa1)-1),diff(pa1),'r*')
```

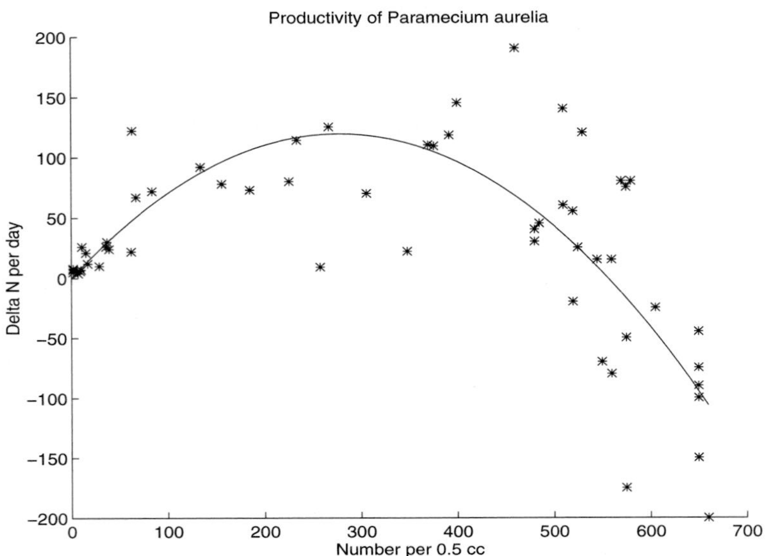

Figure 4.5: The production of *Paramecium aurelia* fitted to a discrete-time logistic model.

```
>> plot(pa2(1:length(pa2)-1),diff(pa2),'g*')
>> plot(pa3(1:length(pa3)-1),diff(pa3),'b*')
>> n = 0:10:max([pa1 pa2 pa3]);
>> r = rpa_ave;
>> k = kpa_ave;
>> plot(n,eval(delta_n_plot),'w')
```

Figure 4.6 is the same type of graph, but for *P. caudatum*; it is produced by our typing

```
>> figure
>> hold on
>> plot(pc1(1:length(pc1)-1),diff(pc1),'c*')
>> plot(pc2(1:length(pc2)-1),diff(pc2),'m*')
>> plot(pc3(1:length(pc3)-1),diff(pc3),'y*')
>> n = 0:10:max([pc1 pc2 pc3]);
>> r = rpc_ave;
>> k = kpc_ave;
>> plot(n,eval(delta_n_plot),'w')
```

There is clearly quite a bit of spread in the data relating ΔN to N, so even the best fit to a logistic model is not fantastic, though it's not terrible either. We say a model is "poorly constrained" if the scatter in the data do not allow us to pin down the model very precisely.

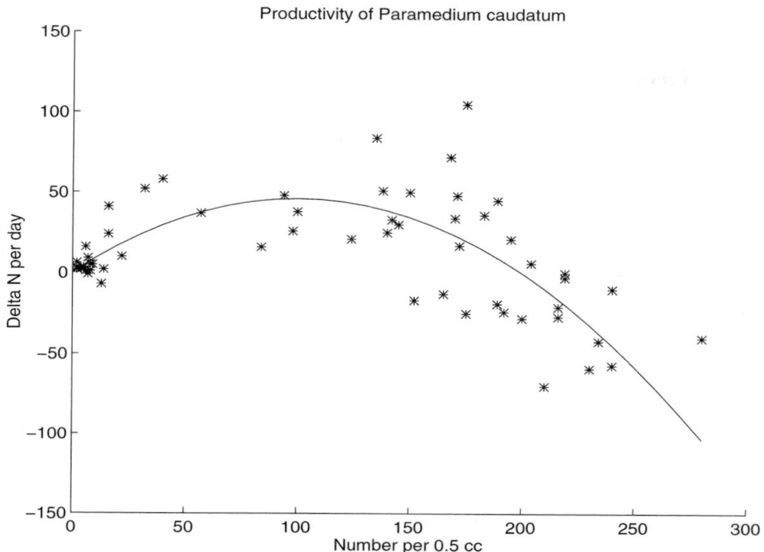

Figure 4.6: The production of *P. caudatum* fitted to a discrete-time logistic model.

The scatter in the data also imply that a model more complicated than the logistic is not justified in this case. The data do not allow us to distinguish a more complicated model from the logistic, and in the absence of other information, the logistic would be preferred because of its simplicity alone.

To see how well the logistic model does in predicting the population census through time, we generate trajectories based on the estimated r's and K's, and compare these with the data. Figure 4.7 illustrates the predicted trajectories for both *P. aurelia* and *P. caudatum*. It is made by our typing

```
>> figure
>> hold on
>> plot(0:length(pa1)-1,pa1,'*g')
>> plot(0:length(pa2)-1,pa2,'*g')
>> plot(0:length(pa3)-1,pa3,'*g')
>> plot(0:length(pc1)-1,pc1,'*m')
>> plot(0:length(pc2)-1,pc2,'*m')
>> plot(0:length(pc3)-1,pc3,'*m')
>> npa = logist_d(rpa_ave,kpa_ave,pa1(1),length(pa1)-1);
>> plot(0:length(pa1)-1,npa,'y')
>> npc = logist_d(rpc_ave,kpc_ave,pc1(1),length(pc1)-1);
>> plot(0:length(pc1)-1,npc,'y')
```

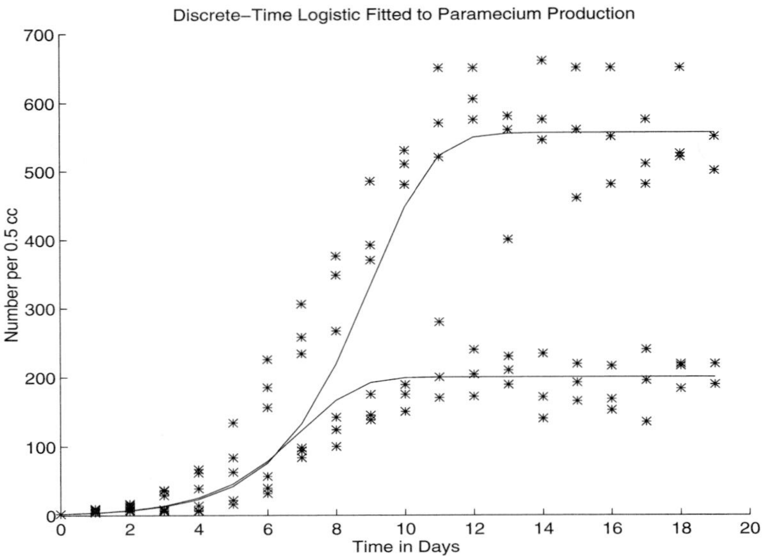

Figure 4.7: Trajectories of *Paramecium aurelia* and *P. caudatum* predicted by a discrete-time logistic model fitted to production functions.

The predicted trajectories don't fit the data astonishingly well. The trajectories do seem to level off at their correct K's, but one trajectory grows too slowly and the other too fast relative to the data. Of course, these discrepancies may not be important to one's purposes, but as we'll now see, we can do better.

4.4.3 Fitting Trajectories Through Time

Instead of fitting the logistic model to data on ΔN vs. N, we can fit to a record of population size through time, N_t. Should we do discrete or continuous? Laboratory *Paramecium* are probably dividing 24 hours a day, so the population growth is probably continuous. But just to be sure, let's estimate the r and K for both discrete and continuous versions of the logistic model, and see if the continuous version really does fit the data better.

Discrete Time

All we need to do is customize a new function to calculate the sum of the squared deviations between the $N(t)$ predicted by the logistic model and the data. Then we can use the function with `fmins` just as before. The function to calculate the sum of the deviations squared is called `rktraj_d`—it should be entered with a text editor and saved in your directory as `rktraj_d.m`.

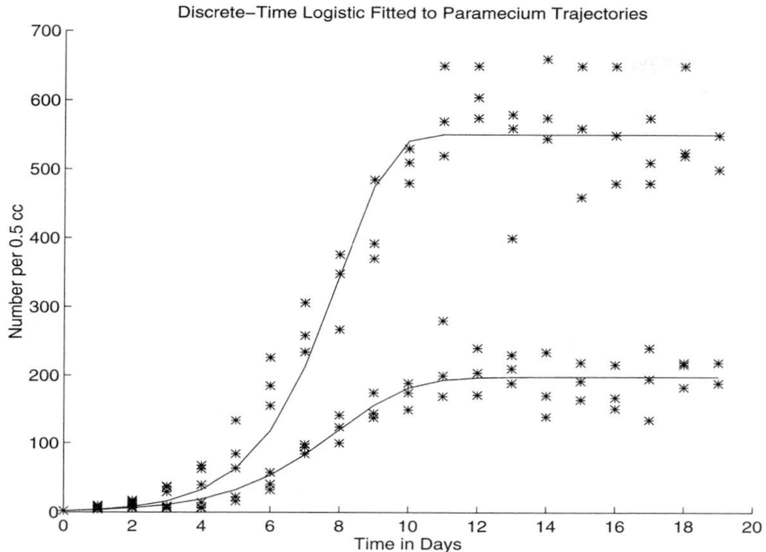

Figure 4.8: Trajectories of *Paramecium aurelia* and *P. caudatum* predicted by a discrete-time logistic model fitted to trajectories.

```
function sumdevsq = rktraj_d(guess,n)
 npredict = [n(1)];
 for t=1:length(n)-1
   nprime = npredict(t)+guess(1)*(1-npredict(t)/guess(2))*npredict(t);
   npredict = [npredict nprime];
 end;
 sumdevsq = sum((n-npredict).^2);
```

`rktraj_d` is called with `guess`, which is a vector whose first element is r and second element is K, and with `n`, whose elements are the sequence of population sizes. `rktraj_d` calculates a predicted sequence of population sizes based on the discrete-time logistic model, in much the same way as the `logist_d` function presented earlier in the chapter. The first line of the function puts the first census datum into the `npredict` vector as the initial condition. Then the loop calculates each successive N_{t+1} as $N_t + r(1 - N(t)/K)N$. At the end of the loop we have a sequence of predicted N_t's to compare with the actual N_t's. The final line of the function calculates the sum of the deviation squared between the actual and the predicted, and this result is returned by the function to the calling program.

Now that we have `rktraj_d`, we just repeat the steps in the last section, but use `rktraj_d` instead of `rkprod`. To obtain the best r and K for the first replicate of *P. aurelia*, type

```
>> best_pa1 = fmins('rktraj_d',[1 500],[],[],pa1)
```

yielding

```
best_pa1 = 0.9559  558.8634
```

The first element is r and the second is K. Now apply `fmins` with `rktraj_d` to the remaining replicates of *P. aurelia* and to the replicates of *P. caudatum*. When the results are collected, and averaged as in the last section, we obtain

```
rpa =   0.9559    0.9919    1.0891
kpa = 558.8634  520.6590  573.4188
rpa_ave =   1.0123
rpa_se =    0.0398
kpa_ave = 550.9804
kpa_se =   15.7322
rpc =   0.8025    0.7664    0.7513
kpc = 197.7185  215.3223  181.6569
rpc_ave =   0.7734
rpc_se =    0.0152
kpc_ave = 198.2326
kpc_se =    9.7218
```

Thus, fitting a discrete-time logistic to N_t vs. t instead of to ΔN vs. N yields slightly different estimates for r and K. To see how well the trajectories of the discrete-time logistic agree with the data based on these new estimates of r and K, make Figure 4.8 by typing exactly the commands used previously for Figure 4.7. Clearly, these new predicted trajectories offer a better fit to the data. But can we do even better?

Continuous Time

To fit a continuous-time logistic model to the data, we tailor still another function to compute the sum of the squared deviations between the model predictions and data. I hope you're getting the hang of this. To fit any ol' model to the data, just tailor an appropriate sum-of-squared-deviations function and then use it with `fmins`. Here we'll use a function called `rktraj_c` that should be entered with a text editor and saved as `rktraj_c.m`. It's patterned after its predecessor of the last section. Here the solution of the logistic differential equation is simply evaluated at successive t's to yield the sequence of predicted N_t's, whereas before we had to iterate.

```
function sumdevsq = rktraj_c(guess,n)
 npredict = [n(1)];
 for t=1:length(n)-1
   nprime = guess(2)/(1+(guess(2)/n(1)-1)*exp(-guess(1)*t));
   npredict = [npredict nprime];
 end;
 sumdevsq = sum((n-npredict).^2);
```

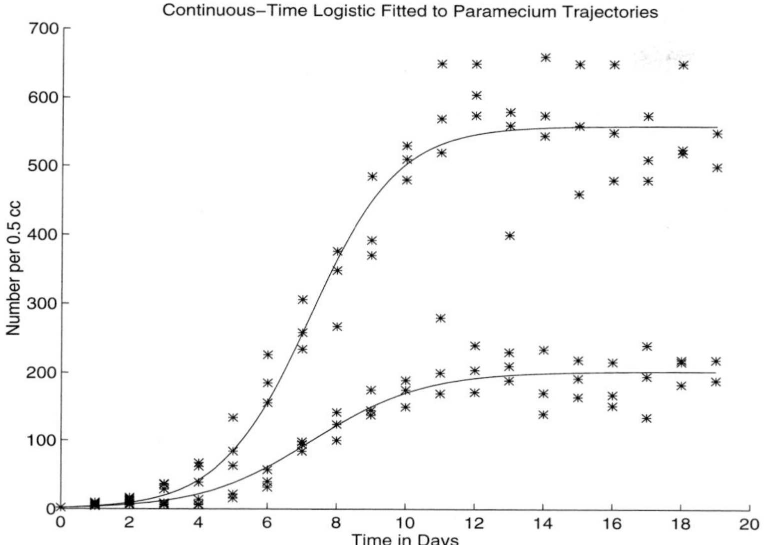

Figure 4.9: Trajectories of *Paramecium aurelia* and *P. caudatum* predicted by a continuous-time logistic model fitted to trajectories.

As before, the function is called with `guess`, which is a vector containing r and K, and with `n`, which is a vector containing the sequence of N_t from the data. The first line of the function's body puts the first data point into the first position of the vector of predicted N_t's as the initial condition. Then the loop simply evaluates the formula for $N(t)$ at integer times points from 1 through the last time in the data. Each N_t is appended to the vector of predicted N_t's. The last line in the function computes the sum of the deviations squared between the `n` and `npredict`.

To find the best fit of a continuous-time logistic model to the trajectory of the first replicate with *P. aurelia*, type

```
>> best_pa1 = fmins('rktraj_c',[1 500],[],[],pa1)
```

yielding

```
best_pa1 = 0.7516   566.0018
```

which means that the continuous-time r is 0.7516 and K is 566.0018. Applying `fmins`, using `rktraj_c` with all the other replicates, and summarizing the results, yields

```
rpa =     0.7516     0.7636     0.8297
kpa = 566.0018  531.0284  582.0278
rpa_ave =    0.7816
```

```
rpa_se =     0.0243
kpa_ave = 559.6860
kpa_se =   15.0571
rpc =    0.6393     0.6313     0.6143
kpc = 203.4624  218.5157  185.5013
rpc_ave =    0.6283
rpc_se =    0.0074
kpc_ave = 202.4931
kpc_se =    9.5428
```

To see how well the continuous-time logistic model compares with the data, produce Figure 4.9 by typing

```
>> figure
>> hold on
>> plot(0:length(pa1)-1,pa1,'*g')
>> plot(0:length(pa2)-1,pa2,'*g')
>> plot(0:length(pa3)-1,pa3,'*g')
>> plot(0:length(pc1)-1,pc1,'*m')
>> plot(0:length(pc2)-1,pc2,'*m')
>> plot(0:length(pc3)-1,pc3,'*m')
>> r = rpa_ave; k = kpa_ave; t = 0:.1:length(pa1)-1;
>> plot(t,eval(n_plot),'y')
>> r = rpc_ave; k = kpc_ave; t = 0:.1:length(pc1)-1;
>> plot(t,eval(n_plot),'y')
```

The agreement is now really good. So it is accurate to say that the growth of Gause's laboratory populations of *Paramecium aurelia* and *P. caudatum* follow the continuous-time logistic model of population growth.

4.5 Comparing the Logistic to Other Models

Is all the world a logistic? You might think so, considering the space devoted to it so far. But there are alternatives you may wish to entertain. An alternative is needed if you try to fit a logistic to data on ΔN vs. N, or to N_t vs. t, as we did above for *Paramecium*; you would discover that the fit is very poor. If so, a more complicated model should be concocted instead. The need for alternatives may also arise if a population model is built up from the underlying processes affecting birth and death rates. If the birth rate decreases linearly with N, and/or the death rate increases linearly with N, then a logistic model emerges with r and K defined in terms of the parameters for the birth and death processes, as we have already seen. But if the birth and death processes don't depend linearly on N, then something else emerges. For example, one may consider how a plant's seed production depends on the number of neighbors it has, and on how much nutrient it can take up, and on how much light it can absorb. Combining this information with specifications about

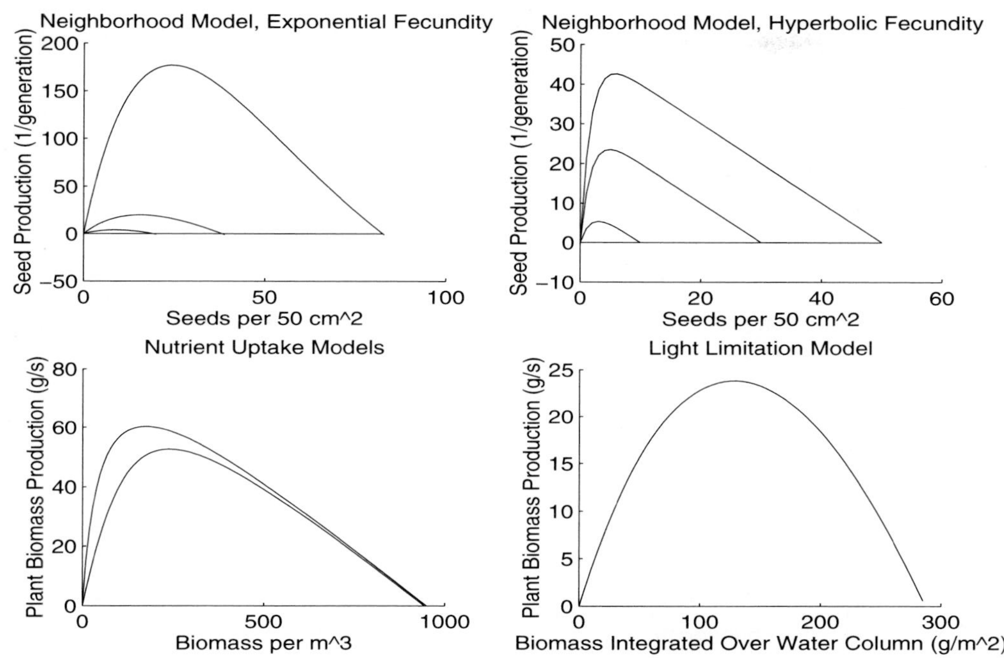

Figure 4.10: Production functions for plant populations. The top left and right are from "neighborhood models" in which an individual's seed production, probability of survival, and distance it was dispersed before germination are assumed to depend on the number of other individuals within some circle around it. The bottom left is from a model in which plant growth is limited by nutrients, and the bottom right from a model in which growth is limited by light.

death rates and perhaps also about spatial distribution then leads to a population-dynamic model. A model of population dynamics built from a description of how the individuals in it function is called an "individual-based model" of population dynamics. Typically, these models do not reduce exactly to a logistic model, although they may be quite close.

Recently, I surveyed the production functions for population-dynamic models developed for organisms extending from plants through invertebrates to vertebrates. The results are reproduced in Figures 4.10–12. The figures all show unimodal downward-opening curves that usually are not as symmetric as the logistic. The details behind each curve vary enormously. Some models are based on two or more state variables, such as the amount of plant biomass together with the amount of nutrient in the soil. With two or more state variables, one may be considered "slow" and the others "fast". For example, nutrients are taken up at a faster rate than leaf- or stem-elongation occurs. In this case, the nutrient concentration can be

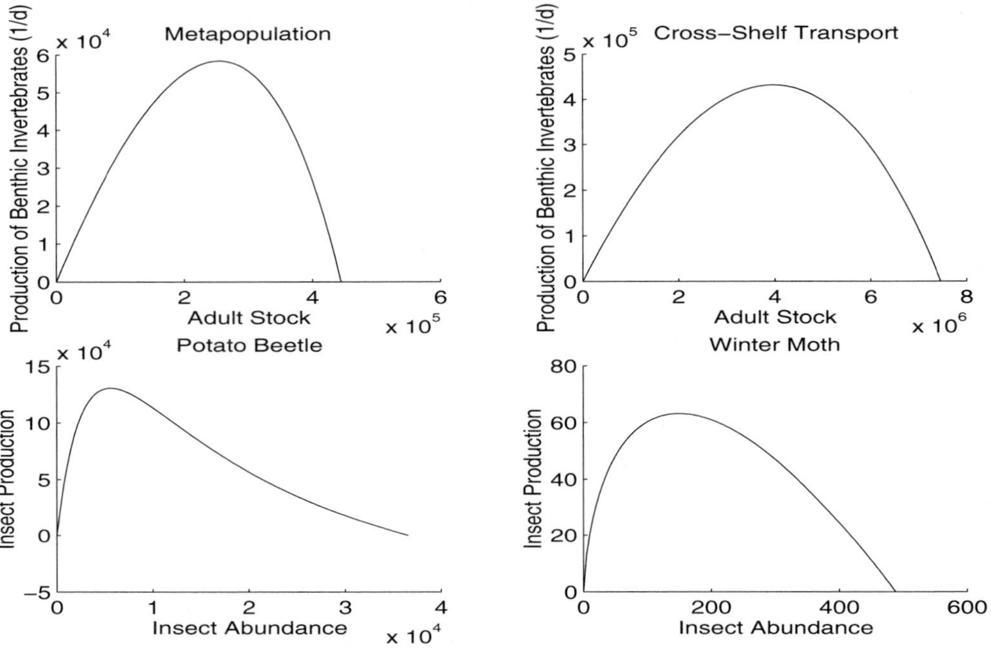

Figure 4.11: Production functions for invertebrate populations. The top left and right are for marine invertebrates with a bottom-dwelling (benthic) adult phase and an ocean-going (pelagic) larval phase. The top left assumes an unstructured "larval pool" and is called a metapopulation model; the top right assumes that the larval pool occupies a band along the coast, within which larvae are moved about by currents and by diffusion. The bottom left and right pertain to insect populations.

assumed to follow, or "track" the pace set by the plant's growth, and the model originally defined with two state variables is then boiled down to a model with just one variable—the slow rate-limiting variable. Tricks like these allow various kinds of models to be compared in a common format, as illustrated in the figures.

Three features of the production functions are worth noting. First, focus on the slope and acceleration where the stock is zero. In the logistic, the slope at $N = 0$ is r and the acceleration there is negative. That is, the first unit of stock earns the highest interest, and each successive unit of stock earns less and less because of the density dependence. (Economists would say that the marginal value of the stock is a decreasing function of stock size.) This is true of all but one of the models of Figures 4.10–4.12 too. The model to the right in Figure 4.12, for a population whose individuals have feeding territories, has a positive slope but no acceleration. What happens is that density-dependence begins only

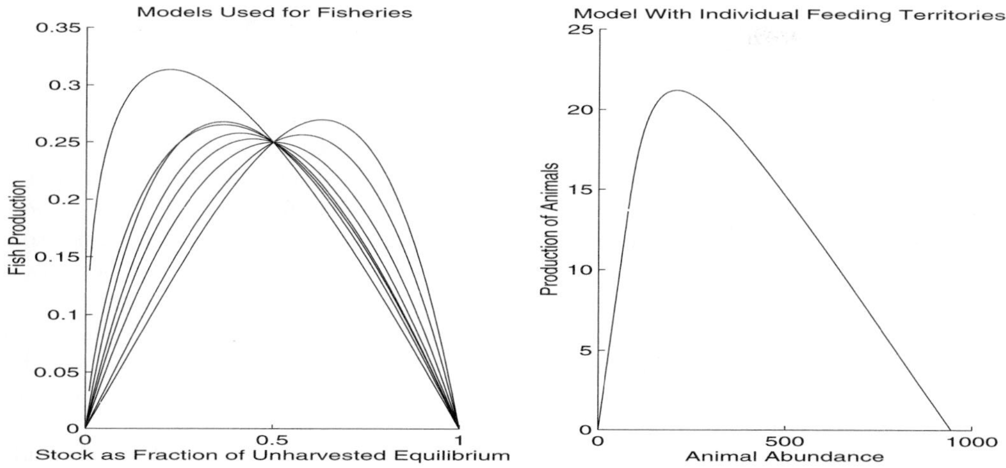

Figure 4.12: Production functions for vertebrate populations. The left illustrates the many forms found in the literature on fisheries management. The right is a neighborhood model for a terrestrial vertebrate where individuals each have feeding territories that become increasingly compressed and less profitable as the population increases. The initial part of the curve is a straight line, where the population dynamics are density-independent. The curve begins to bend when the desired territories of the intersect, and the density dependence begins.

after enough animals have accumulated in the habitat for their territories to intersect. The situation has been termed "density vague" population regulation, and is represented as a production function that begins as a straight line and that bends once the abundance attains the threshold density at which the density dependence occurs.

Although the initial slope of the production functions in Figures 4.10–4.12 are all positive, and all have an initial acceleration that is either negative or zero, ecologists have also discussed the possibility of an initially positive acceleration. An initially positive acceleration is called "positive density dependence" or the "Allee effect," and the idea is that organisms cannot find mates if they are too rare, and perhaps benefit in other ways too from one another's company. If so, a buildup of population initially helps population growth—as though several animals are needed to obtain the critical mass needed for a population explosion. If the initial slope is also positive, coupled with a positive acceleration, then population growth can start from near zero and speeds up faster than exponential growth does. If the initial slope is negative, coupled with a positive acceleration, then the abundance must exceed some threshold for it to start increasing. If the abundance is less than this threshold, then the population goes extinct. Positive density dependence, while possible and plausible, is rarely documented, and so it is almost never built into population-dynamic models, although it could be if needed. The economic significance of positive density dependence is that successive units of stock earn higher rates of return, and thus economically justify more conservation of stock than would otherwise be sought.

Next, focus on the location of the peak to the production function. In the logistic, the peak is midway between 0 and K. By analogy to the logistic, the stock where the production is zero can be called K. The production functions of Figures 4.10-4.12 are skewed either to the right or left of $K/2$ in no particular pattern. The peak production is the maximum sustainable yield (MSY), and the stock leading to MSY is often set as the target stock or goal in resource management. The location of the peak is is relevant to the comparison of this target stock with the unharvested equilibrium, which is at K. Even if another target is chosen, simply knowing where MSY is relative to K is helpful in setting an alternative goal.

Finally, focus on the slope of the production function at K. In some models the production function has a shallow approach to K, and in others it drops sharply. A sharp drop means that the unharvested equilibrium is not very stable, or may even be unstable, depending on how sharp the drop is.

As you can see, the logistic model is a middle-of-the-road model of density-dependent population growth. It's useful to develop theoretical ideas with the logistic; and then, when you want to get real, switch to a process-based model that is special to the kind of system you actually intend to work with.

4.6 The Logistic Weed

A weed is a life style. A weed is a species that lives in temporary habitats spread across a large region, such as sunny gaps in the forest where trees happen to fall. A weed survives by colonizing new pieces of habitat that open up, even as they are excluded from pieces of habitat they have already occupied. A weed species is really a metapopulation, that is, a population of populations. The logistic equation usually refers to a population at a particular site. Yet, despite the conceptual difference between the weed as a metapopulation, and the logistic as representing a single population, a basic model for the dynamics of a weed turns out to mathematically identical to the logistic model.

Here's the idea. The entire region, say 1000 hectares of forest, is assumed to have on the average H pieces of suitable habitat all the time. We let P be the number of the habitats that the weed is occupying, so P can vary between 0 and H. The probability that a weed becomes extinct in a habitat is u, so the total rate at which weeds are being lost from the occupied habitats is uP. The probability that a weed from any given occupied habitat sends a colonist to any particular unoccupied habitat is m. There are $H - P$ unoccupied habitats. So, the rate at which a given occupied habitat is colonizing unoccupied habitats is $m(H - P)$. Because every other occupied habitat is also busy sending out colonists to the unoccupied sites, the total rate at which unoccupied habitats are being colonized by occupied habitats is $m(H - P)P$. The overall change in the number of occupied habitats throughout the forest is then the colonization rate minus the extinction rate, which is

$$\frac{dP}{dt} = m(H - P)P - uP$$

This model is known as the "Levins metapopulation model." Now, the eagle-eyed among you may already have noticed that the right-hand side of this equation is a quadratic, and so can be rearranged to look like the now-familiar logistic model. Specifically, rearrange this to be

$$\frac{dP}{dt} = r(1 - \frac{P}{K})P$$

where r and K are defined in terms of the colonization and extinction parameters as

$$\begin{aligned} r &= mH - u \\ K &= \frac{mH - u}{m} \end{aligned}$$

The Levins metapopulation model when expressed in this form may be called the "logistic weed." Perhaps the most interesting feature of this model is how it highlights the importance of habitat number to the survival of a species with a weedy life style. For the species to persist, the colonization parameter must exceed the extinction parameter, ($mH > u$). If the forest is cut down, or modified through use in such a way that the number of habitats, H, in it is lowered, then mH can drop to a level lower than u. If so, the weedy species goes extinct throughout the entire forest, not only in the part where the forest was cut down.

Thus, there are lots of ways that the logistic model, or logistic-like models, arise in ecology. Therefore, it seems sensible to use the logistic model to illustrate the basic theoretical idea of stability, and to illustrate how we can tell if a system is stable.

4.7 Ecological Stability

Now we come to some simple but *very* basic concepts. Consider a bowl sitting on a table and a marble. Put the marble in the bowl and it eventually comes to rest at the bottom of the bowl. The bottom of the bowl is called an *equilibrium point*—it is called this because once the marble is at the bottom, its position doesn't change any more. The bottom of the bowl is also called a *stable* equilibrium point because, if the marble is started out anywhere away from the bottom, it goes to the bottom anyway. Furthermore, pick up the bowl in your hands and shake it around a bit, not too much, just jiggle it. The marble will then bounce around near the bowl's bottom. This means that the the location of the bottom is a good approximation to where the marble is, even though the bowl, which can be thought of as the marble's environment, is fluctuating.

Next, turn the bowl upside down, making a dome, and place it on the table. Put the marble exactly on its top. It will stay there. The top of the dome is also an *equilibrium point.* Now nudge the marble a bit. It then rolls away, gathering speed as it goes. The top of the dome is called an *unstable* equilibrium point because if the marble is put somewhere near the top, it moves away with increasing speed. Of course, if you pick up the dome and shake it, there's no way a marble will stay near the top. So the dome's top is an area from which a marble is repelled, whereas a bowl's bottom is an area to which a marble is attracted. From a marble's point of view, a hilly landscape is a collection of attracting and repelling domains.

Finally, consider the table itself, and put the marble on the table. It just stays there. Therefore, every point on the table could be called an equilibrium point. But if you touch the marble anywhere, it moves and keeps moving, but at a constant speed. This situation is called *neutral stability*. If you shake the table, the marble wanders off in random directions until it falls off the edge.

What does this all have to do with ecology? It's the force of gravity that moves the ball down into the base of a bowl or from the top of a dome. In ecology, the idea is the same, it's the forces that are different. When a population comes to equilibrium at K, the bowl analogy is appropriate. The forces are exponential growth and density dependence, and these may combine to produce a stable equilibrium at K. As we'll see though, these forces don't *necessarily* combine to produce stability. Instead, an unstable equilibrium may result, as in the dome analogy, and if so, the ecological system is destined for hard times as its marble rolls off the top.

Just to be sure we're clear about all this, remember that the key distinction is between constancy and stability. If a system can persist, in principle, in some state without changing, that state is called an equilibrium. But whether that state is stable is another matter altogether. To be stable, a system has to have forces in it that bring it to the equilibrium if it is not there to begin with[4].

If you read the journals in ecology, you'll encounter a big disconnect between field workers and modelers over the concepts of equilibrium and stability. Modelers are always on the lookout for equilibria, especially stable ones. To predict how a system changes, they find it really useful to know where the equilibria are—it's like going on a hike with a topo map. Knowing where the equilibria are provides a context for how a system changes, just as a topo map does for the ups and downs of a hiking trail. Field workers, for reasons I have never understood, think that modelers studying equilibria are somehow making a philosophical statement that the world is constant. This isn't so. Just as a hiker needs to know where North is, a modeler needs to know where the equilibria are and whether they're stable.

4.7.1 Continuous Time

How can we tell if an ecological system, or any system for that matter, is stable? It's a piece o' cake, almost routine, and here's how. Let's write the production function of a

[4]Some synonyms you may come across: in mathematics, an equilibrium point is called a "fixed point" and also a "stationary state"; in physical chemistry and thermodynamics, an "equilibrium" is distinguished from a "steady state." An equilibrium is a state of the chemical system that doesn't change and for which the input and output of energy and/or material is zero. A steady state is a state that doesn't change and the flow of matter and energy into it is positive and equals the flow out. In ecology, an equilibrium includes both the physical chemist's equilibrium as well as the steady state because, by a physical chemist's definition, all equilibria in biology are really steady states. So for "equilibrium" to have any meaning at all in ecology, it must be broadened to include steady states as well as chemical equilibria. In ecology, the word "resilience" is often used in the same breath as "stability". While stability refers to whether the marble will return to the equilibrium if it is nudged away from it, resilience refers to whether the marble will stay near the bottom if the bowl is shaken. Stability usually implies resilience mathematically.

population-dynamic model for continuous time in the general form

$$\frac{dN}{dt} = F(N)$$

For the logistic, $F(N) = r(1 - N/K)N$. Typically, $F(N)$ is what is called a nonlinear model, which means that the state variable, N, is raised to some power (other than one). The logistic production function, being a downward-opening quadratic, features a term with N^2.

Step 1 is to find all the equilibrium points. This is done by solving for N in

$$F(N) = 0$$

Because the logistic is a quadratic, there are two roots, $N = K$ and $N = 0$. We have to look at *all* the equilibria because that's the only way we can visualize the whole topo map.

Step 2 is to check the stability of each equilibrium found in step 1. We begin by agreeing to focus on the fate of the population when it is near an equilibrium. Let $\hat{N}$ denote an equilibrium (either 0 or K in the logistic). The *deviation* from equilibrium is labeled as n, and defined as $n = N - \hat{N}$. Let's see if we can predict what happens to n near an equilibrium point whose stability we're going to check out. If n is small to begin with, and if it gets smaller through time, then the equilibrium is *stable* because this means the population's deviation from the equilibrium is shrinking through time, so the population is actually moving to the equilibrium point. Conversely, if n is small to begin with, but grows through time, then the equilibrium is *unstable* because the deviation from equilibrium is increasing through time, and the population is moving away from equilibrium. So, let's see how to tell if n decreases or increases through time. Let's substitute $n = N - \hat{N}$ into the population model and see what we can learn. Rearranging, to get $N = n + \hat{N}$, we have

$$\begin{aligned} \frac{dN}{dt} &= F(N) \\ \frac{d(n + \hat{N})}{dt} &= F(n + \hat{N}) \end{aligned}$$

Now $\hat{N}$ is just a number (*it* doesn't change in time, n does), so its derivative is zero, and

$$\frac{dn}{dt} = \frac{d(n + \hat{N})}{dt} = F(n + \hat{N})$$

Fine; at this stage it seems that a differential equation needs to be solved for n, by our integrating the production function $F(n + \hat{N})$ over time. Remember though, that $F(N)$ is typically a nonlinear equation, and we usually can't find an explicit formula to solve this differential equation. So it looks like we're stumped. But, the value to focusing on n instead of N in the first place is that we only care about whether the population goes toward or away from the particular equilibrium point whose stability is being checked. Therefore, we can assume that n is small to begin with, and see what happens in the neighborhood of $\hat{N}$.

If n is small we can develop a power series expansion (Taylor series expansion) around $\hat{N}$ as

$$F(n + \hat{N}) \approx F(\hat{N}) + F'(\hat{N})n$$

where $F'(\hat{N})$ means the derivitive of $F(N)$ with respect to N evaluated at $N = \hat{N}$. This power-series expansion is carried out only to the first order. The terms of second order and beyond are being neglected, because n is assumed to be very small (much less than 1), and so the terms in n^2, n^3 and so forth are tiny in comparison to the term in n. With this assumption, the differential equation for n becomes

$$\frac{dn}{dt} = F(\hat{N}) + F'(\hat{N})n$$

We're almost done. Because $\hat{N}$ is an equilibrium point, $F(\hat{N})$ equals 0 (remember, this is how we found $\hat{N}$ in step 1), so the differential equation for n is

$$\frac{dn}{dt} = F'(\hat{N})n$$

To simplify the notation, let's denote

$$\lambda = F'(\hat{N})$$

so finally we have

$$\frac{dn}{dt} = \lambda n$$

By now, as an expert on exponential growth, you'll recognize this differential equation as none other than exponential growth at rate λ, so the formula for its solution is

$$n(t) = e^{\lambda t} n(0)$$

This formula answers the question of whether an initial deviation from an equilibrium point shrinks or grows through time. *Let λ denote the slope of $F(N)$ at an equilibrium. If λ is negative, any small deviation from that equilibrium shrinks through time, and that equilibrium is stable. If λ is positive, any small deviation from that equilibrium grows through time, and that equilibrium is unstable.* If λ is zero, then we don't know, without doing more work, whether the equilibrium is stable, unstable, or possibly even neutrally stable. λ usually is either positive or negative, and the case where it equals zero doesn't come up very often, so we won't bother with it here. This all may seem a bit much to take in at once, but the bottom line is simple. To tell if an equilibrium is stable or not, calculate the slope of the production function at the equilibrium, $F'(\hat{N})$. If it's negative, the equilibrium is stable; if it's positive, the equilibrium in unstable; and if it's zero, take a coffee break.

4.7.2 The Logistic Model without Harvesting

Let's apply the test for stability to the logistic model—we already know the answer: $\hat{N} = 0$ is unstable and $\hat{N} = K$ is stable, because we already know that the population size moves away from 0 and toward K. So, this gives us a chance to see if the test works. Of course, we'll use MATLAB to do the analysis. So, begin by defining the logistic production function

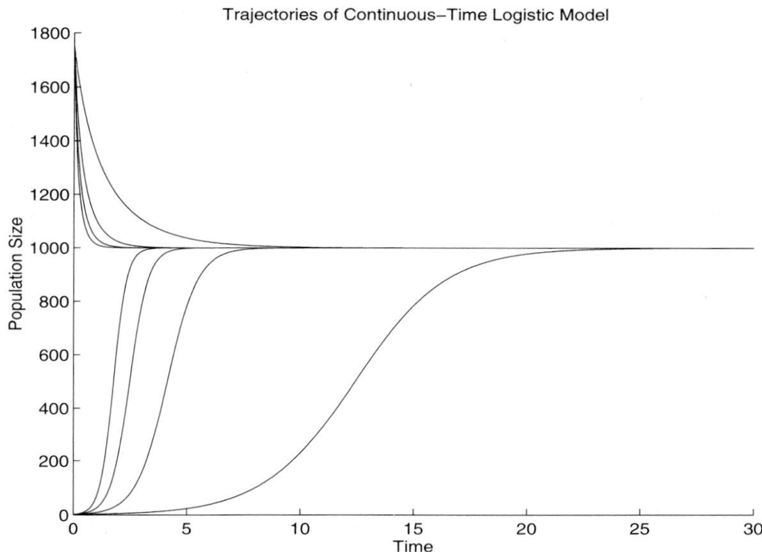

Figure 4.13: Trajectories of the continuous-time logistic model from initial conditions above and below $K = 1000$, and with $r = 0.5, 1.5, 2.5$, and 3.5.

```
>> dn = 'r*n*(k-n)/k'
```

The equilibria are then found with

```
>> nhat = solve(dn,'n')
```

which yields, as we expect

```
nhat = [0]
       [k]
```

Let's assign a separate name to each of the equilibria with

```
>> nhat1 = sym(nhat,1,1)
>> nhat2 = sym(nhat,2,1)
```

At this point, `nhat1` is `0` and `nhat2` is `k`. To see if these are stable, we'll need the formula for the derivative of the logistic production function, which is found as

```
>> ddn = simplify(diff(dn,'n'))
```

yielding

```
ddn = -r*(-k+2*n)/k
```

Now let's evaluate this derivative at each of the two equilibria. For the first equilibrium, we have

```
>> lambda1 = subs(ddn,nhat1,'n')
```

which yields

```
lambda1 = r
```

and for the second equilibrium

```
>> lambda2 = subs(ddn,nhat2,'n')
```

yielding

```
lambda2 = -r
```

We're all done. The λ's at 0 and K are r and $-r$ because the logistic model, being a quadratic equation, is symmetric around its peak at $K/2$. If the λ for an equilibrium is positive, the equilibrium is unstable; and if it's negative, the equilibrium is stable. Here, the λ's turn out to depend only on the intrinsic rate of increase, r. If r itself is positive, then the equilibrium at $\hat{N} = 0$ is unstable and the equilibrium at $\hat{N} = K$ is stable. Of course, you can just look at the graph of the logistic production function and see by inspection that the slope is positive at 0 and negative at K. However, actually calculating the slope at each equilibrium tells you how fast a small deviation from that equilibrium shrinks or expands.

Figure 4.13 illustrates a bunch of trajectories for the logistic model. The initial conditions are below and above K, and the r's are 0.5, 1.5, 2.5, and 3.5. The figure was made with

```
>> n = 'k/(1+(k/no-1)*exp(-r*t))'
>> n_plot = sym2ara(n);
>> figure
>> hold on
>> k = 1000;
>> t=0:.1:30;
>> r = .5;
>> no = 2;
>> plot(t,eval(n_plot),'y')
```

and so on for curves with other selections of `no` and `r`. Notice that all the trajectories converge smoothly to K regardless of whether N is initially above or below K, and of how big r is.

Now that you've cut your teeth on the logistic model, let's move to the stability analysis of a model that's both important and has an answer that's not so obvious in advance.

4.7.3 Harvesting a Logistic Resource

Suppose the logistic model is describing a population that you wish to harvest, such as fish or trees. How much should you take? That is, suppose you can control the harvest rate, h, as so many tons of fish per year, or so many board feet of timber per year. What is the best h?

To begin with, you'll have to take a policy position on what the best h means. One possibility is to aim for the maximum sustainable yield. If the stock size, N, is $K/2$, then the population growth rate is $rK/4$. So if the harvest rate is also $rK/4$, then the harvest balances the population growth, and the population size stays at $K/2$. This is the maximum sustainable yield. Thus, $h = rK/4$ with $N = K/2$ is one possible management target. But there are many other possible sustainable harvest rates too, and they have their pros and cons as well.

Economists have pointed out that harvesting at the maximum sustainable yield may not in fact do the most good. Suppose the resource is owned by a firm, or equivalently, that it is being managed for the public good by a government agency. In an economic context, the goal should be to maximize the revenue derived from the resource, because this is the way either the firm that owns the resource, or the public being served by the governmental agency, can derive the maximum benefit from the resource. To see what the best target is from an economic standpoint, one has to imagine that the stock of organisms is "natural capital." The idea is that each organism is worth some amount, say \$1. This dollar may hypothetically be "invested" as natural capital by its maintenance as an organism. Or, this dollar may be invested in some savings account, where it earns a rate of return called ρ. Abstractly, the problem then is one of best allocating capital between the natural stock of organisms and the financial stock in the savings account. Whereas each dollar in the savings account always earns a return at rate ρ, a dollar invested as an organism earns at the rate described by the production function, $F(N)$, and this rate of return varies because of density dependence. So what is the best allocation between investing dollars in organisms and investing dollars in a savings account?

The answer is found by imagining that you're starting with an empty environment, and then investing each dollar, one by one, either into the natural stock or the savings account. What does the dollar earn if invested in the savings account?—it's ρ. What does the dollar earn if invested as an organism of natural stock? Well, if the present stock is N, the earnings from it are $F(N)$, the natural production function evaluated at N. If the stock is increased by \$1, then the earnings from it become $F(N+1)$. So the change in earnings per change in investment is

$$\frac{F(N+1) - F(N)}{(N+1) - N}$$

In fact, if we were thinking of investments penny by penny rather than dollar by dollar, then a tiny extra investment in stock from N to $N + dN$ would earn extra revenue of

$$\frac{dF(N)}{dN}$$

which is just the derivative of the production function at N. Now for the punch line—as you invest dollar by dollar you put the first dollar where it earns the most, which is, say, into organisms. Then with each additional dollar, the return from the organisms drops because of density dependence, and when you get to the stock size where the return from organisms equals ρ, stop investing into organisms and switch your remaining money into the savings account. In symbols, the best stock size is the N at which

$$\frac{dF(N)}{dN} = \rho$$

Incidentally, economists call $\frac{dF(N)}{dN}$ the "marginal rate of return" of the natural resource. So in words, the best stock size is that at which the marginal rate of return equals the interest rate.

Now let's see what the economically best management target works out to be for a natural stock that grows according to the logistic model. The logistic model with a harvest rate h, is entered in MATLAB as

```
>> dn = 'r*n*(k-n)/k-h'
```

The marginal rate of return from the stock is then

```
>> ddn = simplify(diff(dn,'n'))
```

yielding

```
ddn = -r*(-k+2*n)/k
```

So the economically best stock size is that where the marginal rate of return equals the interest rate, which in MATLAB is found as

```
>> necon = solve(symop(ddn,'-','rho'),'n')
```

which pretty-prints as

```
         (r - rho) k
     1/2 -----------
              r
```

Thus, the economically optimal stock size is less than $K/2$ unless the interest rate ρ happens to be zero. So, the economically optimal stock size is to the left of the peak of the production function. The sustainable harvest rate that this stock supports is

```
>> hecon = simplify(solve(subs(dn,necon,'n'),'h'))
```

which pretty-prints as

```
              2      2
          k (r  - rho )
     1/4 -------------
               r
```

If ρ happens to be zero, the economically optimal harvest happens to be the maximum sustainable yield, otherwise it is less than this.

What, you may be asking, does all this have to do with stability? The concept of an optimal sustainable yield is an equilibrium concept. As you now know, it's vitally important to determine if an equilibrium is stable. Is harvesting at the maximum sustainable yield stable? Is harvesting at the economically best yield stable? Let's find out.

First, to find the equilibria to a logistic model with harvesting, we set the production function to zero and solve for N. When `dn` is defined as above, MATLAB finds the solution upon our typing

```
>> nhat = solve(dn,'n')
```

This yields a symbolic column vector with two equilibria, which are assigned to `nhat1` and `nhat2` with

```
>> nhat1 = simple(sym(nhat,1,1))
>> nhat2 = simple(sym(nhat,2,1))
```

Typing `pretty(nhat1)` then displays

```
                         1/2           1/2
                        k     (r k - 4 h)
         1/2 k - 1/2 -------------------
                               1/2
                              r
```

and similarly, pretty-printing `nhat2`, yields

```
                         1/2           1/2
                        k     (r k - 4 h)
         1/2 k + 1/2 -------------------
                               1/2
                              r
```

The first root is to the left of $K/2$ and the second is to the right of $K/2$. Figure 4.14 illustrates these two equilibria as the points at which the horizontal line for h intersects the logistic production function. For the case in which the harvesting is at the economically best rate, the `hecon` is substituted for `h`. To determine if these equilibria are stable, we evaluate the slope of the right-hand side of the equation for dN/dt, including the harvesting at each of the equilibria. We've already differentiated `dn` with respect to `n`—this is `ddn`. So if we evaluate `ddn` at `n` equal to `nhat1` and `h` equal to `hecon`, by our typing

```
>> lambda1 = simplify(subs(subs(ddn,nhat1,'n'),hecon,'h'))
```

we get

```
lambda1 = rho
```

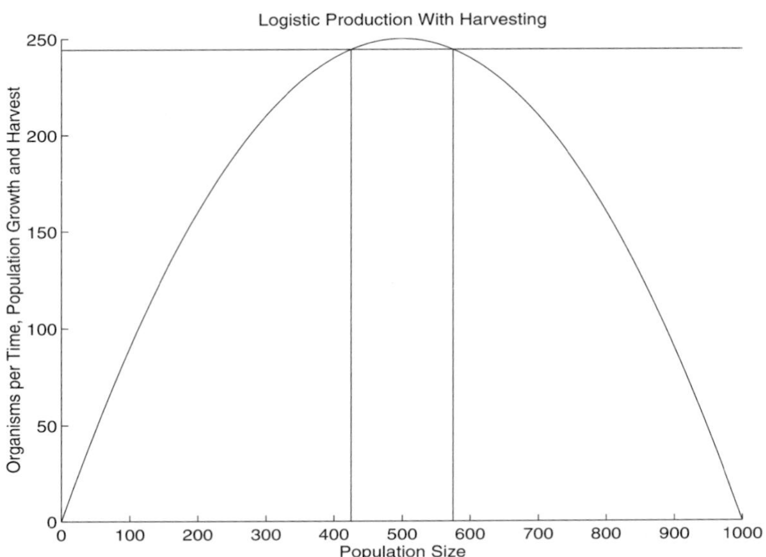

Figure 4.14: Setup for the logistic model with harvesting. The production function is shown as a downward-opening parabola. The harvest rate is a horizontal line. The equilibria for the given harvest rate are the points at which the horizontal line for h intersects the parabola. The economically optimal stock size is the point where the slope of the parabola equals the interest rate, which is the equilibrium to the left of $K/2$.

Similarly, typing

```
>> lambda2 = simplify(subs(subs(ddn,nhat2,'n'),hecon,'h'))
```

yields

```
lambda2 = -rho
```

Therefore, the first equilibrium is unstable, and the second is stable because `lambda1` is positive and `lambda2` is negative. But whoa! The first equilibrium is the economically optimal one—it's the place on the left of $K/2$ where the slope of the production function equals the interest rate.

To see numerical examples of these results, we can generate the setup for the problem in Figure 4.14 with

```
>> dn = 'r*n*(k-n)/k';
>> dn_plot = sym2ara(dn);
>> r = 1;
>> k = 1000;
```

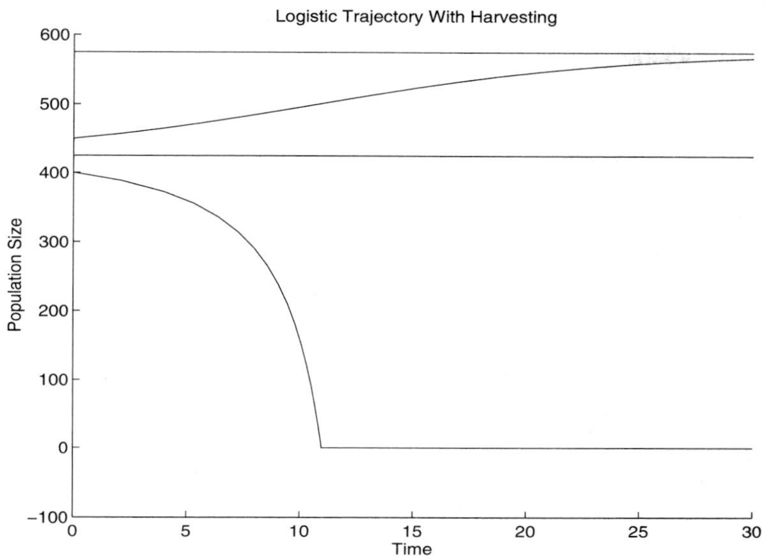

Figure 4.15: The fate of a logistic population harvested at the economically optimal rate. If the population is started at a point below the economically optimal target stock size, it is harvested to extinction. If it is started at a point above the economically optimal target, it grows in size until it reaches the stable equilibrium to the right of $K/2$.

```
>> n = 0:10:k;
>> rho = 0.15;
>> hecon_n = eval(hecon);
>> nhat1_n = eval(subs(nhat1,hecon,'h'));
>> nhat2_n = eval(subs(nhat2,hecon,'h'));
>> figure
>> hold on
>> plot(n,eval(dn_plot),'y')
>> plot([0 k], [hecon_n hecon_n],'c')
>> plot([nhat1_n nhat1_n], [0 hecon_n],'r')
>> plot([nhat2_n nhat2_n], [0 hecon_n],'g')
```

Then Figure 4.15 was produced, to illustrate the fate of a population that is harvested at the economically best rate. To prepare this figure, we must numerically integrate the logistic production function with harvesting, which is a differential equation. As we saw in Chapter 1, numerical integration is done with the `ode23` command in MATLAB. When we use this command, the function to be integrated must be stored as a separate file. So, with a text editor, write the following text into the file `logist_c.m`:

```
function dn = logist_c(t,n)
  global r k h;
  if (n>0)
   dn = r*n*(k-n)/k-h;
  else
   dn = -n;
  end
```

This function simply defines the logistic model with a harvesting term. It is called with the time variable `t`, which is not in fact used, and with the state variable, `n`. The function returns the derivative of `n` with respect to time. The parameters are passed to the function implicitly: when the function is invoked, they are declared as `global` both in the function itself and at the command line in MATLAB. The function, `logist_c`, returns the logistic model when the population size is greater than zero, and returns `-n` if the population size is negative. This return prevents the solution from running away to $-\infty$. Once the file `logist_c.m` has been created, Figure 4.15 is made by our typing

```
>> global r k h;
>> tmax = 30;
>> h = hecon_n;
>> figure
>> hold on
>> [t,n] = ode23('logist_c',0,tmax,nhat1_n+25);
>> plot(t,n)
>> [t,n] = ode23('logist_c',0,tmax,nhat1_n-25);
>> plot(t,n)
>> plot([0 tmax],[nhat1_n nhat1_n],'r')
>> plot([0 tmax],[nhat2_n nhat2_n],'g')
```

The figure illustrates two trajectories, one started above the economically best stock size, and the other below it. The trajectory started below the economically best stock falls to zero, indicating a population harvested to extinction. The trajectory started above the economically best stock rises and converges to the stable equilibrium on the right of $K/2$. These trajectories illustrate that the economically best target harvest and stock size is an unstable equilibrium. The target of a maximum sustainable yield is no better. It too is an unstable equilibrium.

That the economically best target is an unstable equilibrium under constant harvest has serious policy implications. I've suggested that stock at fisheries such as the cod of Newfoundland have become extinct because they have been managed at a target that is an unstable equilibrium. I've also suggested that a stock should be managed for a steady state size of $(3/4)K$, which leads to an sustainable harvest that is also ecologically stable. Over the long run this target is in fact the best economically too, because the loss of income from collapsed fisheries is avoided.

4.7.4 Discrete Time

It is possible that an equilibrium is much larger in a discrete-time model than in a continuous-time model. In the continuous-time logistic, $N = K$ is always stable provided r is positive—it doesn't matter how big r is, just as long as it's positive. In the discrete-time logistic this result is not true, and for $N = K$ to be stable, r must be positive but not too big. To see how this surprising finding comes about, and what it means, let's look into how we can tell if an equilibrium is stable for a discrete-time model.

Although we have been writing a discrete time model in the form

$$\Delta N = F(N)$$

where ΔN is $N_{t+1} - N_t$ and $F(N)$ is the production function, it's more customary in the literature on discrete-time models to express the same model in the form

$$\begin{aligned} N_{t+1} &= N_t + F(N_t) \\ &= G(N_t) \end{aligned}$$

Here, the left-hand side is N_{t+1} and the right-hand side, which is denoted as $G(N_t)$, is simply N_t plus the production function, $F(N_t)$. (I personally prefer the ΔN style, but we follow local custom.) When this style of writing a model is used, an equilibrium is the root of

$$N = G(N)$$

which simply says that at equilibrium, N_{t+1} equals N_t.

OK, so how do we tell if an equilibrium is stable? We'll use the same approach as before. We'll imagine that N_t is very close to an equilibrium, and see if the distance between it and the equilibrium shrinks or grows through time. Let $\hat{N}$ be an equilibrium. The deviation from this equilibrium is $n = N - \hat{N}$. How does this deviation change? Well, according to the model,

$$\begin{aligned} N_{t+1} &= G(N_t) \\ n_{t+1} + \hat{N} &= G(n_t + \hat{N}) \end{aligned}$$

Now we again take advantage of the fact that we are considering only the fate of tiny deviations from equilibrium, and therefore can expand $G(N)$ in a power series to first order, yielding

$$n_{t+1} + \hat{N} = G(\hat{N}) + G'(\hat{N})n_t$$

where $G'(\hat{N})$ means the derivative of $G(N)$ evaluated at $\hat{N}$. Now $\hat{N} = G(\hat{N})$, so we are left with

$$n_{t+1} = G'(\hat{N})n_t$$

To simplify notation we let

$$\lambda = G'(\hat{N})$$

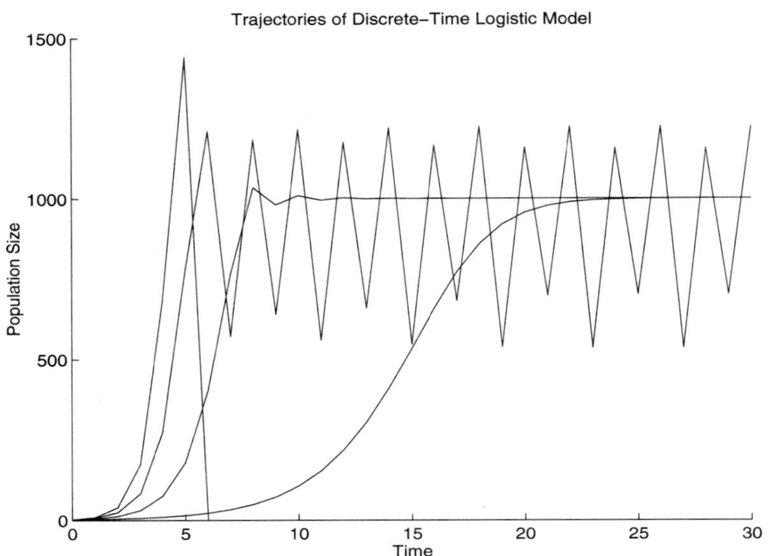

Figure 4.16: Trajectories of the discrete-time logistic model from an initial condition of below $K = 1000$, with $r = 0.5, 1.5, 2.5$, and 3.5.

and then

$$n_{t+1} = \lambda n_t$$

At this point, you'll recognize that we have the discrete-time model for geometric growth. So, if the initial deviation from equilibrium is n_o, the deviation at time t is

$$n_t = \lambda^t n_o$$

The test for stability is now obvious. When does n_t shrink to zero?—when the absolute value of λ is less than one. When does n_t expand through time?—when the absolute value of λ is greater than one. When the absolute value of λ exactly equals one, we don't know whether the equilibrium or stable or not without further work. In summary, *let λ be slope of $G(N)$ at an equilibrium. If $|\lambda| < 1$, any small deviation from that equilibrium shrinks through time, and that equilibrium is stable. If $|\lambda| > 1$, any small deviation from that equilibrium grows through time, and that equilibrium is unstable. If $|\lambda| = 1$, more work is needed to determine if the equilibrium is stable.*

With your experience in stability analysis in continuous time, you can easily apply these criteria to the discrete-time logistic model. First, to define to MATLAB the logistic model in discrete time, type

```
>> ntplusone = 'n+r*n*(k-n)/k'
```

Next, to find the equilibria, type

```
nhat = solve(symop(ntplusone,'-','n'),'n')
```

This yields two roots, 0 and K, which are assigned to `nhat1` and `nhat2`, respectively, by our typing

```
>> nhat1 = sym(nhat,1,1)
>> nhat2 = sym(nhat,2,1)
```

Now calculate $G'(N)$ by typing

```
>> dntplusone = simplify(diff(ntplusone,'n'))
```

The job is completed by evaluations of $G'(N)$ at each of the two equilibria. To evaluate the equilibrium at $N = 0$, type

```
>> lambda1 = simplify(subs(dntplusone,nhat1,'n'))
```

yielding

```
lambda1 = 1+r
```

To evaluate the equilibrium at $N = K$, type

```
>> lambda2 = simplify(subs(dntplusone,nhat2,'n'))
```

which yields

```
lambda2 = 1-r
```

If r is positive, then $N = 0$ is unstable because $1+r$ is greater than one. The equilibrium at $N = K$ is another matter. If r is between 0 and 2 then $|1-r|$ is less than one, and $N = K$ is stable. But if $r > 2$ then $|1-r|$ is greater than one, and $N = K$ is unstable. Thus, in the discrete-time logistic, for $N = K$ to be stable, r can't be too big. If r is too big, then neither equilibrium is stable and the population never settles down to any steady state but keeps fluctuating instead.

Figure 4.16 illustrates trajectories for the discrete-time logistic; these can be compared with those of the continuous-time logistic shown in Figure 4.13. The discrete-time version of the logistic is decidedly less prone to smooth stable trajectories than the continuous-time version. Figure 4.16 was generated by our typing

```
>> figure
>> hold on
>> k = 1000; runlen = 30; no = 2;
>> r = .5;
>> n = logist_d(r,k,no,runlen);
>> plot(0:runlen,n,'y')
```

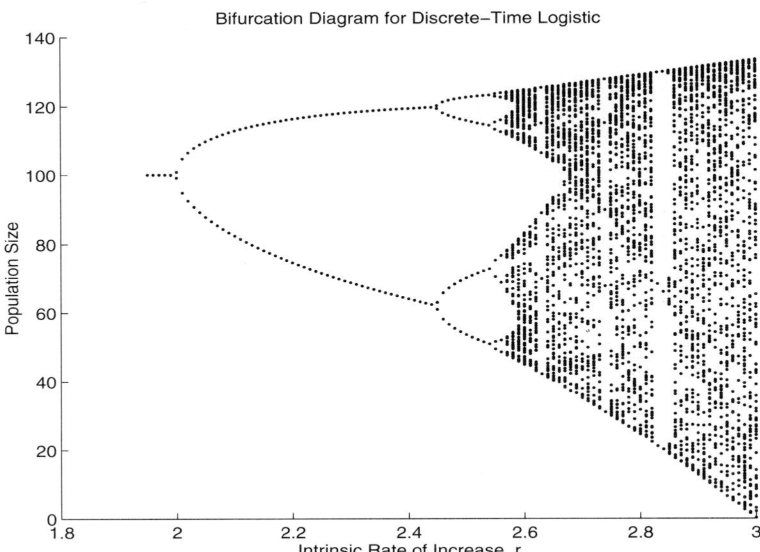

Figure 4.17: The asymptotic population sizes occupied in the discrete-time logistic model as a function of the intrinsic rate of increase.

and so on for other values of r. The general lesson from this exploration of the discrete-time logistic model is that density dependence is not necessarily a stabilizing mechanism. To bring about stability, the density-dependent feedback on population growth must act quickly and accurately. If not, the density dependent feedback can cause the population to overshoot the equilibrium, and if the overshoots are big enough, the population never converges to equilibrium and instead continues to fluctuate forever or to go extinct.

4.8 Ecological Chaos

What happens when r is greater than 2 in the discrete-time logistic? Figure 4.16 shows an oscillation between 4 distinct population sizes when r is 2.5. Is this a special case? What happens for other values of r? To find out, let's run the function `logist_d` for a range of values between 2.0 and 3.0. In fact, let's start at $r = 1.95$, because we know that K is a stable equilibrium for this r, and it's always a good idea to start at a point where we already know the answer and work out from there. Furthermore, we want to look at the population size that occurs after the effect of the initial condition has disappeared. So, let's iterate the logistic model for 500 time steps and display the population sizes that occur in the last 100 of these time steps.

```
>> figure
```

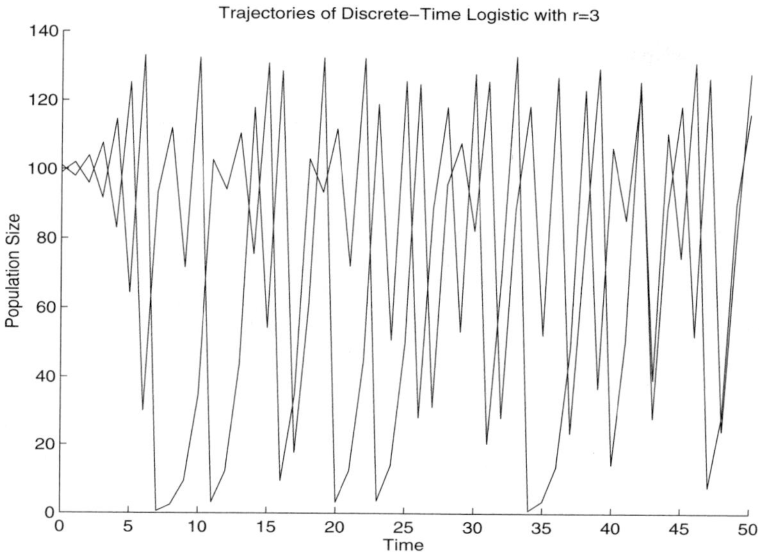

Figure 4.18: The fate of two chaotic discrete-time logistic populations started very close to each other.

```
>> hold on
>> for r=1.95:.01:3
>>   n = logist_d(r,100,101,500);
>>   rx = r*ones(1,501);
>>   plot(rx(401:501),n(401:501),'.')
>> end
```

These commands set up a loop, with `r` running from 1.95 to 3 in steps of 0.01. Each time the loop is carried out, the discrete-time logistic model is iterated by `logist_d` for 500 steps; the present `r` and a `k` of 100 are used, and the starting population size is 101. The sequence of population sizes is returned as the vector `n`. We want to plot the last 100 of these as dots in the y-direction, and use the present value of `r` as the x-coordinate. The command `r*ones(1,501)` gives us a vector of 500 numbers, each of which is `r`. Then, `plot(rx(401:501),n(401:501),'.')` plots the last 100 of the population sizes, and uses the population size as the y-coordinate and `r` as the x-coordinate. The loop then repeats for another `r`. The resulting graph appears as Figure 4.17, and is called a "bifurcation diagram."

When r is less than 2, the only population size that occurs is at K, because K is a stable equilibrium. As r is varied above 2, the population size oscillates between two values, one above K and the other below K. This oscillation is called a "two-point cycle." Varying a parameter is sometimes called "tuning," by analogy with tuning the dial on a radio. As r

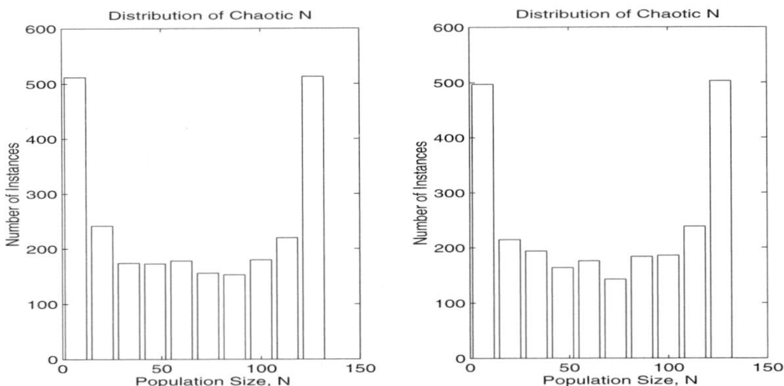

Figure 4.19: The statistical distribution of population sizes for two chaotic discrete-time logistic populations started with different initial conditions.

is tuned above 2, the two values between which the population oscillates, appear to "grow" out from the equilibrium at K. This is called a "bifurcation." As r is tuned still higher, the oscillation between two values turns into an oscillation between four values. In fact, an r of 2.5, as illustrated in Figure 4.16, is in the interval of r's in which the population oscillates in a four-point cycle. As r is tuned even higher, the number of points in the oscillations generally increases, and when r equals 3, something special happens—the population is said to be "chaotic."

Chaos in mathematical models is all the rage recently—it's even been mentioned in the movie "Jurassic Park." Here's what chaos is all about. When r=3, the population generally doesn't settle down to any one steady state or to a particular oscillation between some fixed number of states. Instead, it wanders around the interval between 0 and $(4/3)K$, and comes arbitrarily close to any point within that interval. Furthermore, if two populations start out near to one another, they don't stay together, but deviate from one another so completely that after a while one cannot tell that they were ever near one another to begin with.

Figure 4.18 illustrates this idea. Two populations with $r = 3$ and $K = 100$ are started close to each other, one at $N = 101$ the other at $N = 99$. They are each run for 2500 iterations. Here's how the runs are generated:

```
>> n1 = logist_d(3.0,100,101,2500);
>> n2 = logist_d(3.0,100, 99,2500);
```

Figure 4.18 shows the first 50 time steps in the total iteration. It is made by our typing

```
>> figure
>> hold on
>> plot(0:50,n1(1:51),'y')
>> plot(0:50,n2(1:51),'r')
```

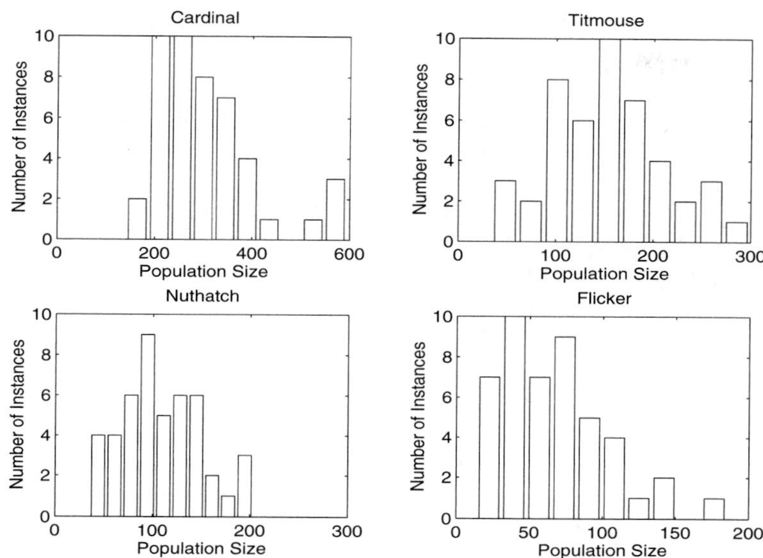

Figure 4.20: The statistical distribution of population sizes for birds from Buckeye Lake Park, Ohio.

Although both populations start out near one another, soon each has wandered far away from the other.

Because two populations that start near one another completely lose their closeness through time, in practice one cannot hope to predict the long-term fate of a population if its trajectory is chaotic. One cannot in practice ever know the initial condition exactly, and any error in the initial condition is magnified through time, so that one cannot predict what the population's size will be far off into the future. This inability to predict the future is true even though the model has no randomness. But don't despair. It's not hard at all to predict the fate of a chaotic population in a statistical sense, nor is it hard to falsify the claim that a real population is exhibiting the dynamics of discrete-time logistic chaos. Although two chaotic populations that are initially close to each other do lose their closeness, they also both converge to the same statistical distribution of population sizes through time. We can make a histogram of the population sizes through time for both of the chaotic populations we just computed. The histograms appear in Figure 4.19; to make them, we type

```
>> figure
>> subplot(1,2,1)
>> hist(n1)
>> subplot(1,2,2)
>> hist(n2)
```

Both distributions are nearly the same, and in fact are identical in the limit as time becomes very large. The distributions are ∪-shaped, with peaks at 0 and $(4/3)K$; this shape indicates that the populations spend a lot of time above K, then drop to near 0 and then bounce back up over K again. This statistical regularity is quite predictable, even though the detailed trajectory of a specific population is not predictable unless the initial condition is known exactly.

To close this chapter, let's return to the birds from Buckeye Lake Park in Ohio, where we began the chapter. Recall that we did find density dependence in these bird populations. To see if their fluctuations through time could be the result of chaotic dynamics as represented by the discrete-time logistic model, first estimate the r and K for these populations. For example, to find the r and K for the cardinal, type

```
>> rk_card = fmins('rkprod',[.5 300],[],[],card);
>> rcard = rk_card(1)
>> kcard = rk_card(2)
```

from which we learn that r and K for the cardinal are 0.4793 and 332.3017 respectively. Folowing the same steps for the other birds, we find that

```
species    r       k
card     0.4793 332.3017
tmouse   0.5043 162.3266
nhatch   0.7871 121.5122
flick    0.4601  79.0966
```

It's obvious that these r's are low. These r's should cause K to be a stable equilibrium. The r must be greater than 2 for oscillatory phenomena to emerge, and near 3 for the oscillations to be complex enough to appear chaotic. The real r's for the Buckeye Lake birds are not consistent with chaos. Furthermore, we can plot a histogram of the population sizes of the Buckeye Lake birds. Figure 4.20 illustrates these, which are made when we type

```
>> figure
>> subplot(2,2,1)
>> hist(card)
>> subplot(2,2,2)
>> hist(tmouse)
>> subplot(2,2,3)
>> hist(nhatch)
>> subplot(2,2,4)
>> hist(flick)
```

The histograms are all ∩-shaped, not ∪-shaped as predicted by the dynamics of discrete-time logistic chaos. Recall that fluctuations in population size may be caused by demographic and environmental stochasticity, as well as by chaotic dynamics. Environmental stochasticity, whereby both r and K vary somewhat from year to year, is almost surely the main cause of the fluctuations in the abundance of the Buckeye Lake birds.

Although the Buckeye Lake birds may not be exhibiting ecological chaos, the jury is still out on how often ecological systems fluctuate chaotically. Chaos is turbulence. Fluids readily show chaotic dynamics, as storms, breaking waves, and waterfalls illustrate. Ecological chaos is ecological turbulence. An ecological system that's sitting in a chaotic physical environment acquires a signature of chaos from the environment. But do ecological interactions generate chaos, turbulence, on their own, without help from the environment? Although such a scenario is possible, I doubt it's very common, but the jury's still out.

About now you may be asking, "Where do populations get their r and K to begin with?" The answer is, "From evolution." And that brings us to the next chapter.

4.9 Further Readings

Alexander, S. and Roughgarden, J. (1996). Larval transport and population dynamics of intertidal barnacles: a coupled benthic/oceanic model. *Ecological Monographs*, 66:259–275.

Bishop, Y. M. M., Fienberg, S. E., and Holland, P. W. (1975). *Discrete Multivariate Analysis: Theory and Practice.* The MIT Press.

Gause, G. F. (1934). *The Struggle for Existence.* Hafner Press (Reprint 1964).

Gilpin, M. E. and Ayala, F. J. (1973). Global models of growth and competition. *Proc. Nat. Acad. Sci. (USA)*, 70:3590–3593.

Hassell, M. P., Lawton, J. H., and May, R. M. (1976). Patterns of dynamical behavior in single-species populations. *Journal of Animal Ecology*, 45:471–486.

Hirsch, M. W. and Smale, S. (1974). *Differential Equations, Dynamical Systems, and Linear Algebra.* Academic Press.

Huisman, J. and Weissing, F. J. (1994). Light-limited growth and competition for light in well-mixed aquatic environments: an elementary model. *Ecology*, 75:507–520.

Levins, R. (1969). Some demographic and genetic consequences of environmental heterogeneity for biological control. *Bulletin of the Entomological Society of America*, 15:237–240.

Lotka, A. J. (1925). *Elements of Physical Biology.* Williams and Wilkins (Reprinted 1956, Elements of Mathematical Biology, Dover).

May, R. M. (1976). Simple mathematical models with very complicated dynamics. *Nature*, 261:459–467.

May, R. M. (1977). Thresholds and breakpoints in ecosystems with a multiplicity of stable states. *Nature*, 269:471–477.

May, R. M., Beddington, J. R., Horwood, J. W., and Sheperd, J. G. (1978). Exploiting natural populations in an uncertain world. *Mathematical Biosciences*, 42:219–252.

Pacala, S. W. and Silander, J. A. (1990). Field tests of neighborhood population dynamic models of two annual weed species. *Ecological Monographs*, 60:113–134.

Roughgarden, J. (1997). Production functions from ecological populations: a survey with emphasis on spatially implicit models. In Tilman, D. and Kareiva, P., editors, *Spatial Ecology: The Role of Space in Population Dynamics and Interspecific Competition*, page in press. Princeton University Press.

Roughgarden, J. and Iwasa, Y. (1986). Dynamics of a metapopulation with space-limited subpopulations. *Theoretical Population Biology*, 29:235–261.

Roughgarden, J. and Smith, F. (1996). Why fisheries collapse and what to do about it. *Proc. Nat. Acad. Sci. (USA)*, 93:5078–5083.

Schaffer, W. M. (1985). Order and chaos in ecological systems. *Ecology*, 66:93–106.

Schoener, T. W. (1973). Population growth regulated by intraspecific competition for energy or time: some simple representations. *Theoretical Population Biology*, 4:56–84.

Sugihara, G. and May, R. M. (1990). Nonlinear forcasting as a way of distinguishing chaos from measurement error in time series. *Nature*, 344:734–741.

Tilman, D. (1977). Resource competition between planktonic algae: an experimental and theoretical approach. *Ecology*, 58:338–348.

4.10 Application: Avoiding the Collapse of Wildebeest

In this chapter, several different models of population growth with resource limitation have been discussed[5]. All of the models have considered the carrying capacity as a constant. In reality of course, carrying capacity varies from year to year, because it depends on factors such as rainfall, frequency of frosts, and so forth. This year-to-year weather-driven variation does not undermine the basic principles of resource-limited population growth, although it may obscure the clarity of density-dependent trends.

A simple, but useful, extension of logistic population growth modeled with weather-driven variation has been applied to the management of Serengeti wildebeest populations. Wildebeests are regularly poached, and because their populations are currently quite large, Serengeti park officials would like to legalize a limited amount of harvesting. In allowing harvesting, park officials seek some compromise between maximizing harvest and minimizing the risk of wildebeest extinction (or herd collapse).

The problem is complicated by the fact that wildebeest demography is tightly connected to rainfall (because rainfall determines the amount of plant material produced each year as food), and rainfall varies unpredictably among years. A simple way of incorporating this complication is to define the wildebeests' carrying capacity as a function of rainfall; in

[5]This section is contributed by Peter Kareiva.

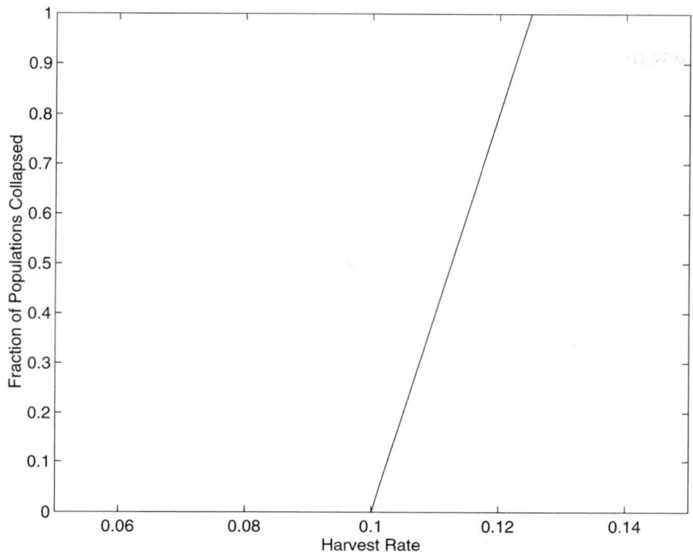

Figure 4.21: The results of a stochastic model simulating wildebeest population levels. This graph shows what fraction of the runs will result in the collapse of the wildebeest population, for a given harvest rate.

particular, the relation `K = rainfall*20748` has been shown to describe the dynamics of past wildebeest populations exceedingly well.

This `K` is plugged into what is called the Beverton-Holt model for resource limited growth. The Beverton-Holt model is but one of a wide variety of different equations that qualitatively behave like the logistic equation, although their details vary. That is, at low population densities, the per-capita reproduction rate is the highest; and when carrying capacity is exceeded, the per-capita net reproduction rate falls to below 1 (so that the population shrinks). The Beverton-Holt equation for resource-limited growth is

$$N_{t+1} = \frac{RN_tK}{K + (R-1)N_t}$$

where R is the maximum geometric growth factor achieved as N goes to zero and K is the carrying capacity.

We can use MATLAB to apply this equation to the management issue of harvesting in a variable environment. We now treat K simply as a variable governed by the equation `K=20748*rainfall`, with `rainfall` drawn as a random variable from a list of past rainfalls recorded for the Serengeti. The idea then is to simulate future scenarios with different degrees of harvesting, where h represents the fraction of population harvested. The complete model thus is

$$N_{t+1} = (1-h)\frac{RN_tK}{K+(R-1)N_t}$$

To implement this in MATLAB, we first create an array of actual observed rainfalls from the Serengeti,

```
>> rain=[100,38,100,104,167,167,165,79,91,77,134,192,235,159,...
        211,257,204,300,187,84,99,163,97,228,208,83,44,112,191,...
        202,137,150,158,20];
```

and then generate the carrying capacities associated with each rainfall with

```
>> global K;
>> K= 20748*rain;
```

We will consider five different harvest rates,

```
>> h=[.05 .075 .1 .125 .15];
```

and use a function `four1` that iterates wildebeest populations from one year to the next, according to the Beverton-Holt model. It uses random draws of rainfall (where `No` is the initial population size, `h` is the fraction of population that is harvested, and `runlen` is how many years the simulation will be run):

```
function [collapse] = four1(No,h,runlen)
 global K;
 R = 1.1323;
 n = [];
 clps = 0;
 for s = 1:100
   n = [No];
   for t = 1:runlen
     i = randperm(33);
     k = K(i(1));
     a = n(t)* ( (R*k) / (k+(R-1.0)*n(t)) ) * (1-h);
     if a < 150000
        if a ~= 0.0
        clps = clps+1;
        end;
        a=0.0;
     end
     n =[n a];
   end
 end
 collapse=clps/100;
```

The above function uses a `R` of 1.1323 (which has been estimated from field data on population growth for wildebeests), and sets the threshold of 150,000 wildebeests as a collapsed herd. (This number is nearly the record low for the Serengeti herd.) After you've saved the function, you can simulate the fate of the wildebeest for each harvest level. Type

```
>> clps=[];
>> for i=1:5;
>>    clps(i)=four1(250000,h(i),100);
>> end;
>> figure; plot(h,clps)
```

The output of this model is a graph that plots the fraction of 100 stochastic runs in which the population collapses during a 100-year simulation period (Figure 4.21). Clearly, as the rate of harvest increases, there is an increase in the likelihood of collapse, although the pattern is clearly nonlinear. The details of this relationship will depend on the time horizon over which the simulation is run (50 years? 200 years?) and the threshold population below which a herd is categorized as collapsed. But in all cases there is a tradeoff between harvesting a large number of wildebeests and risk of population collapse.

The key reference for this application is

Pascual, M. and Hilborn, R. 1995. Conservation of harvested populations in fluctuating environments: the case of the Serengeti wildebeest. *Journal of Applied Ecology* 32: 468-480.

Chapter 5

Ecology Evolving

A species is putty to the hand of natural selection. Natural selection picks the colors, sculpts the shape, and modulates the behavior of organisms to fit their environment. Or does it? This golden metaphor of natural selection as a sculptor fashioning organisms increasingly well-adapted to their environment expresses an enticing ideal. Examining its validity is the major topic of this chapter. Because we use some form of this principle all the time, we need to examine whether natural selection increases adaptation to the environment. When we say an organism does X to survive or to reproduce, we are implicitly affirming that natural selection brings about traits *because* they improve survival and reproduction. Discovering the adaptive function of a trait explains why it exists. We need to know if this line of reasoning is OK. The remaining topics consider how ecology affects evolution and vice versa. This chapter extracts from evolutionary biology what is needed for ecology. Along the way it treats some of the most interesting conceptual topics in science today, including why we age as fast as we do and why we reproduce when we do.

Evolutionary biology started with Darwin's observations of plants and animals on the Galapagos Islands and his comparisons of them with those he had recently seen on the mainland of South America. At Darwin's time a biological species was thought to be the biological equivalent of a chemical species, such as water, salt, oxygen and any other compound. Wherever you go, water is water and salt is salt. You don't visit a foreign country and find that ice melts at a higher temperature there, or that salt doesn't dissolve in water there. It's absolutely fundamental that a chemical species represents the same thing everywhere and for all time. Biological species were supposed to be this way too. A robin is a robin, a blue jay is a blue jay, everywhere and for all time—but no, this isn't true. Darwin observed that the birds of the Galapagos were modified versions of those he had already met in South America. It gradually dawned on him that, unlike chemical species, biological species *are* changeable. Darwin went on to hypothesize that natural selection, the natural counterpart to people's breeding of horses, dogs, and cattle, is what caused species to be modified. Since the 1920's biologists have investigated whether Darwin's hypothesis that natural selection causes evolution is correct. It's important to remember though, that the most fundamental of Darwin's discoveries is simply that biological species *change* through time, in contrast to chemical species that are forever.

Biologists define evolution as a change through time in the proportion of genes for various traits. Today, in the human species as a whole more people have brown or black hair than blond hair. If, several generations from now, more people have blond hair than brown or black, then evolution will have taken place. As we will see, there are many reasons why the proportion of genes in a species change through time. Each mechanism that can change the proportion of genes in a population is called a "force of evolution." By far the most important force is called "natural selection," about which we say much more later. As an ecologist, you need to know the menu of possible causes of evolutionary change, and how natural selection compares with the other forces. The place to begin is with the genetic variation itself.

5.1 Genetic Variation

Suppose you're interested in the evolutionary fate of a trait such as tolerance to high air temperature, a trait of increasing importance as global warming continues. Suppose caterpillars die when the air temperature exceeds 100°F in the shade. Where I presently live, it's rare for summer temperatures to exceed 100°F, but according to projections of global warming, temperatures in excess of 110°F will be commonplace in several decades. Will the butterflies where I live, and where you live as well, go extinct as a result of global warming? To answer this question using contemporary evolutionary biology, one must first hunt around for genetic variation among butterflies in the ability to survive at high temperatures. We need to find the one in a hundred, or one in a thousand, that actually can make it through a day at 110°F. On balance, this rare form probably can't do something else that the common form can do, because otherwise it would already be the common form itself. But the environment of the future is projected to be lethal for the presently common type, and so the future of the species rests on whether the heat-tolerant form will spread through the species in time to save it from extinction.

When you have located some rare individuals who can make it through a hot day, the second step is to carry out some crossing experiments to determine the genetics of how heat tolerance is inherited. Now we come to an Achilles' heel of evolutionary biology today. It's expensive to determine the genetics of ecologically interesting traits, and you can bet that no one knows the genetics of heat tolerance for the butterflies in my backyard, or yours either[1]. By now, a theoretician's remedy to this lack of knowledge will come as no surprise. We'll just assume that the genetics of heat tolerance is determined by the simplest possible genetic system, namely, one locus with two alleles[2]. That is, we'll pretend

[1]Actually, a colleague, Ward Watt, knows the genetics of heat tolerance in the sulphur butterflies of *his* backyard at the Rocky Mountain Biological Laboratory in Colorado—so it is *possible* to know the genetics of temperature tolerance, it's just expensive.

[2]For our purposes, a "locus" is a spot on a chromosome. Two different genes that can occupy the same spot on a chromosome are called "alleles." Typically, an organism has one chromosome from its father and a matching chromosome from its mother. Therefore, it has two alleles at each locus, and the pair of alleles

that there are two alleles at some locus, A_1 and A_2, such that A_1A_1 is the genotype of the presently common form, and A_1A_2 is the genotype of the presently rare heat-tolerant form. We'll also pretend that the other possible homozygote, A_2A_2, is even more tolerant of high temperatures than the heterozygote, A_1A_2, is. At the moment there aren't any A_2A_2 individuals because the A_2 gene is rare, so when it does occur it winds up being paired with an A_1 allele. Assuming that this simple genetic basis is true, we'll then figure out how evolution by natural selection works. Of course, if a more accurate prediction is needed, then the time and expense of determining the actual genetic basis must be invested. Luckily though, predictions developed from a one-locus two-allele genetic assumption are quite robust, so that the time and expense of determining the actual genetic basis of a trait are rarely worthwhile for ecological purposes.

We'll call the "gene pool" the collection of all the alleles at the A locus. If there are 1,000 caterpillars, then there are 2,000 alleles in the gene pool at this locus, because each caterpillar has 2 alleles in it. Let p stand for the fraction of the gene pool that is A_1, and q the fraction that is A_2[3]. If 1 in 1,000 caterpillars is heat-tolerant with genotype A_1A_2 and everyone else is A_1A_1, then p is 1999/2000, which is 0.9995, and q is 1/2000, which is 0.0005. Successful evolution of tolerance to heat means that A_2 must come to replace A_1 in the gene pool, which means that over time p must go from 0.9995 down to 0.0005 and q must go from 0.0005 up to 0.9995. If this happens fast enough, the species will have evolved heat tolerance and avoided extinction. Evolutionary theory in principle can tell us how fast evolution goes, and whether the butterflies will win the race to adapt to global warming.

5.1.1 The Hardy-Weinberg Law

Although one can directly count the number of A_1A_1, A_1A_2, and A_2A_2 genotypes, as we imagined we did above when we found 999 A_1A_1 caterpillars, 1 $A_1A_2A_2$ caterpillar, and 0 A_2A_2 caterpillars, usually there's a simple relation between the numbers of each genotype and the gene-pool fractions, p and q. By themselves, the gene-pool fractions don't say how the genes are organized into genotypes, but if we know something about how the mating takes place, then we can connect p and q with the genotype numbers. By definition, the mating is said to be "random union of gametes" with respect to the A locus, if the probability that an A_1 allele is paired with an A_2 allele is what would be expected if these alleles just collided with one another at random. This is not as fanciful as it might seem. Sea urchins, for example, really do release sperm and eggs into the sea where they're mixed around by the ocean waves, and fertilization takes place by random collision of gametes in the water. Now think about the pattern of genotypes that will emerge from this process. p of the gametes carry A_1 and q of the gametes carry A_2. Make a table of the possibilities:

---

at a locus is called its "genotype" at that locus. The organism's "phenotype" is the trait that these alleles together express, such as heat tolerance. If both an organism's alleles at a locus are the same it is called a "homozygote," otherwise it is a "heterozygote."

[3] p are q are also called "allele frequencies" or "gene frequencies."

Sperm	Egg
A_1: p	A_1: p
A_1: p	A_2: q
A_2: q	A_1: p
A_2: q	A_2: q

The probability that the sperm has A_1 is p, and the probability that it combines with an egg carrying A_1 too is also p, so the probability that a zygote has two A_1's is p^2. Similarly, the probability that a zygote winds up with two A_2's is q^2. The probability that the zygote has an A_1 and an A_2 is $2pq$ because the probability that the sperm brings the A_1 and the egg brings the A_2 is pq *plus* the probability the sperm is A_2 and egg is A_1 is qp—either way the zygote winds up with an A_1 and an A_2, so the total probability of this is $2pq$. Thus, if we know p and q, and if we also know that the mating is by random union of gametes, the fraction of the population in each genotype is

Genotype	Fraction
A_1A_1	p^2
A_1A_2	$2pq$
A_2A_2	q^2

This ratio of genotypes is called the *Hardy-Weinberg* ratio. It is formed after just one generation of random union of gametes[4]. Returning to our caterpillars, 999 of them were A_1A_1 and one was A_1A_2. Why didn't we see a A_2A_2 caterpillar too? Because under random union of gametes, the anticipated ratio of genotypes is 0.9995×0.9995, $2 \times 0.9995 \times 0.0005$, and 0.0005×0.0005. This works out to be 0.99900025, 0.00099950, and 0.00000025[5]. These ratios imply that a population of 1000 caterpillars would have 999 individuals of A_1A_1, and 1 individual of A_1A_2, as postulated. The population is too small to have any A_2A_2 individuals. If the population size were 100 million caterpillars, then 25 A_2A_2 individuals would be expected.

The Hardy-Weinbery law is a special case of Hardy-Weinberg ratios. If it turns out that p doesn't change through time, then the genotypic proportions don't either. The Hardy-Weinberg law states that if p is constant through time, then the proportions of A_1A_1, A_1A_2, and A_2A_2 also remain constant through time. If p does change through time, the Hardy-Weinberg *ratios* are always resynthesized anew each generation by random union of gametes using the current value of p. The numerical value of the proportions of A_1A_1,

[4]If the egg and sperm don't have the same fractions of both alleles to begin with, one extra generation is needed for both egg and sperm to come to have the same p and q, provided the A locus is not sex-linked. Thus, if p and q are not the same in both egg and sperm, it takes two, rather than one, generations for the Hardy-Weinberg ratios to form.

[5]To calculate these numbers with the variable precision arithmetic capability of MATLAB, type

```
>> p = '1999/2000'; q = '1/2000';
>> vpa(symop(p,'*',p))
>> vpa(symop('2','*',p,'*',q))
>> vpa(symop(q,'*',q))
```

A_1A_2, and A_2A_2 will change through time if p changes through time, but they will always be in Hardy-Weinberg *ratios* to one another.

The practical use of the Hardy-Weinberg relations today is to connect the gene pool ratios with the genotype ratios, and incidentally to explain why a rare allele is present only in a heterozygous individual unless the population is very large. The Hardy-Weinberg ratios are also used as a basis for comparison with other genotypic ratios that emerge when the mating is not according to random union of gametes. If organisms mate with their relatives more often than expected by random collision, then the mating is said to involve "inbreeding." When there is inbreeding, the ratio of the heterozygote to homozygote categories is lowered relative to the Hardy-Weinberg proportions. If a normally random-mating population is inbred, perhaps because of low population size, then deleterious recessive alleles are expressed because there are now relatively more homozygotes. These deleterious recessive alleles were hidden in heterozygotes when there was random mating. The overall negative effect of exposing the hidden deleterious alleles is called an "inbreeding depression." A naturally inbreeding species doesn't suffer an inbreeding depression because its deleterious recessive alleles have already been lost. If organisms mate with others having the same traits as they do more often than expected by random collision, the mating is called "positive assortative" mating. This has many of the same effects as inbreeding.

The Hardy-Weinberg law, that the genotype ratios don't change if p doesn't change, seems rather obvious, but not in historical context. The alternative to Mendelian inheritance is called "blending inheritance." According to this idea, offspring are the average of their parents. For example, consider fur thickness, and suppose one parent's fur is 1 cm thick and the other's is 2 cm. Then the fur thickness of the offspring is 1.5 cm if the inheritance is blending. Now consider all the matings in the population. The net effect of blending inheritance throughout the whole population is to lose variability in fur thickness because the children of each mating cannot have fur as thick or as thin as their parents, but wind up with something in between. So, the traits of thick and thin fur thickness are gradually lost, and every one ends up the same, with an intermediate fur thickness. Thus, with blending inheritance, the mating process itself, carried out through time, causes variation to decline. In contrast, with Mendelian inheritance, the mating process itself, carried out through time, doesn't change the degree of variation in the population, and this message is what the Hardy-Weinberg law stated for the first time.

The Hardy-Weinberg law was also exceedingly important in further clarifying the difference between a biological species and a chemical species. A chemical species can exist in pure form, such as a pure crystal of salt or diamond. Any color in a crystal of salt or diamond is an imperfection. In contrast, variation is part of the essential nature of a biological species—it's because of genetic variation that a species can change, and what variation it has is organized in a regular pattern, namely, in the Hardy-Weinberg ratios or its equivalent for other mating systems. There is no such thing as a pure robin or pure blue jay, or pure cockerspaniel or pure poodle. Some of the worst injustices of human history have come from attempts at ethnic cleansing to obtain racial purity. In fact, a biological population without genetic variation is impossible naturally, and if somehow created artificially, would soon be maladapted to its environment and destined for rapid extinction.

5.2 Natural Selection

Artificial selection dates to the time when people originally domesticated wild plants and animals. Natural selection has been around even longer, since the beginning of life. Natural selection takes place when the individuals of the different genotypes either have different probabilities of survival or produce different numbers of offspring, or both. Therefore, some genotypes spread in the population relative to the others. If the population size is growing, all genotypes could be expanding in absolute numbers even as their relative proportions are changing. If the population size is more or less constant, then the spread of some genotypes means that others are declining.

5.2.1 Predicting What Natural Selection Does

It's easy to develop a formula for how genes spread because of natural selection. We simply keep track, like accounting, of how many of each genotype there are from beginning to end of the life cycle. To illustrate, let's use the simplest life cycle around—all the organisms are synchronized in their life stages, and they reproduce only once and then die. We'll use discrete time with the time step being one generation. At the start of the generation all the organisms are zygotes, as indicated in the "Zygotes" column in the table below:

Genotype	Zygotes	Adults	Gametes
A_1A_1	p^2N	$l_{11}p^2N$	$2m_{11}l_{11}p^2N$
A_1A_2	$2pqN$	$l_{12}2pqN$	$2m_{12}l_{12}2pqN$
A_2A_2	q^2N	$l_{22}q^2N$	$2m_{22}l_{22}q^2N$

The zygotes are made by a random union of gametes, according to the p and q at the beginning of the generation. The zygotes therefore are in the Hardy-Weinberg ratios of p_t^2, $2p_tq_t$, and q_t^2, where p_t is the fraction of A_1 in the gene pool at the start of generation t. If there are N_t zygotes in total, then $p_t^2N_t$ is the number of A_1A_1 genotypes, $2p_tq_tN_t$ is the number of A_1A_2 genotypes, and $q_t^2N_t$ the number of A_2A_2 genotypes, at the beginning of generation t. We now label the probability that an individual of genotype A_1A_1 survives to adulthood as l_{11}, and similarly for l_{12} and l_{22}. Now, if we multiply the number of zygotes of any genotype by the probability of their surviving to adulthood, we obtain the number of adults of that genotype, as indicated in the "Adults" column in the diagram. Moving right along, we now check out what happens after the adults mate. Let m_{11} label one-half the number of gametes that an individual of genotype A_1A_1 makes that are actually incorporated into the zygotes that start the next generation. OK, I know that sounds weird, but we do this to make the units work out right. For example, if an individual of genotype A_1A_1 typically leaves 100 gametes that are actually incorporated into the zygotes at $t+1$, then m_{11} is 50. Similarly, m_{12} and m_{22} are each one half of the number of gametes produced by an individual of A_1A_2 and of A_2A_2 respectively, that are actually incorporated into the zygotes of time $t+1$. This is all summarized in the "gametes" column of the diagram. Now we're ready to compute the fraction in the gene pool at time $t+1$. By definition,

$$p_{t+1} = \frac{A_1 \text{ alleles at } t+1}{\text{total alleles at } t+1}$$

To fill in the needed information, we ask, How many A_1 alleles are there at $t+1$? It's all the gametes made from parents of genotype A_1A_1 plus one-half of the gametes made from parents of genotype A_1A_2. What's the total number of gametes? It's the sum of the "gametes" column. Plugging this in, we get

$$p_{t+1} = \frac{2m_{11}l_{11}p_t^2N_t + (1/2)2m_{12}l_{12}2p_tq_tN_t}{2m_{11}l_{11}p_t^2N_t + 2m_{12}l_{12}2p_tq_tN_t + 2m_{22}l_{22}q_t^2N_t}$$

Finally, some stuff divides out, the N_t and a "2"; the "(1/2)" takes out a "2," and we can pull the p_t out of the numerator leaving a more manageable equation:

$$p_{t+1} = \frac{m_{11}l_{11}p_t + m_{12}l_{12}q_t}{(m_{11}l_{11}p_t^2 + m_{12}l_{12}2p_tq_t + m_{22}l_{22}q_t^2)}p_t$$

We're nearly done. In evolutionary biology it's customary to lump the survival and reproduction into a single number whenever possible. Here we can lump the product of m and l into a single coefficients, usually denoted as a W and called *selective values*, according to

$$\begin{aligned} W_{11} &= m_{11}l_{11} \\ W_{12} &= m_{12}l_{12} \\ W_{22} &= m_{22}l_{22} \end{aligned}$$

where upon the equation for the state of the gene pool at $t+1$ is predicted from that at time t by

$$p_{t+1} = \frac{W_{11}p_t + W_{12}q_t}{(W_{11}p_t^2 + W_{12}2p_tq_t + W_{22}q_t^2)}p_t$$

This is a fundamental equation in evolutionary biology, and versions of it were originally derived by R. A. Fisher, J. B. S. Haldane, and S. Wright during the 1920's. It predicts the course of evolution by natural selection at a particular locus from any initial condition, when the selective values for the genotypes at that locus are known.

The selective value is a measure of what biologists call the "fitness" of an individual. An individual's fitness is the number of copies of its genes that it puts into the next generation—it's the individual's impact on the next generation's gene pool. The W_{ij}'s are the formulas used to measure genetic fitness for this particular model of evolution based on discrete nonoverlapping generations.

A companion equation predicts the population size at $t+1$. If you think about it, the information that goes into the W's, the probability of survival and the number of successful gametes for each genotype, is precisely what is needed to predict the population size at the next generation. So why not do both?—that is, use the W's to predict both the gene pool and population size at $t+1$, even though our main interest is in the gene pool to begin with. By definition

$$N_{t+1} = (1/2)(\text{total alleles at } t+1)$$

because each individual has two alleles at the locus, and the number of individuals is one-half the total number of alleles. So, summing the "gamete" column from the diagram above yields

$$N_{t+1} = (1/2)(2m_{11}l_{11}p_t^2 N_t + 2m_{12}l_{12}2p_t q_t N_t + 2m_{22}l_{22}q_t^2 N_t)$$

Then factoring out the N_t, and using W's instead of ml's, we obtain

$$N_{t+1} = (W_{11}p_t^2 + W_{12}2p_t q_t + W_{22}q_t^2)N_t$$

The equations for p_{t+1} and N_{t+1} can be used together to predict how the gene pool *and* the population size change through time from any initial condition for these state variables, when the three W's are known. Evolutionists tend to focus only on the equation for p_{t+1}, while ecologists, being fundamentally interested in abundance as well as in traits, use both equations.

The expression in parentheses above has a very important meaning of its own. Because p_t^2, $2p_t q_t$, and q_t^2 are the ratios of the three genotypes, the average selective value in the population is precisely this expression, and is labeled as $\bar{W}$. The average selective value is also called the "average fitness," or mean fitness, because the W's are measures of fitness, as previously mentioned. From an evolutionary standpoint, the average fitness measures the average degree of adaptation to the environment among the members of the population. From an ecological standpoint, the average fitness is a geometric growth factor. Because $N_{t+1} = \bar{W}N_t$, we see that $\bar{W}$ is just like the "big R", of previous chapters—the only difference is that the time step must be one generation here, whereas we used "big R" for the geometric increase over any time step in previous chapters. So, to summarize, the equations that are used together to predict both the change in the gene pool and the change in population size from one generation to the next, are

$$\begin{aligned} p_{t+1} &= \frac{p_t W_{11} + (1-p_t)W_{12}}{\bar{W}_t} p_t \\ N_{t+1} &= \bar{W}_t N_t \end{aligned}$$

where the average fitness, or factor of geometric increase over one generation, is

$$\bar{W}_t = p_t^2 W_{11} + 2p_t(1-p_t)W_{12} + (1-p_t)^2 W_{22}$$

Notice, we've used $(1-p_t)$ instead of q_t to emphasize that there are only two variables in this system of equations, p_t and N_t.

The area of biology that treats evolution from a genetic standpoint is called "population genetics" and was founded by R. A. Fisher, J. B. S. Haldane, and S. Wright, who introduced the equation for p_{t+1} presented above. Originally, Darwin described natural selection in terms of exponential growth bringing about competition for limiting resources. It's often not realized that Darwin's original presentation of natural selection is unnecessarily restrictive, and has not been used in biology since the founding of population genetics. The derivation of the equation for p_{t+1} makes no mention at all of competition for resources or of limits to growth. Natural selection as understood in population genetics is all about differences between net rates of reproduction. Even if the population as a whole is growing without limits, the components in it that reproduce faster than others grow in numerical proportion,

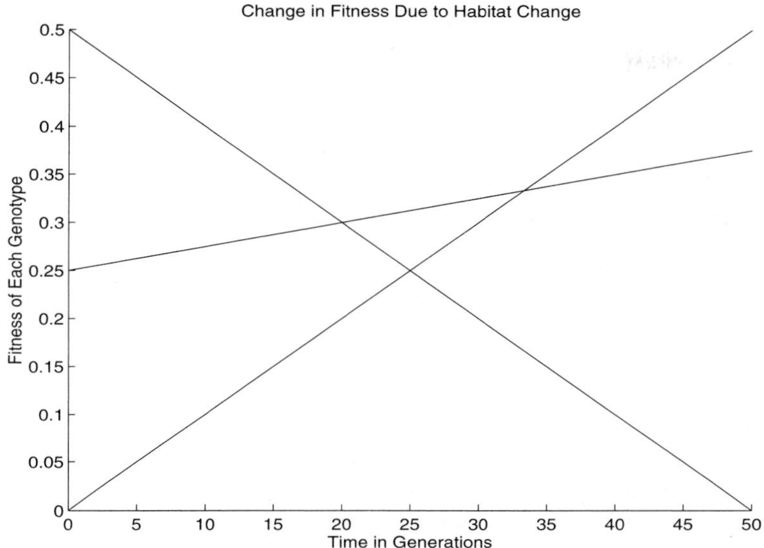

Figure 5.1: The probability of survival for each of three genotypes in a hypothetical butterfly population during global warming through the next 50 years, assuming one generation per year. A_1A_1 starts at 1/2, A_1A_2 at 1/4, and A_2A_2 at 0.

and on a percentage basis, eventually come to constitute nearly the entire species. Of course, there *might* be limiting resources as well, and if so, an important special case of natural selection called "density-dependent selection" is all about this. Natural selection itself is a much more general concept. Once again, natural selection is the differential net production of offspring by different genotypes.

5.2.2 Evolutionary Scenario for Global Warming

Returning to the butterflies endangered by global warming, what do the formulas we have developed say about their chances of surviving the next 50 years or so? The scenario we would like to model is that the survivorship of the A_1A_1 genotype progressively declines through time, while that of the A_1A_2 and A_2A_2 genotypes progressively improve through time. The fertility, m_{ij}, is assumed to be the same for all genotypes, and not to change with global warming, although of course, this could be modeled too if desired. A possible setup for this scenario is to imagine that the survivorships of the A_1A_1, A_1A_2, and A_2A_2 genotypes change linearly in time over the next 50 years according to

```
>> l11 = '0.5*(1-t/50)';
>> l12 = '0.25+t/400';
>> l22 = 't/100';
```

Figure 5.1 illustrates each genotype's probability of survival in future years, and is made by typing

```
>> t=0:50;
>> figure
>> hold on
>> plot(t,eval(l11),'g');
>> plot(t,eval(l12),'y');
>> plot(t,eval(l22),'r');
```

The probability that an A_1A_1 individual survives from birth to reproduction starts out at 1/2, and declines linearly to 0 after 50 years. This decline indicates that it cannot tolerate the hot temperatures of the future. Conversely, the probability of survival for the A_2A_2 genotype increases linearly with time from 0 to 1/2, so that it is the mirror image of the A_1A_1 genotype. The heterozygote starts out at 1/4 and increases linearly to 3/8 to indicate that it is better adapted to warm temperatures than it is to present conditions, but is not as well adapted to the present conditions as A_1A_1 is, nor to the conditions expected in 50 years as A_2A_2 is; it is the best during some of the intervening years.

Next, set the fertility to a common value of 2. Type

```
>> m = 2;
```

This fertility implies that the presently most common genotype, A_1A_1, has a fitness of 1 now, and that the fitness of A_2A_2, which is now 0, will be 1 in 50 years when the fitness of A_1A_1 will have dropped to 0.

We will write a short program to iterate the equation for p_{t+1} over the next 50 years; it will assume that there is one generation of butterflies each year. The vector p will store the gene pool fraction of A_1 through time, and is initialized with

```
>> p = [1999/2000];
```

The initial condition, corresponding to $t = 0$, is referenced in the vector p as p(1). Thus p_t is referenced as p(t+1) in the vector p. The vector n will store the population size through time, and is initialized with

```
>> n = [2000];
```

The vector w will store the average fitness at each generation, and is initialized with

```
>> w =[1];
```

Then a loop to iterate the equation for 50 generations is started with

```
>> for t=1:50
```

Within the loop, the fitnesses are calculated from the survivorship and fertility of each generation with

```
>>    w11 = m*eval(l11);
>>    w12 = m*eval(l12);
>>    w22 = m*eval(l22);
```

Then the average fitness at time t is calculated with

```
>>    wbar = p(t)^2*w11+2*p(t)*(1-p(t))*w12+(1-p(t))^2*w22;
```

p_{t+1} is called pprime and is calculated as

```
>>    pprime = (p(t)*w11 + (1-p(t))*w12)*p(t)/wbar;
```

N_{t+1} is called nprime and is calculated as

```
>>    nprime = wbar*n(t);
```

Now that we have p_{t+1}, N_{t+1}, and $\bar{W}$, we'll store them by appending them to the p, n, and w vectors with

```
>>    p = [p pprime];
>>    n = [n nprime];
>>    w = [w wbar];
```

and the loop ends with

```
>> end
```

Once you've typed the lines above, you now have the p, n, and w vectors that contain the projected course of evolution for the butterfly species. Let's graph these results. p_t is graphed with

```
>> figure
>> plot(0:50,p(1:51))
```

and appears as Figure 5.2. Notice that the major evolutionary change occurs near 40 years. N_t is graphed with

```
>> figure
>> plot(0:50,n(1:51))
```

and appears as Figure 5.3. The figure shows that the population quickly approaches extinction. In the model, N_t is not 0, but it is less than 1, implying extinction in practice. To understand why the population goes extinct before evolution can come to its rescue, look at the average fitness during the 50 generations. You do this by typing

```
>> figure
>> plot(0:50,w(1:51))
```

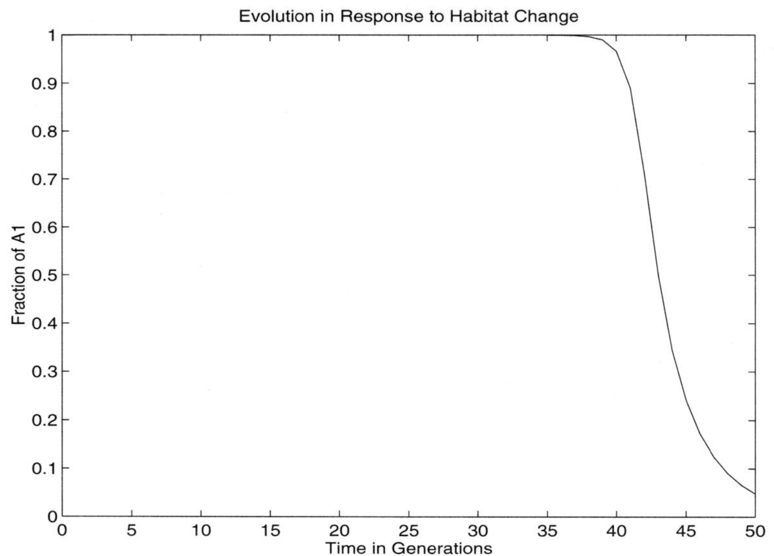

Figure 5.2: Allele A_1's fraction in the gene pool, p_t, through the next 50 years. A_1 confers adaptation to cool conditions, and A_2 to hot conditions. As p_t drops from near 1 to near 0, the population evolves from being cool-adapted to being warm-adapted.

which appears as Figure 5.4. The average fitness, which can be thought of as the geometric growth factor for each generation, is less than 1 except at the very beginning. This is because at the very beginning there is only one genotype with a fitness of 1, which is what is needed for direct replacement. Because the other genotypes have a fitness less than 1, after the beginning, the average fitness is less than 1, and the population declines each generation. So, if the population starts out with an average fitness of 1, then matters will only get worse during the course of its evolution, and it will become extinct before the evolution of genotypes adapted to new conditions has been completed.

The way around this dilemma, of the population becoming extinct before it has time to evolve to new conditions, is to hope that it starts out with an average fitness exceeding 1, so that it is actually expanding. This present-day potential for expansion provides a margin of safety to protect the population as it evolves to new conditions. To determine the minimum margin of safety, compute the geometric mean of the average fitness over the course of the projected evolution. Type `prod(w(2:51))^(1/50)`. This calculates the product of the last 50 entries of `w`, and raises this product to the $(1/50)^{\text{th}}$ power. This number is less than 1, indicating that on the average, the population declines each generation by this factor. So, to compensate for this decline, we must boost the fitness of each genotype by the reciprocal of this number. Specifically, compute a according to

```
>> a = 1/(prod(w(2:51))^(1/50))
```

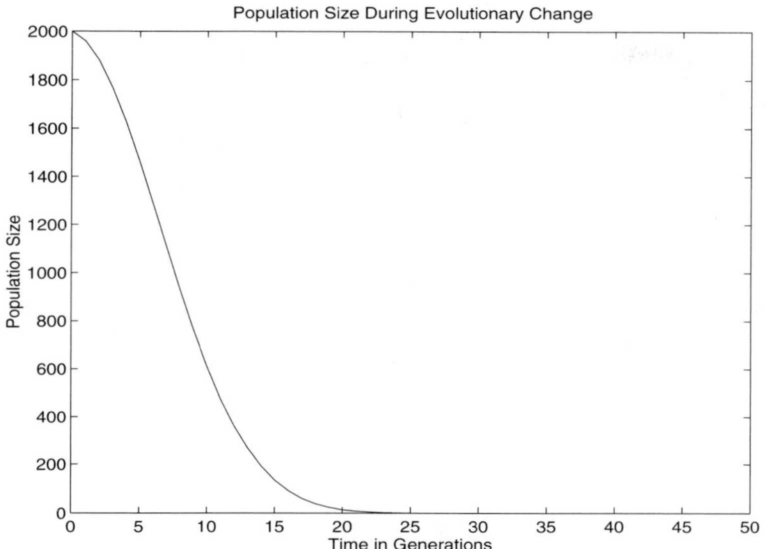

Figure 5.3: The population size, N_t, during 50 years of evolution. The population effectively becomes extinct before completing its evolution from cool-adapted to warm-adapted.

which yields

```
a = 1.8115
```

This is the margin of safety needed by this population to survive the 50 generations while it evolves to the new conditions brought about by global change. If we multiply each fitness by this number, the geometric mean over the entire 50 generations will equal 1, implying that the population will then get back to the size of 2000 that it started with. So, if we repeat the iteration of p_{t+1}, but now compute the fitnesses each generation with

```
>>  w11 = a*m*eval(l11);
>>  w12 = a*m*eval(l12);
>>  w22 = a*m*eval(l22);
```

we end up with revised `p`, `n`, and `w` vectors. The revised `p` is the same as the original `p`, so the genetic course of evolution over the 50 years is the same as in Figure 5.2. The revised `w` vector is just a times the original one, so its graph looks like Figure 5.4 except with different axis calibrations. What's interesting is the revised `n` vector. Upon graphing it with

```
>> figure
>> semilogy(0:50,n(1:51))
```

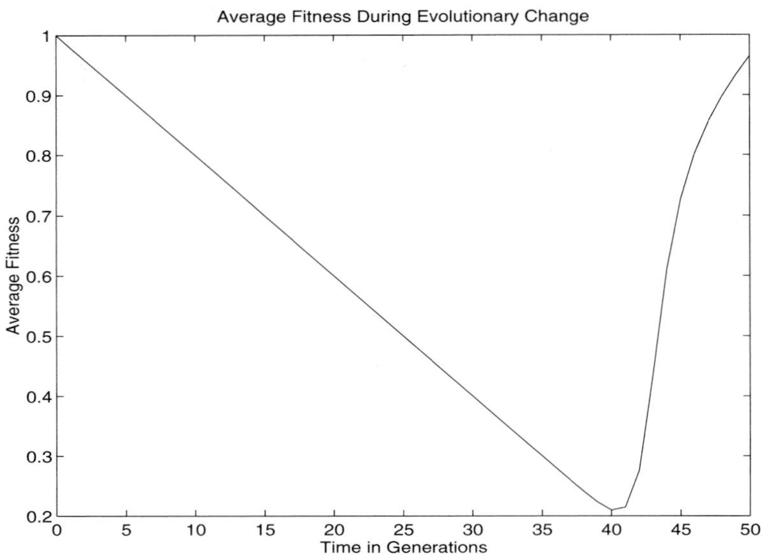

Figure 5.4: The average fitness, $\bar{W}_t$, during 50 years of evolution. The average fitness each year is less than 1, indicating that the population declines for each generation.

we obtain Figure 5.5. The population size after 50 years is 2000, just where it started out. Along the way, the population initially increased because its average fitness was still greater than 1, then it started to die back as its average fitness fell below 1, and finally resumed its increase as the population successfully completed its evolution.

Thus, some populations probably will evolve successfully to global climate change, but others are certain to become extinct before they have time to evolve the adaptations necessary to survive in the new conditions. The loss in average fitness during the evolution of new adaptations is called a "genetic load" or "substitutional load" by population geneticists. Here, we see that the ecological implication of a genetic load is that the population's geometric growth factor at the start of the evolution must be sufficiently greater than 1 for it to avoid extinction while the evolution is taking place. Of course, this discussion assumes that the population is remaining in the affected habitat. Another possibility is that the butterflies move north. A species doesn't move by a mass migration. Instead, the northern edge of the species boundary expands while the southern boundary dies back, so that the center of the species shows a net shift to the north. One might guess that most species will respond to global warming by shifting their location, but those that don't will have to evolve new capabilities to persist.

This example illustrates that unlike Achilles, population genetics has two vulnerable heels. Its first heel is that the genetic basis of the trait of interest must be determined if accurate predictions are needed, although pretending that the trait is inherited with two

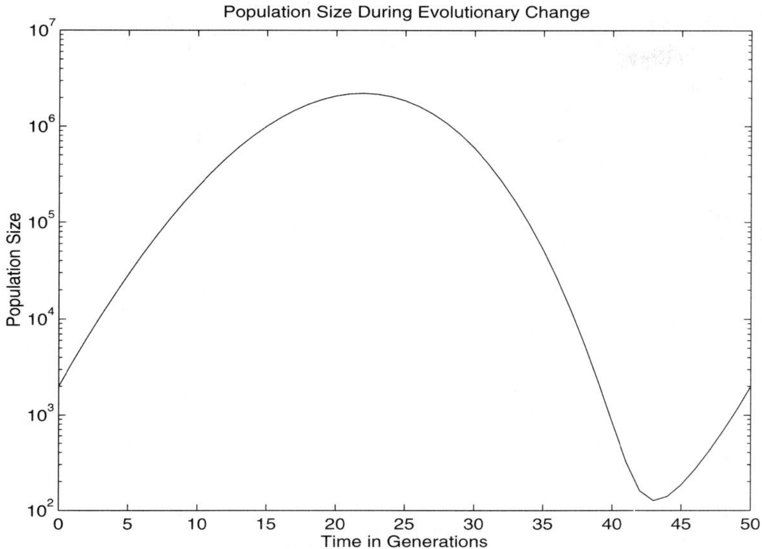

Figure 5.5: The population size, N_t, during 50 years of evolution. Each fitness has been multiplied by the reciprocal of the geometric mean of the average fitnesses from the preceding figure. This boost provides the population with a margin of safety needed for it to survive its evolution. The population size at the beginning and end of the 50-year period is 2000.

alleles at one locus probably provides a good approximation regardless of the true genetic basis. The second Achilles heel is that the survival and fertility of each genotype must be measured. Measuring these for a population is hard enough, as we've seen in the previous chapters. But for accurate evolutionary predictions we need to measure these for each genotype in the population as well. This seems like cruel and unusual punishment. So, in practice, population-genetic theory is used mostly to construct scenarios, like that above. It gives us an idea of what could happen, and if we develop models for the most plausible possibilities, we can plan accordingly.

The population-genetic theory of natural selection predicts how a species evolves, not how a new species is formed. "Speciation" is the name for the processes that make species, and evolutionary biology does not yet have mathematical models to describe how speciation occurs. Presently two definitions of a species are in wide use. By the biological definition, a species is a set of organisms who can potentially mate with one another—they are the organisms who share a common gene pool. By the cladistic definition, a species is a set of organisms who are recognizably distinct. The biological definition is favored by population geneticists, the cladistic definition by scientists actually involved in classifying species and determining their relationships to one another. It's fair to say that by either definition, speciation usually occurs slowly, because geologic changes to the habitat are usually involved.

Therefore, ecologists take as a given the number of species in the world, and study whether particular species will become modified by natural selection in various ecological contexts—just as we have done in our modeling of the evolutionary response to global change.

Let's now look in more detail at how natural selection brings about evolution. What controls the speed of evolution, and what does the gene pool look like when evolution is finished?

5.2.3 Classification of Evolutionary Outcomes

If the gene pool at a locus contains more than one allele, say both A_1 and A_2, the population is said to be "polymorphic" at that locus. If the gene pool at a locus contains only one allele, say A_1, the population is said to be "monomorphic" or "fixed" for that allele. Be sure to see the contrast between heterozygote and homozygote, which refer to an *individual*, whereas "polymorphism" and "monomorphism" or "fixed" refer to a *population*. As we will now see, natural selection can cause the gene pool to be either polymorphic or fixed, according to the fitnesses of each genotype.

When focusing only on how natural selection affects the gene pool, and ignoring its effect on population size, we may use measures of relative fitness that we have scaled for our own convenience. In the equation for p_{t+1}, the W_{ij} are in both the numerator and denominator. Therefore, we may multiply all the W_{ij}'s by any constant we wish, and the constant will divide out, leaving the equation unchanged. One convention is to divide all the W_{ij}'s by the biggest W_{ij}. In symbols, let $W_{\max}$ denote the biggest of the W_{ij}'s. The "relative selective values" or "relative fitnesses" are denoted with lower case letters, and defined as

$$w_{ij} = \frac{W_{ij}}{W_{\max}}$$

By this convention, the biggest *relative* selective value is 1, and the other relative selective values are less than 1. Another convention is to divide all the W_{ij}'s by W_{12}. This scales the *relative* selective value of the heterozygote, w_{12}, to 1, and makes the other relative selective values greater or less than 1, according to whether they were greater or less than that of the heterozygote to begin with. Still other scaling conventions turn up now and then—the point is that if one is focusing only on the gene pool and not the population size, any scaling convention will work.

How can we determine the various possible outcomes of natural selection? In the last chapter we learned that we can predict where the trajectories of a model are leading: we determine where all the equilibria are and analyze whether they are stable or not. This provides the "topo map" on which the various trajectories of evolution will go, according to their starting points. So, let's begin by checking out where the equilibria are. You can think of the remainder of this section as a clinic on how to use the techniques of stability analysis that we developed in the previous chapter.

Let's begin by telling MATLAB about the equation for p_{t+1}. We're using MATLAB's symbolic side now, not its numerical side as we did when we were simulating the evolutionary response to global change. Again we'll let `pprime` stand for p_{t+1}, so type

```
>>pprime = '(p*w11 + (1-p)*w12)*p/(p^2*w11 + 2*p*(1-p)*w12 + (1-p)^2*w22)';
```

To find the equilibria, we find the values of p for which $p_{t+1} = p_t$; this is requested of MATLAB by our typing

```
>> phat = simplify(solve(symop(pprime,'-','p'),'p'))
```

and MATLAB replies with

```
phat = [                          0]
       [                          1]
       [-(w12-w22)/(w11-2*w12+w22)]
```

Thus, there are three possible equilibria. $p = 0$ means the gene pool is fixed for A_2, $p = 1$ means the gene pool is fixed for A_1, and the third equilibrium is potentially a polymorphic equilibrium; that is, an equilibrium in which both A_1 and A_2 are present. An equilibrium where the gene pool is fixed is also called a "boundary equilibrium." A polymorphic equilibrium is also called an "interior equilibrium." Whether this third solution really does represent a polymorphic equilibrium depends on whether it evaluates to a number between 0 and 1; otherwise it doesn't matter. For later convenience, let's assign each of these possible equilibria to a specific symbol. We type

```
>> phat1 = sym(phat,1,1)
>> phat2 = sym(phat,2,1)
>> phat3 = sym(phat,3,1)
```

MATLAB's reply (with a little editing from me) is

```
phat1 = 0
phat2 = 1
phat3 = (w12-w22)/((w12-w11)+(w12-w22))
```

Let's reformat `phat3` to see it more clearly, and label it as $\hat{p}$:

$$\hat{p} = \frac{1}{\frac{w_{12}-w_{11}}{w_{12}-w_{22}} + 1}$$

If you look closely at the formula for $\hat{p}$, you'll see that it evaluates to a number between 0 and 1 only if both $(w_{12} - w_{11})$ and $(w_{12} - w_{22})$ have the same sign. This can happen in only two ways: either (a) the heterozygote has a higher fitness than both homozygotes, or (b) the heterozygote has a lower fitness than both homozygotes. Thus, the third equilibrium is valid and needs to be considered only in these circumstances; otherwise there are just two equilibria.

Now we come to stability. When are these equilibria stable? To find out, we differentiate p_{t+1} with respect to p_t and evaluate the derivative at each equilibrium. It's most convenient to start with the first two equilibria, leaving the third to later. MATLAB computes the derivative, called here `dpprime`, when we type

```
>> dpprime = simplify(diff(pprime,'p'));
```

Now, to evaluate this derivative at `phat1`, to be called `lambda1`, we type

```
>> lambda1 = simplify(subs(dpprime,phat1,'p'))
```

and MATLAB replies with

```
lambda1 = w12/w22
```

The equilibrium at which A_2 is fixed is stable if $w_{22} > w_{12}$ because then `lambda1` is less than 1, and is unstable if $w_{12} > w_{22}$ because `lambda1` is then greater than 1. Recall from the previous chapter that an equilibrium is stable if lambda is between -1 and 1. Well, the w_{ij}'s are positive, so the requirement at -1 doesn't come up. But the requirement at 1 does. If the equilibrium where A_2 is fixed is unstable, then A_1 can increase within the population—this makes sense because the fitness of the heterozygote containing A_1 is higher than the fitness of the prevailing A_2 homozygote. Similarly, to evaluate the derivative at `phat2`, we type

```
>> lambda2 = simplify(subs(dpprime,phat2,'p'))
```

with MATLAB replying (plus a little editing by me)

```
lambda2 = w12/w11
```

The equilibrium at which A_1 is fixed is stable if $w_{11} > w_{21}$ because then `lambda2` is less than 1, and is unstable if $w_{12} > w_{11}$ because `lambda2` is then greater than 1. If the equilibrium where A_1 is fixed is unstable, then A_2 can increase within the population—this again makes sense because the fitness of the heterozygote containing A_2 is higher than the fitness of the prevailing A_1 homozygote.

We can summarize these results as four cases:

1. The more A_1 the better—called "directional selection" for A_1. In this case $w_{11} > w_{12}$ and $w_{12} > w_{22}$. Therefore, the equilibrium at $p = 0$ is unstable, the equilibrium at $p = 1$ is stable, and $\hat{p}$ doesn't matter because it's not between 0 and 1. Putting this together, we see that trajectories lead away from $p = 0$ and wind up at $p = 1$. Diagrammatically, we may write

$$0 \longrightarrow\longrightarrow\longrightarrow\longrightarrow 1$$

2. The more A_2 the better—directional selection for A_2. Here $w_{22} > w_{12}$ and $w_{12} > w_{11}$. This is obviously the reverse of the preceding case, so its diagram is

$$0 \longleftarrow\longleftarrow\longleftarrow\longleftarrow 1$$

3. The heterozygote is best. Here $w_{12} > w_{11}$ and $w_{12} > w_{22}$. Therefore, both $p = 0$ and $p = 1$ are unstable, and trajectories lead away from both 0 and 1 into the interior

of the interval. Furthermore, $\hat{p}$ is between 0 and 1, and so it potentially matters. Diagrammatically, we have

$$0 \longrightarrow\longrightarrow \hat{p} \longleftarrow\longleftarrow 1$$

Thus, a polymorphic condition is indicated. A "protected polymorphism" is defined as the condition in which all the monomorphic equilibria are unstable; the presence of this condition implies that some sort of polymorphism must exist. At this stage in the discussion we haven't analyzed whether $\hat{p}$ is a stable equilibrium, so in principle the protected polymorphism could represent an equilibrium *point*, if $\hat{p}$ is indeed stable, or the protected polymorphism could represent an oscillation of p through time. Which of these is really happening we'll look into further below. The bottom line is that some sort of polymorphic condition definitely results if the heterozygote is the best.

4. The heterozygote is worst. Here $w_{12} < w_{11}$ and $w_{12} < w_{22}$. Therefore, both $p = 0$ and $p = 1$ are stable, and trajectories lead away toward both 0 and 1 from the interior of the interval. Also, $\hat{p}$ is between 0 and 1, and again it potentially matters. Diagrammatically, we have

$$0 \longleftarrow\longleftarrow \hat{p} \longrightarrow\longrightarrow 1$$

The bottom line in this case is that the gene pool comes to a monomorphic equilibrium at either 0 or 1 depending on the initial condition. If the initial p is sufficiently close to 0 to begin with, it winds up at 0; and if it's sufficiently close to 1 to begin with, it winds up at 1. The set of initial conditions that wind up at an equilibrium is called its "domain of attraction." A good guess at this stage in the discussion is that the demarcation point separating the domain of attraction of $p = 0$ from $p = 1$ is $\hat{p}$, which is perhaps an unstable equilibrium point in this circumstance.

We've offered some guesses about the stability of $\hat{p}$, so let's see if we're correct. The λ for this equilibrium is generated with

```
>> lambda3 = simplify(subs(dpprime,phat3,'p'))
```

yielding

```
lambda3 = -(w12*w22+w11*w12-2*w11*w22)/(w11*w22-w12^2)
```

Because we're interested in this equilibrium only when the heterozygote is either better or worse than both homozygotes, this is a good time to use relative selective values that are scaled to the heterozygote. If the selective values are all divided by the selective value of the heterozygote, the relative selective value of the heterozygote is 1. Using this convention, we can simplify λ_3 by putting $w_{12} = 1$ and typing

```
>> lambda3 = simplify(subs(lambda3,'1','w12'))
```

yielding

```
lambda3 = (-w22-w11+2*w11*w22)/(w11*w22-1)
```

This is a little better but its meaning is still not transparent. So, upon doodling around on a sheet of paper, I rearranged the expression into an alternative form

$$\lambda_3 = 1 - \frac{(1 - w_{11})(1 - w_{22})}{1 - w_{11}w_{22}}$$

To confirm that this formula is the same as `lambda3`, you can type

```
>> lambda3alt = simplify('1 - (1-w11)*(1-w22)/(1-w11*w22)')
```

and observe that `lambda3` and `lambda3alt` are identical. Now the meaning of λ_3 is transparent. (a) If both w_{11} and w_{22} are less than 1, indicating a heterozygote better than both homozygotes, then λ_3 is less than 1 because $\frac{(1-w_{11})(1-w_{22})}{1-w_{11}w_{22}}$ is positive. It is also greater than or equal to 0 and only equals 0 in the extreme case where both w_{11} and w_{22} are 0. Hence, the equilibrium at $\hat{p}$ is stable and the approach to this equilibrium is not oscillatory. (b) If both w_{11} and w_{22} are greater than 1, indicating a heterozygote that is worse than both homozygotes, then λ_3 is greater than 1 because $\frac{(1-w_{11})(1-w_{22})}{1-w_{11}w_{22}}$ is negative. Hence, the equilibrium at $\hat{p}$ is unstable in this circumstance.

In summary then, if the heterozygote is better than the homozygotes, the equilibria at $p = 0$ and $p = 1$ are unstable, and the equilibrium at $\hat{p}$ is between 0 and 1, and it is stable. In this circumstance, natural selection causes the population to evolve a stable polymorphism. But if the heterozygote is worse than the homozygotes, then the equilibria at $p = 0$ and $p = 1$ are stable, the equilibrium at $\hat{p}$ is between 0 and 1, and it is unstable. The unstable $\hat{p}$ marks the boundary between the the domains of attraction of the stable equilibria at $p = 0$ and $p = 1$.

5.2.4 Illustration of Evolutionary Outcomes

After grinding through the stability analysis of the equation for p_{t+1}, we now know what to expect of natural selection. Let's generate some illustrations of the various possibilities. To do this, let's package the commands we previously used when we simulated the evolutionary response to global change into an external function. With a text editor, create a file called `natsel.m` as follows:

```
function [p,w] = natsel(w11,w12,w22,po,runlen)
 p = [po];
 w = [];
 for t=1:runlen
  wbar = p(t)^2*w11 + 2*p(t)*(1-p(t))*w12 + (1-p(t))^2*w22;
  pprime = (p(t)*w11 + (1-p(t))*w12)*p(t)/wbar;
  p = [p pprime];
  w = [w wbar];
 end
```

This function accepts as inputs the three fitnesses, the initial value of p, and the number of generations for the iteration, and it returns a vector of p_t and $\bar{w}_t$. We're not bothering

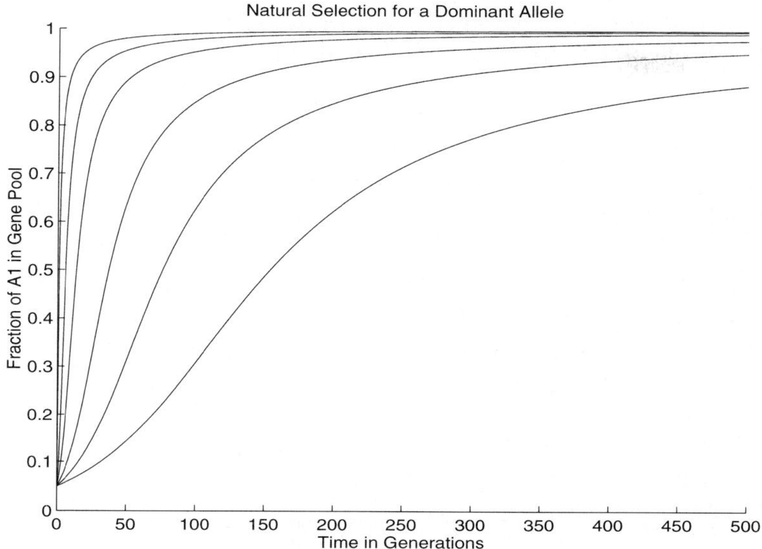

Figure 5.6: Directional selection in favor of A_1, made with the assumption that A_1 is dominant and A_2 recessive. The relative fitnesses are 1, 1, and $1 - s$, with $s = 1$, 0.5, 0.25, 0.1, 0.05, and 0.025.

to record N_t here because we're focusing only on the gene pool. Once this has been saved in your directory or folder, it's ready to use. In the following, we'll label the A_1 allele that being favored by selection, and the A_2 allele that being opposed by selection, unless heterozygote superiority or inferiority is involved.

Suppose that in our first illustration, the A_1 allele is dominant and the A_2 allele is recessive. In this case, the fitness of the A_1A_1 and A_1A_2 genotypes are identical. A figure with six illustrations is generated by our typing

```
>> figure
>> hold on
>> for s=[1 .5 .25 .1 .05 .025]
>>   [p w] = natsel(1,1,1-s,.05,500);
>>   plot(0:500,p);
>> end;
```

The result appears as Figure 5.6. The initial value of p is taken as 0.05, and the iteration runs for 500 generations. The parameter `s` measures the strength of natural selection, because it is the difference between the relative fitness of the type being favored by natural selection and the type being opposed by natural selection. Varying `s` from 1 to 0.025 means varying the strength of natural selection from strong to weak. Obviously, the stronger the natural

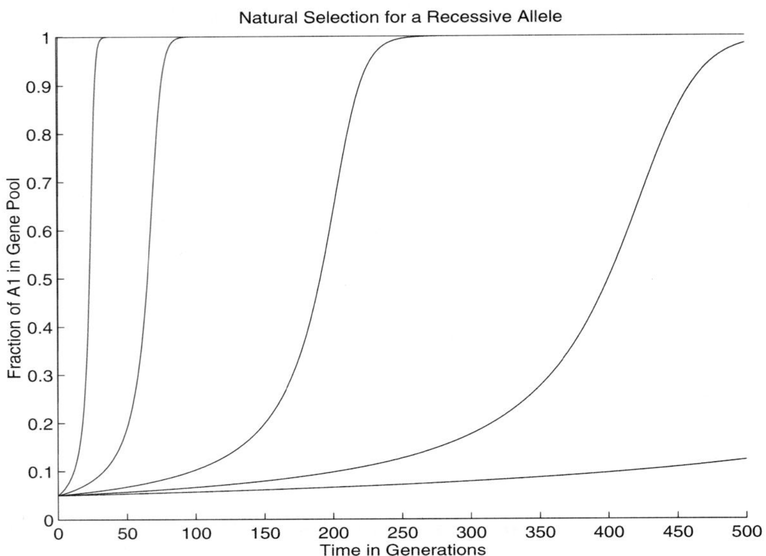

Figure 5.7: Directional selection in favor of A_1, made with the assumption that A_1 is recessive and A_2 dominant. The relative fitnesses are 1, $1-s$, and $1-s$, with $s = 1$, 0.5, 0.25, 0.1, 0.05, and 0.025.

selection, the faster the evolution. The overall pattern of evolutionary change shows a fast start leading to a slowdown as p reaches about 0.9 or so. The slowdown occurs because, as it becomes increasingly rare, the A_2 allele is increasingly present only in heterozygotes. It therefore is not expressed. If A_2 is not expressed in the phenotype, it is not exposed to selection. Thus, selection for a dominant allele, and against a recessive allele, starts out fast but slows to a snail's crawl.

Our second illustration considers the reverse, where the A_1 allele is recessive and the A_2 is dominant. To generate this case, type

```
>> figure
>> hold on
>> for s=[1 .5 .25 .1 .05 .025]
>>   [p w] = natsel(1,1-s,1-s,.05,500);
>>   plot(0:500,p);
>> end;
```

and the result appears in Figure 5.7. The overall pattern shows a slow start leading to a fast ending—the opposite of the preceding figure. Here the slow start is because the favored allele, A_1, which is recessive, is initially mostly in heterozygotes, and so is not expressed and is not exposed to the natural selection in its favor.

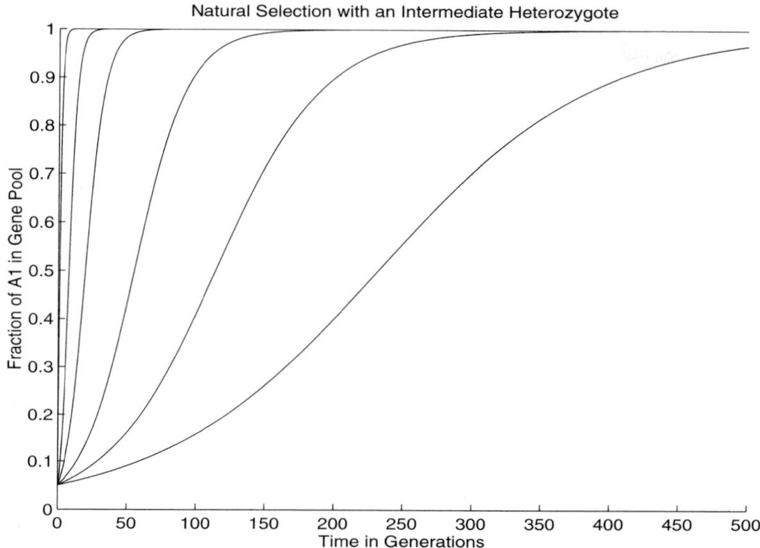

Figure 5.8: Directional selection in favor of A_1, made with the assumption that A_1 and A_2 are codominant. The relative fitnesses are 1, $1 - s/2$, and $1 - s$, with $s = 1$, 0.5, 0.25, 0.1, 0.05, and 0.025.

A third illustration presents the intermediate case of directional selection, where A_1 and A_2 are equally dominant, called "codominance," resulting in a heterozygote whose fitness is intermediate between that of the two homozygotes. Figure 5.8 is generated by typing

```
>> figure
>> hold on
>> for s=[1 .5 .25 .1 .05 .025]
>>  [p w] = natsel(1,1-s/2,1-s,.05,500);
>>  plot(0:500,p);
>> end;
```

Evolution in this case shows a more regular pace from beginning to end, and may be taken as the paradigm for directional selection.

The fourth illustration presents selection in favor of the heterozygote. Type

```
>> figure
>> hold on
>> for s=[1 .5 .25 .1 .05 .025]
>>  [p w] = natsel(1-s,1,1-s,.05,500);
>>  plot(0:500,p);
```

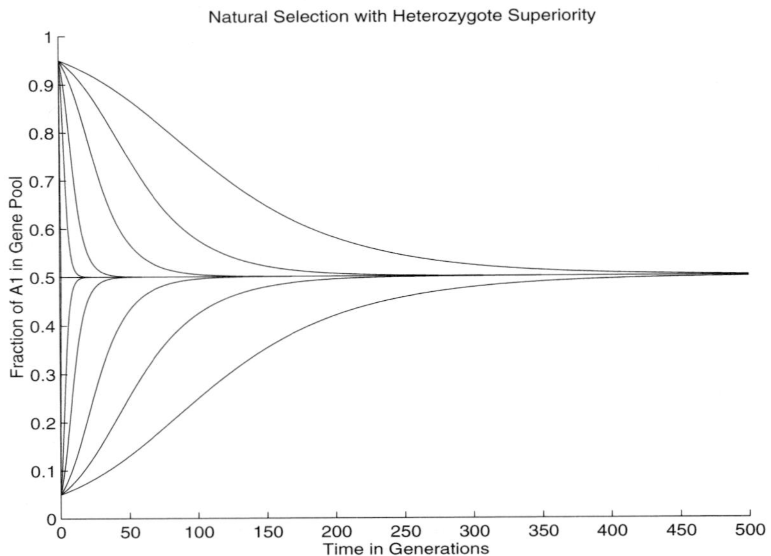

Figure 5.9: Natural selection favoring the heterozygote. The relative fitnesses are $1 - s$, 1, and $1 - s$, with $s = 1$, 0.5, 0.25, 0.1, 0.05, and 0.025.

```
>>  [p w] = natsel(1-s,1,1-s,.95,500);
>>  plot(0:500,p);
>> end;
```

and Figure 5.9 appears. Evolution clearly leads to a polymorphism wherein both A_1 and A_2 are retained in the population. The trajectories lead to the same polymorphism from any initial condition—the initial conditions illustrated are 0.05 and 0.95. The equilibrium polymorphism is at $\hat{p}$, whose formula was presented earlier.

The final illustration presents selection in opposition to the heterozygote. Typing

```
>> figure
>> hold on
>> for s=[1 .5 .25 .1 .05 .025]
>>  [p w] = natsel(1,1-s,1,.45,500);
>>  plot(0:500,p);
>>  [p w] = natsel(1,1-s,1,.55,500);
>>  plot(0:500,p);
>> end;
```

yields Figure 5.10. Evolution leads to a monomorphic equilibrium in all cases. Initial conditions less than $\hat{p}$, which is 0.5 in this example, wind up at $p = 0$; and initial conditions greater than $\hat{p}$ wind up at $p = 1$.

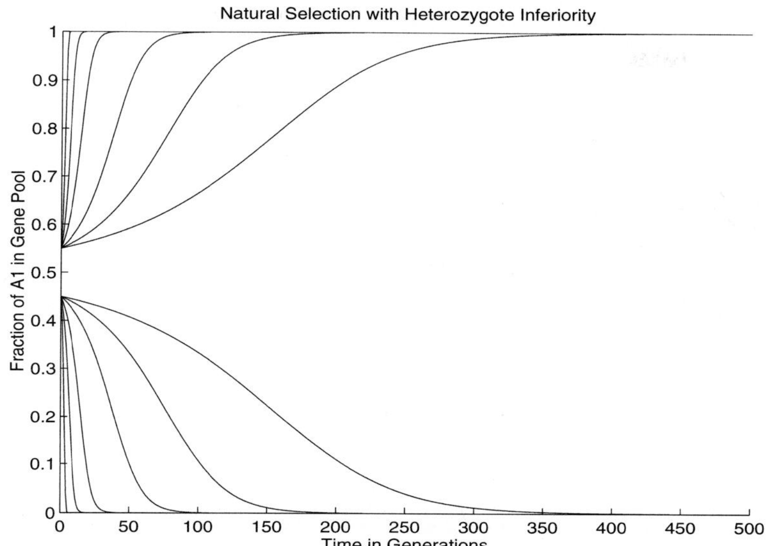

Figure 5.10: Natural selection opposing the heterozygote. The relative fitnesses are 1, $1 - s$, and 1, with $s = 1$, 0.5, 0.25, 0.1, 0.05, and 0.025.

5.2.5 The Fundamental Theorem of Natural Selection

The typical reaction of an ecologist to all the cases of natural selection illustrated in Figures 5.6–5.10 is, "Yuchh! Isn't there something more interesting one can say about natural selection than a catechism of all possibilities?" Yes, there is. Return then to the golden metaphor of natural selection as a sculptor. Does natural selection mold a population to fit its environment? Well, how would we know? What could we use as a general measure of how well a population fits its environment? And if we can find a general measure of adaptation, does natural selection increase it through time?

The most obvious general measure of adaptation in a population is the average *relative* fitness, $\bar{w}$. If this goes up through time, then on the average, the individuals in the population are better able to reproduce and to survive than before, and if so, we could reasonably say that the population was getting a better fit to its environment. So let's see what happened to the average relative fitness, $\bar{w}$, through time in some of the illustrations of evolution by natural selection from Figures 5.6-5.10.

For purposes of comparison, the strength of selection is set to a common value by our typing

```
>> s = .05;
```

Now let's begin with directional selection. In the preceding figures, we've illustrated directional selection for a dominant, codominant, and recessive allele, and curves for each of

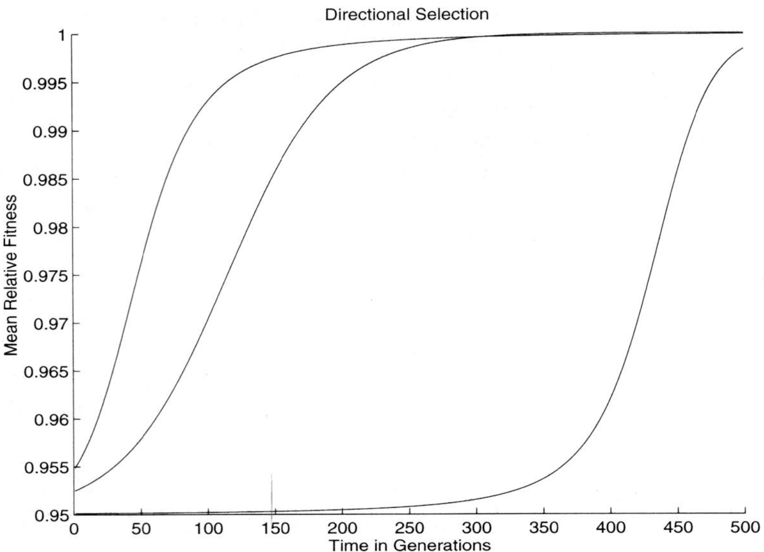

Figure 5.11: The average relative fitness, $\bar{w}$, through time for three examples of directional selection. The curves from left to right involve selection for a dominant allele, a codominant allele, and for a recessive allele, respectively.

these possibilities are now placed in Figure 5.11 by our typing

```
>> figure
>> hold on
>> [p w] = natsel(1,1,1-s,.05,500);
>> plot(1:500,w,'y');
>> [p w] = natsel(1,1-s/2,1-s,.05,500);
>> plot(1:500,w,'r');
>> [p w] = natsel(1,1-s,1-s,.05,500);
>> plot(1:500,w,'g');
```

In all three curves, the average relative fitness, $\bar{w}$, clearly increases through time. This increase is consistent with the idea that natural selection is continuously molding the population to fit the environment better and better. Next, we'll check out what happens to $\bar{w}$ through time with selection for a heterozygote; this condition leads to a polymorphism. Type

```
>> figure
>> [p w] = natsel(1-s,1,1-s,.05,250);
>> plot(1:250,w);
```

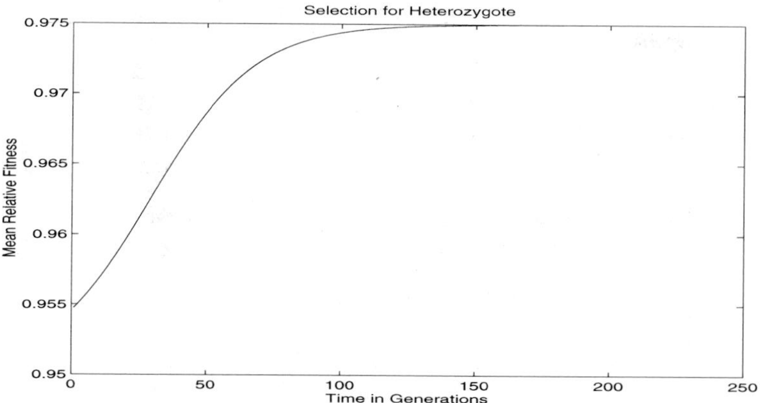

Figure 5.12: The average relative fitness, $\bar{w}$, through time, for selection in favor of the heterozygote.

Figure 5.12 shows that again the average fitness, $\bar{w}$, increases through time. $\bar{w}$ never levels off at 1 though, because only the heterozygote has a fitness of 1, and at the stable polymorphic equilibrium, the population maintains steady-state levels of the homozygotes too, both of which have fitnesses less than 1. Therefore, the population average of w_{ij} is less than 1.

Last, let's look into selection against the heterozygote. We type

```
>> figure
>> [p w] = natsel(1,1-s,1,.55,250);
>> plot(1:250,w);
```

The average fitness increases through time also in Figure 5.13. $\bar{w}$ levels off at the fitness of the homozygote for the allele that becomes fixed.

In the example of Figure 5.13, both homozygotes have a fitness of 1, so $\bar{w}$ has to level off at 1. But an example that is not so symmetric could be made by the plotting of $\bar{w}$ through time from fitnesses of (1, 0.950, 0.975). One would again see that $\bar{w}$ always increases through time, but levels off at either 1 or 0.975, depending on whether the equilibrium that was eventually reached was either $p = 1$ or $p = 0$. Thus, $\bar{w}$ does go up from generation to generation, but does not necessarily wind up at the highest possible value, because evolution may start out in the domain of attraction of the equilibrium for the less fit of the two homozygotes.

All in all, the prospects for the golden metaphor look promising at this stage. In all the examples we have investigated, the average relative fitness, $\bar{w}$, which can be viewed as a measure of how well the population is fitting the environment, increases through time as a result of natural selection. The final value of $\bar{w}$ actually attained along a particular trajectory may not be the best of all possible final values, but the final value of $\bar{w}$ is always better than where it started out.

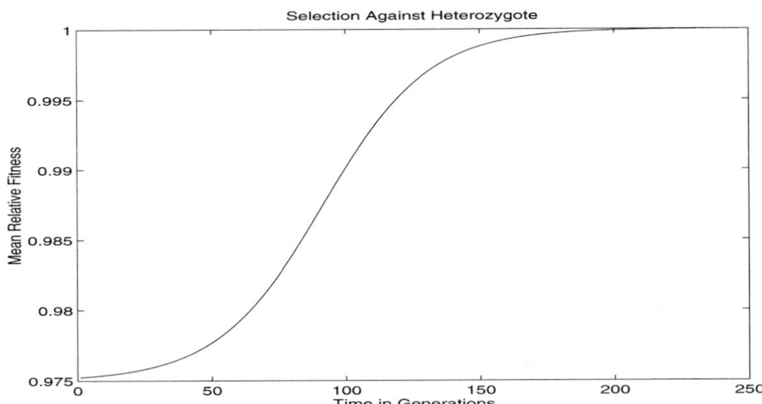

Figure 5.13: The average relative fitness, $\bar{w}$, through time, for selection opposing the heterozygote.

The golden metaphor of natural selection as a sculptor who continuously improves adaptation has been concisely expressed in population genetics in what is known as The Fundamental Theorem of Natural Selection. There are several variants of this theorem, and the most picturesque is Sewall Wright's version, called the "adaptive topography." The idea is to show that a population "climbs" up a hill, represented by $\bar{w}$, to an ever higher degree of adaptedness. Here's how this theorem is derived with MATLAB.

To start, of course, remind MATLAB of the formulas for p_{t+1} and $\bar{w}$. Type

```
>> pprime = '(p*w11+(1-p)*w12)*p/(p^2*w11+2*p*(1-p)*w12+(1-p)^2*w22)';
>> wbar = 'p^2*w11+2*p*(1-p)*w12+(1-p)^2*w22';
```

Then we develop the formula for Δp, which is $p_{t+1} - p_t$. We type

```
>> delta_p = factor(symop(pprime,'-','p'))
```

which yields (I've split the output over two lines)

```
delta_p = -p*(-1+p)*(p*w11-2*w12*p+w22*p+w12-w22)/
          (p^2*w11+2*w12*p-2*w12*p^2+w22-2*w22*p+w22*p^2)
```

This formula tells us how the gene pool changes for each generation. Now the goal is to connect this formula to the slope of $\bar{w}$—to show that Δp climbs $\bar{w}$. That is, if the slope of $\bar{w}$ is positive, so that climbing means moving to the right, then Δp must be positive too, and vice-versa if the slope of $\bar{w}$ is negative—a negative slope means p must move to the left if it is to climb to a higher $\bar{w}$. So let's see if we can connect the formula for Δp to the formula for the slope of $\bar{w}$. MATLAB generates for formula for the slope of $\bar{w}$ when we type

```
>> dwbar = diff(wbar,'p')
```

yielding

```
dwbar = 2*p*w11+2*(1-p)*w12-2*w12*p-2*(1-p)*w22
```

when we assume that the relative fitnesses, w_{ij}, are independent of p. For clarity, we can remove the 2 in the formula if we type

```
>> dwbar_over2 = factor(symop(dwbar,'/','2'))
```

yielding

```
dwbar_over2 = p*w11-2*w12*p+w22*p+w12-w22
```

At this point, notice that `dwbar_over2` is the same as the second expression in parentheses in the numerator of `delta_p`. Therefore, we *have* connected Δp with the slope of $\bar{w}$, and the formula for Δp may be written as

$$\Delta p = \frac{p(1-p)}{\bar{w}} \frac{1}{2} \frac{d\bar{w}}{dp}$$

To confirm that this rearrangement is correct, type

```
>> delta_palt = simplify(symop('p*(1-p)','*',dwbar_over2,'/',wbar))
```

which yields (the output is split over two lines)

```
delta_palt = -p*(-1+p)*(p*w11-2*w12*p+w22*p+w12-w22)/
             (p^2*w11+2*w12*p-2*w12*p^2+w22-2*w22*p+w22*p^2)
```

This is the same as the formula for `delta_p` above. Therefore, if the relative fitnesses don't depend on p, the gene-pool changes do represent the population climbing to higher and higher points on the $\bar{w}$ curve. In this context, the $\bar{w}$ curve is called the "adaptive topography" and natural selection is viewed as causing the population to climb to the nearest peak on the adaptive topography.

To illustrate how natural selection causes the population to climb to the nearest peak of the $\bar{w}$ curve, an animation can be helpful and even fun to develop. With a text editor create the following function, to be saved in a file called `climb_ns.m`. This function first draws the graph of $\bar{w}$ and then waits for you to input an initial value for p. You do this interactively with the mouse. Using the mouse, you move the cross-hairs to the spot on the $\bar{w}$ curve at which you want the evolution to begin. Then click a button on the mouse. The function proceeds to iterate the equation for p_{t+1} and to graph the population's value of $\bar{w}$ at each time as an asterisk. The function then pauses until you use the mouse to enter another initial value of p. Then the function iterates and graphs the equation for p_{t+1} again. Here then is how the function works in detail. It is called with the three relative fitnesses and the number of generations for the function to iterate:

```
function climb_ns(w11,w12,w22,runlen)
```

The graph of $\bar{w}$ is drawn in yellow with the commands

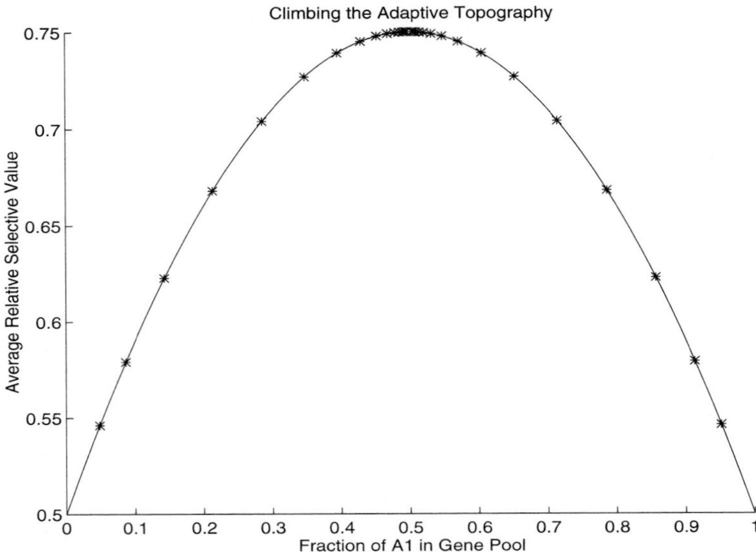

Figure 5.14: An illustration of natural selection causing a population to climb to the nearest peak of the $\bar{w}$ curve.

```
figure
hold on
p = 0:.01:1;
wbar = 'p^2*w11 + 2*p*(1-p)*w12 + (1-p)^2*w22';
plot(p,eval(sym2ara(wbar)),'y')
```

The loop for the two instances in which p_{t+1} is iterated, is initialized with

```
for trial = 1:2
```

The position of the mouse is detected by our calling the `ginput(1)`; this command requests one data point and returns its x and y coordinates as `p` and `w`. We actually use only `p`, but we need to accept the value for `w` anyway.

```
  [p w] = ginput(1);
```

Now that we have an initial value for p_t, we start the iteration with

```
  for t=1:runlen
```

Within the iteration, at each time we first calculate the present value of $\bar{w}$ and graph it as a green asterisk with

```
    wbar = p^2*w11 + 2*p*(1-p)*w12 + (1-p)^2*w22;
    plot(p,wbar,'g*')
```

Then we calculate p_{t+1}, called pprime, with

```
    pprime = (p*w11 + (1-p)*w12)*p/wbar;
```

and then reassign p_t to p_{t+1} with

```
    p = pprime;
```

Both loops are now ended with

```
  end
end
```

Once this function has been saved under the name climb_ns.m, it is ready to be used. You can call it from MATLAB's command prompt.

Figure 5.14 offers an example of an animation of climbing the adaptive topography. It was made by my typing at the command prompt,

```
>> climb_ns(0.5,1,0.5,15)
```

This example is for heterozygote superiority, and the peak to $\bar{w}$ occurs at 0.5, where a stable polymorphism is predicted to result. Then I clicked the mouse to input an initial p of about 0.05, and dots appeared. They indicated that the population was moving to the right and coming to equilibrium at $p = 0.5$. Then I clicked the mouse to input an initial p at about 0.95, and dots appeared indicating the population was moving to the left, again coming to equilibrium at $p = 0.5$. Go ahead and try this for other choices of relative selective values, such as (1,0.5,0.5), (1,0.5,1), and (0.75,1,0.5). In this way, you'll see as many illustrations as you wish of the most picturesque way to present the golden metaphor of evolutionary biology—that natural selection continuously improves a species' adaptation to the environment until no further improvement is possible with its genetic variation.

The formula for Δp showing that the gene pool changes by climbing the $\bar{w}$ curve is one of several theorems all of which I'm referring to as a Fundamental Theorem of Natural Selection. The version that visualizes evolution as climbing the $\bar{w}$ curve was developed in great detail by S. Wright. Another version is that $\Delta\bar{w} \geq 0$; that is, the average relative fitness increases with each generation until equilibrium is reached. This version, derived by J. Kingman, is particularly useful in mathematical analysis[6]. A third version, developed by R. A. Fisher, is that $\Delta\bar{w}$ is proportional to a measure of the amount of genetic variability in the population. The idea is that evolution is fast when there's a lot of variability, and is slow when all the organisms are nearly identical to one another. Strictly speaking, this particular version should perhaps be the only theorem called the fundamental theorem because Fisher coined the name. Still, I prefer to call all of these theorems a fundamental theorem of natural selection because they all express essentially the same principle.

[6]In mathematics, $\bar{w}$ is said to be a "global Liapanov function," which is a function of the state variables that increases through time until an equilibrium is reached.

All versions of the fundamental theorem of natural selection are based on assumptions, and some of these may be relaxed without changing the result, whereas others are crucial. Much research in theoretical population genetics has been aimed at destroying or rescuing the fundamental theorems in the face of changed assumptions. The main assumption that can be completely relaxed is that of two alleles at one locus. The fundamental theorems can be generalized to any number of alleles at a single locus. But, if the relative selective values are not constants, then modifications are needed. If the w_{ij} depend on p, then we have what is called "frequency-dependent selection." For example, the fitness of an A_1A_1 individual may be helped or harmed by the number of A_2A_2 individuals present. If so, the w_{ij} depend on p and the fundamental theorems must either be greatly modified or even junked. If the w_{ij} depend on the population size, N, we have what is called "density-dependent selection." In this case, the fundamental theorems can be modified without too much alteration in spirit. Similarly, if age is involved, then versions of a fundamental theorem can be tailored appropriately. But if forces of evolution other than natural selection are added to the mix, then they can seriously intervene in the course of evolution, and prevent maximal adaptation from being achieved.

So, we can't leave matters where they now stand. If we're gonna buy into the principle that natural selection continuously improves a species' adaptation to its environment, we need to check out how fragile this attractive idea really is. Therefore, let's take a look at another force of evolution, called genetic drift, and see what it does, and see how its strength compares with that of natural selection.

5.3 Genetic Drift

5.3.1 Sampling Error

The gene-pool composition may be affected by what might be thought of as "population transmission noise," particularly if the population is tiny. This noise arises because of what is called "sampling error" in statistics. Here's the idea. Suppose a tiny population consists of only 4 individuals, so the gene pool is 8 genes, and pretend for the moment that both A_1 and A_2 have the same effect on fitness, so natural selection doesn't prefer one over the other. All that's going on is a tiny population reproducing itself through time. At the beginning of a generation, a total of 8 gametes, out of perhaps thousands to millions of gametes that the four parents produced, are lucky enough to be fertilized. Suppose also that the gene pool among the parents is 50% A_1 and 50% A_2, so that p is $1/2$. Then p will also be $1/2$ among the thousands of gametes. So let's grab 8 of these gametes at random, the 8 that will actually become the next generation. Well, we *could* wind up with four gametes carrying A_1 and four gametes with A_2. Or, we might not. We might also wind up, by chance, with three A_1 and five A_2, or vice versa. Indeed, if we happen to wind up with any set of gametes other than exactly four A_1 and four A_2, then p changes slightly. If we happen to wind up with, say, three A_1 and five A_2, then p will have dropped from $1/2$ to $3/8$ just by accident. Such an accident is called a sampling error—it's when the gametes that are actually incorporated into zygotes have a gene-pool composition different than the set of all gametes from which they were drawn[7]. The mere fact that so few gametes are actually used to make up a tiny population's generation makes sampling error inevitable. The gene pool changes because of sampling error, so we must consider sampling error as a force of evolution. The cumulative impact of sampling error over many generations is called "genetic drift."

Using MATLAB we can simulate how genetic drift happens. With a text editor, make the following function, to be saved as `drift.m`. The function takes as input the initial fraction of A_1 alleles, `po`, the population size, `n`, and the number of iterations, `runlen`, and returns a column vector, `p`, of the fraction of A_1 in the gene pool through time. The function is declared by

```
function [p] = drift(po,n,runlen)
```

The initial condition is placed into the first position of the `p` column vector with

```
p = [po];
```

Then the loop to carry out the iterations is initialized with

```
for t = 1:runlen
```

Within the loop, the variable `a1` will be used to keep a count of the number of A_1 alleles in the gene pool at time `t`. It is initialized to 0 with

```
  a1 = 0;
```

Now we'll draw the $2N$ genes that we need to make up the new gene pool. To do this we set up a loop to run from the first to the $2N^{\text{th}}$ gamete; we type

```
  for i=1:(2*n)
```

For each allele we call the random-number generator `rand`, and if the random number is less than the current p, then the allele we've drawn is an A_1. Otherwise it's an A_2. So if the random number is less than the current p, we increment `a1` by 1, according to

```
    if rand < p(t)
      a1 = a1 + 1;
    end
```

To now end the loop to draw all $2N$ alleles, we type

```
  end
```

Now that we've synthesized a new gene pool, we calculate the fraction of A_1 alleles in it, and append it as the next row in the `p` column vector. To do this, use

```
  p = [p; (a1/(2*n))];
```

The loop for the main iteration then ends with

```
end
```

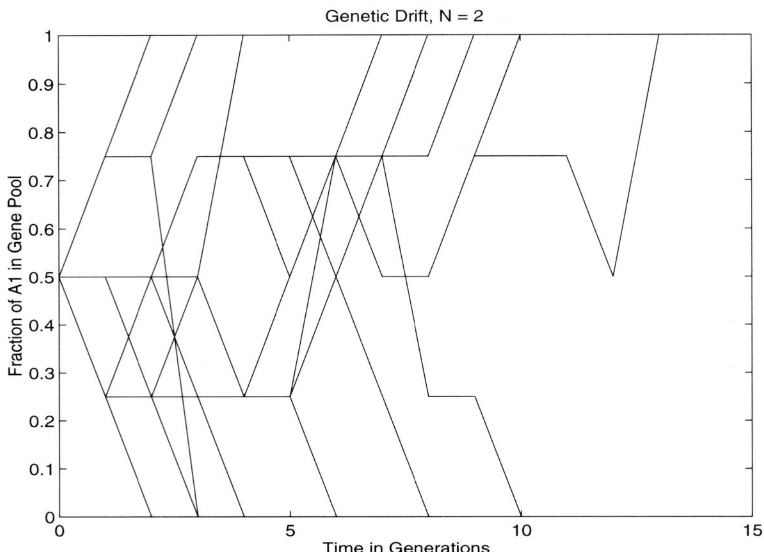

Figure 5.15: The trajectories from an ensemble of 18 replicates of genetic drift in a population of 2 individuals. The starting p, the fraction of A_1 in the gene pool, is 1/2.

Once this has been saved in your directory as `drift.m`, it's ready to be used from MATLAB's command window.

To illustrate genetic drift, let's start with a very tiny population, one whose size is just 2 individuals. Let's run 18 replicates. The choice is 18 simply because MATLAB has six colors to graph with, and we can use each color three times in the same graph without its getting too confusing. The 18 replicates are called an "ensemble." We will use a matrix `e` to store each replicate as a column vector. To start off, we initialize the random-number generator and the `e` matrix, by our typing

```
>> rand('seed',0);
>> e = [];
```

The 18 replicates of genetic drift are each going to start with a p of 1/2, use a population size of 2, and run for 15 generations. The replicates are generated and stored in `e` by our typing

```
>> for i=1:18
>>  e(:,i) = drift(0.5,2,15);
>> end
```

[7] In statistics, sampling error is the difference between the composition of a sample and the composition of the population from which it was drawn.

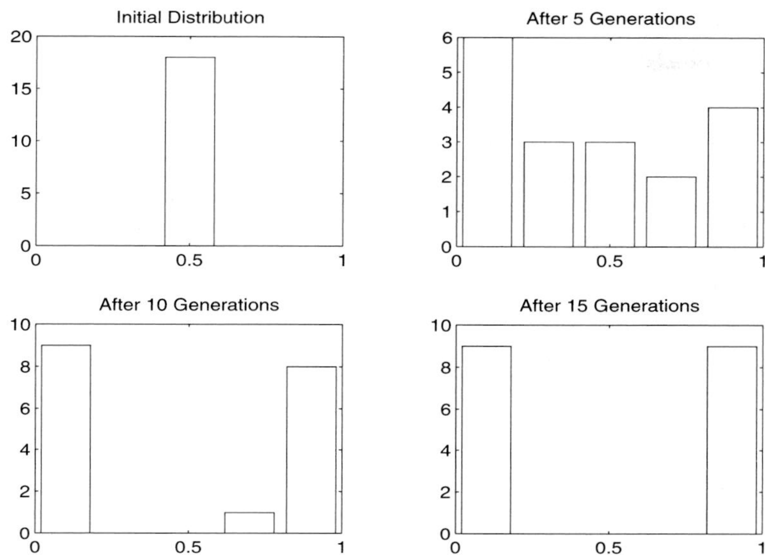

Figure 5.16: The distribution of p, the fraction of A_1 in the gene pool, in an ensemble of 18 replicates of genetic drift in populations of 2 individuals each, at various times, starting with all replicates having $p = 1/2$.

To graph the trajectories from the ensemble of replicates, we then type

```
>> figure
>> plot(0:15,e)
```

The result is illustrated in Figure 5.15. All the trajectories start out at $p = 1/2$ and all end up at either $p = 0$ or $p = 1$. Thus, the initial polymorphic condition is gradually lost through time, and the population ends up being monomorphic, with a gene pool fixed for either A_1 or A_2.

Another way to present the results simulating genetic drift is to provide a histogram of p across the replicates for various times. The `hist` command in MATLAB plots histograms from a vector of data. Here we have to specify explicitly what bins to use, because the default histogram of 10 bins doesn't look very good. The bins to be used for a population size of `n` are `(1/2)*(1/(2*n+1)):1/(2*n+1):1-(1/2)*(1/(2*n+1))`, meaning that the center of the first bin is at $\frac{1}{2}\frac{1}{2N+1}$, the step size to the centers of consecutive bins is $\frac{1}{2N+1}$, and the center of the last bin is $1 - \frac{1}{2}\frac{1}{2N+1}$. A figure with four histograms for the times 0, 5, 10, and 15 is then made if we type

```
>> bins = .1:.2:.9;
>> figure
```

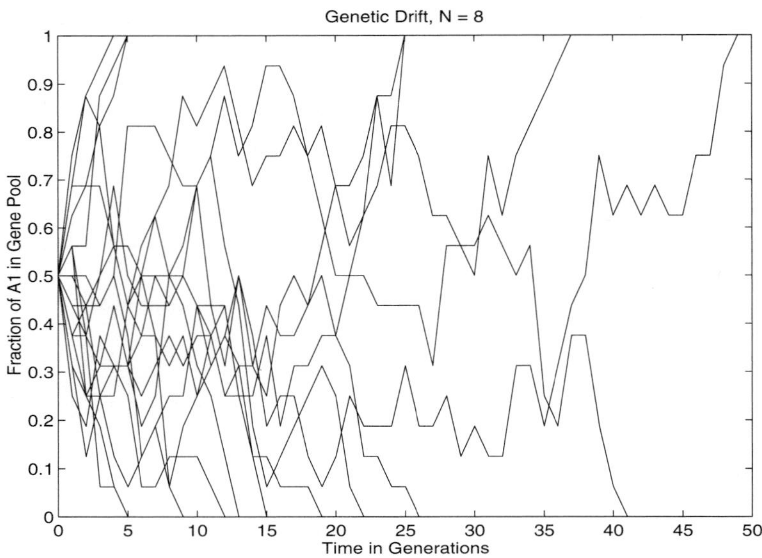

Figure 5.17: The trajectories from an ensemble of 18 replicates of genetic drift in a population of 8 individuals. The starting p, the fraction of A_1 in the gene pool, is 1/2.

```
>> subplot(2,2,1)
>> hist(e(1,:),bins)
>> subplot(2,2,2)
>> hist(e(6,:),bins)
>> subplot(2,2,3)
>> hist(e(11,:),bins)
>> subplot(2,2,4)
>> hist(e(16,:),bins)
```

The result is illustrated in Figure 5.16. The figure shows that the distribution of p is unimodal at time 0, when all the replicates have $p = 1/2$, and gradually becomes flat, then U-shaped, and finally consists of only two bars corresponding to $p = 0$ and $p = 1$. This figure is important because it shows that the detailed trajectory of any one population cannot be predicted through time, but the histogram of how an ensemble of populations develops through time can be predicted.

Intuitively, it would seem that the existence of genetic drift depends on a tiny population size. Suppose that the population size is large, say one million. Then two million gametes drawn from the total pool of gametes are actually incorporated into zygotes. Well, what's the chance that a sample of two million gametes misrepresents the composition of the gamete pool from which it was drawn? Pretty low. Sampling error depends on a small sample size,

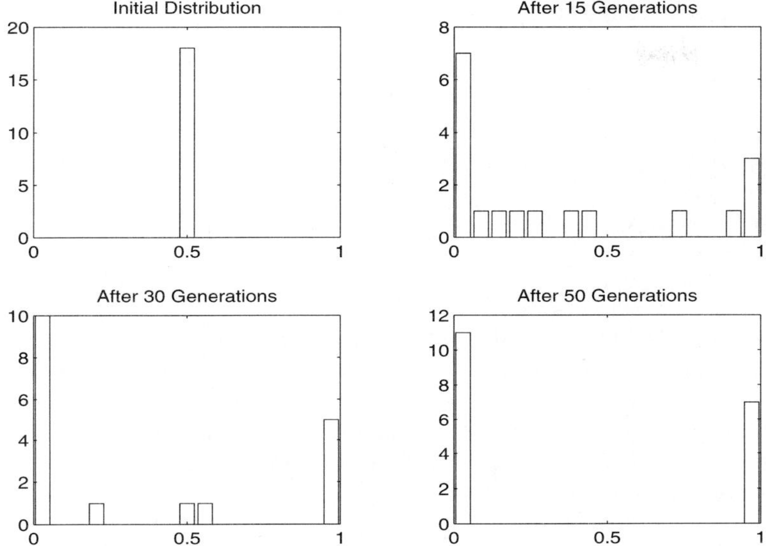

Figure 5.18: The distribution of p, the fraction of A_1 in the gene pool, in an ensemble of 18 replicates of genetic drift in populations of 8 individuals each, at various times, starting with all replicates having $p = 1/2$.

and the principle of genetic drift depends on sampling error. Therefore, a large population shouldn't show much genetic drift. To see if this intuition is correct, let's construct an ensemble of 18 replicates with a larger population size than before—let's try a population size of 8. An ensemble of 18 replicates, each starting with a p of 1/2, a population size of 8 individuals, and running for 50 generations, is constructed by our typing

```
>> rand('seed',0);
>> e = [];
>> for i=1:18
>>   e(:,i) = drift(0.5,8,50);
>> end
```

The trajectories are then graphed with

```
>> figure
>> plot(0:50,e)
```

which results in Figure 5.17. The overall result is the same as before, in the sense that all the trajectories end up at either $p = 0$ or $p = 1$. What's different is that the process takes much longer. Similarly, we may also plot the distribution of p in the ensemble at various times. We use

```
>> bins = 1/34:1/17:33/34;
>> figure
>> subplot(2,2,1)
>> hist(e(1,:),bins)
>> subplot(2,2,2)
>> hist(e(16,:),bins)
>> subplot(2,2,3)
>> hist(e(31,:),bins)
>> subplot(2,2,4)
>> hist(e(51,:),bins)
```

This yields Figure 5.18. The overall pattern is again one of starting out with a unimodal distribution with all populations at $p = 1/2$ that gradually transforms into a bimodal distribution with only two bars at $p = 0$ and $p = 1$. But again, what's different is that the process takes much longer. So, it's true that the bigger the population, the slower its genetic drift. As we'll now see, a slower genetic drift also means a weaker genetic drift, when it is pitted against natural selection.

5.3.2 Drift Mixed with Selection

Using MATLAB, we can simulate what happens when both natural selection and genetic drift are happening. To do this, let's modify `drift.m` by adding natural selection after the genetic drift has taken place. Specifically, with a text editor, create a file called `drift_ns.m` as follows. The function accepts as inputs the relative fitnesses of the A_1A_1, A_1A_2, and A_2A_2 genotypes, together with the initial fraction of A_1 in the gene pool, the population size, and the number of generations for the iteration. It returns a column vector of the gene-pool fraction of A_1 through time.

```
function [p] = drift_ns(w11,w12,w22,po,n,runlen)
```

The body of the function begins with the same commands used in `drift.m` to carry out the genetic drift:

```
p = [po];
for t = 1:runlen
  a1 = 0;
  for i=1:(2*n)
    if rand < p(t)
      a1 = a1 + 1;
    end
  end
```

So, the p after drift, called `pd`, is

```
  pd = a1/(2*n);
```

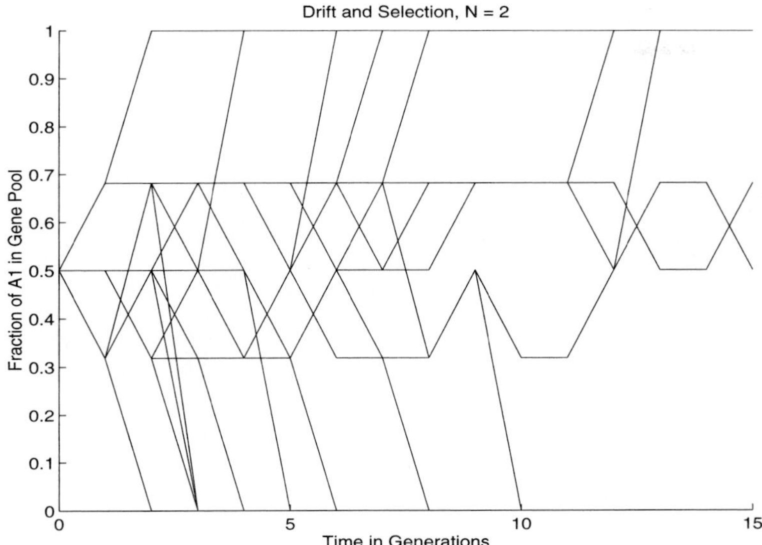

Figure 5.19: The trajectories from an ensemble of 18 replicates of a natural selection in favor of a heterozygote, combined with genetic drift, in a population of 2 individuals. The starting p, the fraction of A_1 in the gene pool, is 1/2. The relative fitnesses are $w_{11} = 0.5$, $w_{12} = 1$, and $w_{22} = 0.5$.

Now for the natural selection. `pd` is p_t and we compute p_{t+1}, called `pprime`, with the now-familiar formulas

```
  wbar = pd^2*w11 + 2*pd*(1-pd)*w12 + (1-pd)^2*w22;
  pprime = (pd*w11 + (1-pd)*w12)*pd/wbar;
```

Finally, we append `pprime` to the bottom of the `p` column vector, and end the loop,

```
  p = [p; pprime];
end
```

So, once this function has been saved as `drift_ns.m`, we can simulate what happens when both genetic drift and natural selection occur together.

The most interesting situation to be investigated is the one in which the natural selection favors the heterozygote, because here natural selection and genetic drift are pulling against each other—natural selection leads to polymorphism and genetic drift leads to monomorphism. Let's then generate an ensemble of replicates of genetic drift with natural selection and compare them with the ensembles we previously made of genetic drift alone. The ensemble of 18 replicates based on a population size of $N = 2$, and with relative selective values of 1/2, 1, and 1/2, is made as before, by our typing

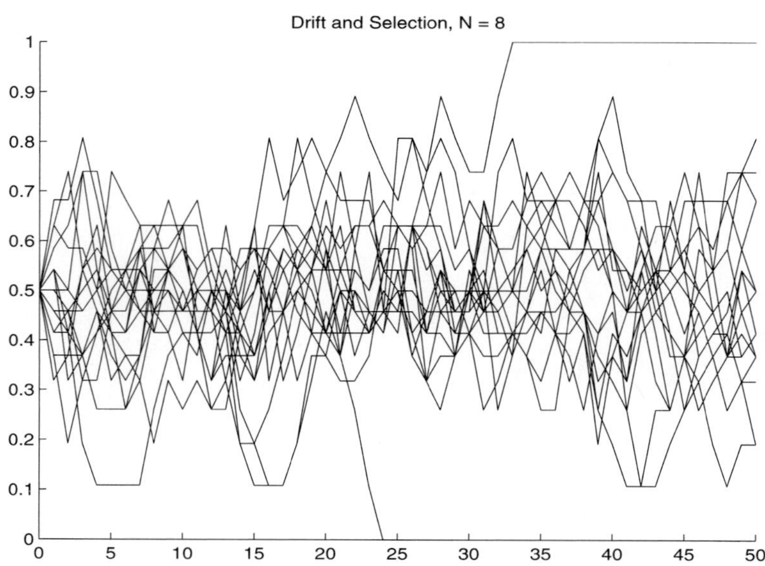

Figure 5.20: The trajectories from an ensemble of 18 replicates of a natural selection in favor of a heterozygote, combined with genetic drift, in a population of 8 individuals. The starting p, the fraction of A_1 in the gene pool, is 1/2. The relative fitnesses are $w_{11} = 0.5$, $w_{12} = 1$, and $w_{22} = 0.5$.

```
>> rand('seed',0);
>> e = [];
>> for i=1:18
>>  e(:,i) = drift_ns(0.5,1,0.5,0.5,2,15);
>> end
```

To plot the trajectories, type

```
>> figure
>> plot(0:15,e)
```

This yields Figure 5.19. The figure shows that, in 15 generations, the drift eventually overpowers natural selection, and all but two of the populations eventually become monomorphic. Next, let's try a population size of $N = 8$. The ensemble of replicates, and the figure of trajectories are obtained from

```
>> rand('seed',0);
>> e = [];
>> for i=1:18
>>  e(:,i) = drift_ns(0.5,1,0.5,0.5,8,50);
```

```
>> end
>> figure
>> plot(0:50,e)
```

As Figure 5.20 shows, now natural selection has the upper hand—it preserves polymorphism in 16 of the 18 replicates in the face of drift over the same time period, during which all the replicates became monomorphic in the absence of selection. The general conclusion is that the larger the population size, the weaker is genetic drift compared to given degree of natural selection.

In Figures 5.19 and 5.20, natural selection and genetic drift were pushing the gene pool in opposite directions. Of course, other cases of selection favor monomorphism; e.g., selection to fix the A_1 allele. In such a case, drift and selection may pull in the same direction, and together may produce the final evolutionary outcome more quickly than would either force alone.

Theoretical-population geneticists have studied the comparative strengths of natural selection and genetic drift is great detail over the last 50 years. They have produced an approximate guide to cases in which one or the other force can be ignored. Scale the relative fitness such that the biggest is 1. Then let s be the difference between the smallest relative fitness and 1, (i.e., $s = 1 - w_{\min}$) and N be the population size. The approximate guide is that

$$\begin{array}{lll} s & >> & \frac{1}{N} \quad \text{Selection Important, Drift Ignored} \\ s & \approx & \frac{1}{N} \quad \text{Both Selection and Drift Important} \\ s & << & \frac{1}{N} \quad \text{Selection Ignored, Drift Important} \end{array}$$

Here the symbols $>>$ mean, greater than by a factor of ten or more; $<<$ means, less than by a factor of ten or more; and $\approx$ means, within a factor of ten. In Figures 5.19-20, $s = 0.5$. When $N = 2$, s equals $1/N$ so both selection and drift must be considered—in fact, drift was the more powerful. When $N = 8$, s is greater than $1/N$, but not by a factor of ten or more, and again selection and drift must both be considered. The population would have to be greater than 20 for drift to be weak enough to be ignored relative to a strength of selection of $s = 0.5$.

Thus, genetic drift is an important force of evolution in two circumstances. First, if the population size is tiny, then even rather strong selection may be impeded by genetic drift. Second, if the two alleles are nearly identical in terms of their biochemical function, then they might not have much effect on an organism's fitness. If s is 10^{-6} for example, the population size would have to be greater than ten million for this rather weak selection to make a difference in the gene pool. Instead, even if the population size were 10^5, which is quite large for a vertebrate population, the gene pool at this locus would be governed by genetic drift. Genetic drift sets a threshold as to how strong the natural selection must be, for it to have an impact on the gene pool. The strength of selection, s, must be greater than the reciprocal of the population size, $1/N$, by a factor of ten or more before it can be regarded as evolutionarily important. Two alleles whose effects on fitness are less than this threshold are said to be "selectively neutral."

5.4 Mutation and Recombination

Mutation is the conversion of one allele to another by an "error" in the organism's DNA replication in its germ line; i.e., in the line of cells that leads to gamete production. If the allele that arises by mutation is not presently in the population, it is said to be a "novel mutation." Mutation in this sense is the ultimate source of all genetic variation. But mutation doesn't happen just once and then stop. Mutation of alleles into other alleles that are already in the population is called "recurrent mutation." Recurrent mutation by itself influences the proportion in the gene pool of the various alleles, and therefore must be considered still another force of evolution, in addition to natural selection and genetic drift as discussed previously.

The effect of recurrent mutation on the gene pool is, happily, much easier to work out than natural selection and genetic drift. So here it goes. Suppose u is the probability that an A_1 allele mutates to A_2. A typical value for u is 1 in a million, 10^{-6}. Similarly, let v be the probability that an A_2 allele mutates to A_1. Well then, the fraction of the gene pool that is A_1 at time $t+1$ is the sum of the A_1 alleles that didn't mutate into A_2 plus the alleles that did mutate from A_2 into A_1. In symbols, this is

$$p_{t+1} = (1-u)p_t + v(1-p_t)$$

That's all there is to it. So as always, we'll start by seeing if there are any equilibria associated with this process, and if so, whether they're stable. Let's have MATLAB do the work, we'll just ask the questions. To tell MATLAB about this model, let `pprime_m` stand for p_{t+1}, and type

```
>> pprime_m = '(1 - u)*p + v*(1-p)';
```

Then have MATLAB find the equilibria, which are the p such that $p_{t+1} = p_t$. Type

```
>> phat_m = simplify(solve(symop(pprime_m,'-','p'),'p'))
```

whereupon MATLAB replies with

```
phat_m = v/(u+v)
```

There is just one equilibrium, and it is always between 0 and 1. So far so good. Now we'll find out, is it stable? Take the derivative of p_{t+1} with respect to p_t, evaluate it at the equilibrium, and see if it is between -1 and 1. So, MATLAB computes the derivative upon our typing

```
>> dpprime_m = simplify(diff(pprime_m,'p'))
```

and tells us that

```
dpprime_m = 1-u-v
```

This doesn't even depend on p[8]. Therefore, the derivative evaluated at equilibrium is

$$\lambda_m = 1 - u - v$$

Because u and v are usually something like 10^{-6} or smaller, λ_m is usually a little less than 1, indicating that the equilibrium is stable.

Recall, though, that the λ is important because it indicates the fate of a small deviation from equilibrium according to

$$(\text{Deviation from Equilibrium})_{t+1} = \lambda(\text{Deviation from Equilibrium})_t$$

Therefore, if λ is less than 1, but still very close to 1, then the equilibrium is stable, *but* deviations from equilibrium decay away very slowly. In this case, λ_m will be something like $1 - 10^{-6}$ with real mutation rates, so the equilibrium, though stable, is approached very very slowly, over a million or so generations. For example, if

```
>> u = 0.000002;
>> v = 0.000001;
```

the equilibrium is at

```
>> eval(phat_m)
ans = 0.3333
```

and λ_m, written to enough decimal places, is

```
>> format long
>> eval(lambda_m)
ans = 0.99999700000000
>> format short
```

The very, very slow speed of evolution by recurrent mutation puts it in the same class of importance as genetic drift in large populations. If selection is exceedingly weak, say with a strength of selection, s, of 10^{-6} or less, and the population is big, say 10^6 or more, then the gene pool is influenced by both drift and recurrent mutation acting together. Otherwise, forget recurrent mutation as a force of evolution. Of course, novel mutation remains important, by definition, as a source of new genetic variation.

5.4.1 Recurrent Mutation Mixed with Selection

Recurrent mutation, although it is not important in its own right unless both selection is weak and the population size large, illustrates how the genetic system can intervene in evolution: the genetic system can prevent natural selection from raising the average adaptation in a population to the highest level actually provided by the genetic variation.

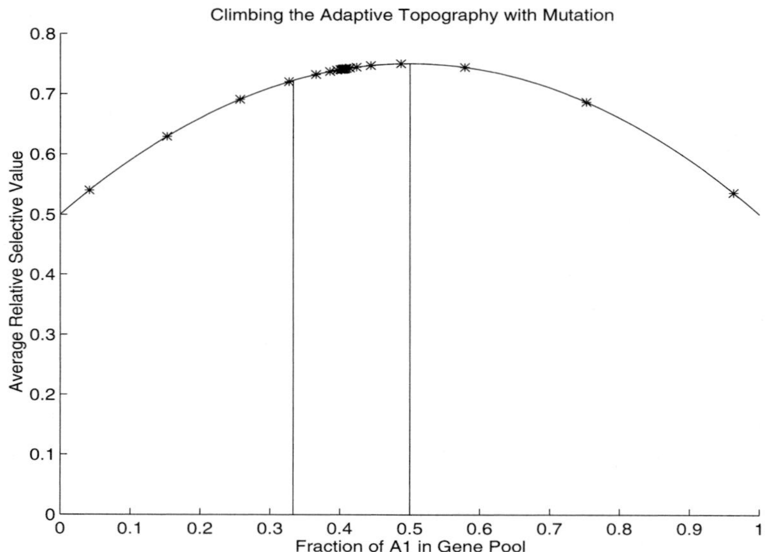

Figure 5.21: Illustration of recurrent mutation preventing natural selection from causing a population to climb to the nearest peak of the $\bar{w}$ curve.

Recurrent mutation messes up the golden metaphor of natural selection, and seeing how the mess comes about is a tip-off to other ways the genetic system can mess up evolution too.

By adding recurrent mutation to the animation we developed earlier, we can show how natural selection progressively increases the average fitness, $\bar{w}$, through time. Recall the function `climb_ns.m` that was used to produce Figure 5.14. Let's add recurrent mutation to this. The function will be saved as a file named `climb_mu.m`. It is called with the three fitnesses, the mutation rates, and the number of iterations to be carried out,

```
function climb_mu(w11,w12,w22,u,v,runlen)
```

The function then initiates the figure, and draws the $\bar{w}$ curve in yellow with

```
figure
hold on
p = 0:.01:1;
wbar = 'p^2*w11 + 2*p*(1-p)*w12 + (1-p)^2*w22';
plot(p,eval(sym2ara(wbar)),'y')
```

Then a red vertical line is drawn from the horizontal axis to the spot on $\bar{w}$ at which the equilibrium under natural selection alone would occur.

[8] dp_{t+1}/dp_t doesn't depend on p_t because p_{t+1} is linear in p_t.

```
p = (w12-w22)/((w12-w11)+(w12-w22));
plot([p p],[0 eval(wbar)],'r')
```

Next, a cyan vertical line is drawn from the horizontal axis to the spot on $\bar{w}$ at which the equilibrium from recurrent mutation alone would occur.

```
p = v/(u+v);
plot([p p],[0 eval(wbar)],'c')
```

Then the function carries out two trials, and in each asks the mouse for input to determine the initial p. As before, the present $\bar{w}$ is calculated first and plotted with a green asterisk. Then the gene-pool fraction after the natural selection has taken place is computed, which is called pprime_s:

```
for trial = 1:2
  [p w] = ginput(1);
  for t=1:runlen
   wbar = p^2*w11 + 2*p*(1-p)*w12 + (1-p)^2*w22;
   plot(p,wbar,'g*')
   pprime_s = (p*w11 + (1-p)*w12)*p/wbar;
```

What's new is the next line. After the natural selection, the recurrent mutation takes place. The gene-pool fraction is now computed taking the recurrent mutation into account, and is called pprime_m,

```
   pprime_m = (1-u)*pprime_s + v*(1-pprime_s);
```

Finally, p_t is reassigned to pprime_m, and the loops are terminated with

```
   p = pprime_m;
  end
end
```

So, type all this in, and save it as a file named climb_mu.m.

To see what happens when recurrent mutation is mixed in with natural selection, let's use huge mutation rates of $u = 0.2$ and $v = 0.1$, which are 10^5 greater than real rates. As before, let the fitnesses be 0.5, 1, and 0.5. Upon typing

```
>> climb_mu(0.5,1,0.5,0.2,0.1,15)
```

we see in Figure 5.21 the curve of $\bar{w}$ and vertical lines indicating where the mutation-alone and selection-alone equilibria are—at $p = 1/3$ and $p = 1/2$, respectively. Now put the cross-hairs at $p = 0.05$ and click. Green asterisks appear. These show the population climbing past the mutation-alone equilibrium and stopping at a p between the mutation-alone and the selection-alone equilibria. This equilibrium is *not* at the peak of $\bar{w}$, because that is the point of the selection-alone equilibrium. Next put the cross-hairs at $p = 0.95$ and click. Green asterisks show the population climbing to the peak of $\bar{w}$, and then continuing on past, until stopping at the equilibrium between the mutation-alone and selection-alone equilibria. This animation demonstrates that recurrent mutation mixed in with natural selection prevents the population from attaining the peak of $\bar{w}$. Thus, the fundamental theorem of natural selection, and the golden metaphor of evolution, are false in these circumstances.

At this point you might say, well, the example used huge mutation rates, and in practice, with real mutation rates of 10^{-6} or less, the stopping point for mutation mixed with selection will be very close to the selection-alone equilibrium, which is the peak of $\bar{w}$. That's true, but don't rejoice yet. Another genetic process, called "recombination" is analogous to recurrent mutation, and its rates are as high as 0.5 each generation. Recombination can mess up the fundamental theorem and the golden metaphor every bit as much as the example of Figure 5.21, and be completely realistic as well.

So, what is recombination and how can it mess up evolutionary progress? Recombination pertains to "between-locus" rearrangements of genes. Suppose a trait is determined by two loci, with two alleles at each. Then the genotype within a gamete is one of four possibilities

$$\underline{A_1B_1} \quad \underline{A_1B_2} \quad \underline{A_2B_1} \quad \underline{A_2B_2}$$

where an underline (or overline) indicates alleles within the same gamete, and possibly even on the same chromosome. These gamete genotypes can be combined into the following ten zygotic genotypes

$$\frac{A_1B_1}{\overline{A_1B_1}} \quad \frac{A_1B_1}{\overline{A_1B_2}} \quad \frac{A_1B_1}{\overline{A_2B_1}} \quad \frac{A_1B_1}{\overline{A_2B_2}} \quad \frac{A_1B_2}{\overline{A_1B_2}} \quad \frac{A_1B_2}{\overline{A_2B_1}} \quad \frac{A_1B_2}{\overline{A_2B_2}} \quad \frac{A_2B_1}{\overline{A_2B_1}} \quad \frac{A_2B_1}{\overline{A_2B_2}} \quad \frac{A_2B_2}{\overline{A_2B_2}}$$

Now focus on the gametes made by the two possible double-heterozygotes,

$$\frac{A_1B_1}{\overline{A_2B_2}} \longrightarrow \begin{cases} \underline{A_1B_1} \\ \underline{A_1B_2} \\ \underline{A_2B_1} \\ \underline{A_2B_2} \end{cases}$$

and

$$\frac{A_1B_2}{\overline{A_2B_1}} \longrightarrow \begin{cases} \underline{A_1B_2} \\ \underline{A_1B_1} \\ \underline{A_2B_2} \\ \underline{A_2B_1} \end{cases}$$

By definition, "recombination" is the appearance of gametes from a parent in which the alleles at the two loci are interchanged. For example, the recombinant gametes from a parent whose genotype is $\frac{A_1B_1}{\overline{A_2B_2}}$ are the $\underline{A_1B_2}$ and $\underline{A_2B_1}$ gametes, if any. If all the gametes from this parent are $\underline{A_1B_1}$ or $\underline{A_2B_2}$, then recombination has not occurred. If some of the gametes are $\underline{A_1B_2}$ and $\underline{A_2B_1}$, then recombination has occurred. The proportion of recombinant gametes out of the total number of gametes is called the "recombination fraction," conventionally denoted as r. The recombination fraction is zero if there is no recombination, and in this case the loci are said to be perfectly "linked." Recombination can occur if the two loci are on the same chromosome and "crossing over" occurs, which means two chromosomes break and the fragments of different chromosomes reconnect; or if the two loci are on different chromosomes, and the chromosomes assort independently of each other at meiosis. Regardless of details about how the recombination occurs, the recombination fraction, r, varies from near zero (if the loci are tightly linked), to 1/2 (if the loci are at opposite ends of the same chromosome or on different chromosomes).

Recombination is analogous to recurrent mutation. In the limit that the two loci are perfectly linked, with no recombination between them, then each gamete genotype, $\underline{A_iB_j}$, is in effect an "allele" itself. That is, $\underline{A_1B_1}$, $\underline{A_1B_2}$, $\underline{A_2B_1}$, and $\underline{A_2B_2}$ are in effect four different alleles at one locus. Indeed, we could define a C_i locus with alleles C_1 being $\underline{A_1B_1}$, C_2 being $\underline{A_1B_2}$, C_3 being $\underline{A_2B_1}$, and C_4 being $\underline{A_2B_2}$. The C locus with four alleles would be formally identical to the A and B loci taken together with perfect linkage between them. The genotypes at this composite locus are then C_iC_j, with $i = 1\ldots4$, and $j = i\ldots4$. Recombination is then a special form of recurrent mutation in which a parent with genotype C_1C_4 produces gametes of type C_2 and C_3 in addition to C_1 and C_4. Similarly, a C_2C_3 parent produces C_1 and C_4 gametes by recombination in addition to the nonrecombinant gametes of C_2 and C_3. The analogue of the mutation rate is the recombination fraction, r, which as we have seen, may be as high as 1/2 and often is.

The reason recombination can be blamed for messing up the fundamental theorem of natural selection is that the fundamental theorem is valid for *one* locus with any number of alleles, when there is assumed to be no recurrent mutation or genetic drift, and that the relative fitnesses are constants. We've illustrated the fundamental theorem with one locus and only two alleles, but it works just fine with three or more alleles, and it would work just fine for the C locus above with four alleles, provided r is zero. But recombination, $r > 0$, throws a monkey wrench into the evolutionary process because it prevents the fundamental theorem from being realized at the C locus. Thus, recombination, like recurrent mutation, may prevent the population from attaining the highest degree of average adaptation possible with the genetic variation that is actually present in the population. But the rate of recombination, unlike mutation, is not tiny, and cannot be ignored. The fundamental theorem of natural selection, and the golden metaphor of evolution, are false in these circumstances too.

So, what's to be made of this unhappy finding?

5.5 Does Evolution Optimize?

Ecologists want evolutionary biology to license optimality as an approach to our understanding why organisms have the phenotypes that they do. To illustrate, suppose an organism can spend an amount of time during the day, τ, in foraging, which provides nutrients and energy to make eggs. Say the number of eggs an organism produces, m, increases with τ as $m = e\tau$ where e is the number of eggs produced per unit time foraging. Suppose also that an animal who forages is vulnerable to predation while it is foraging, so that its survival drops the more it forages. Say that the probability of surviving, l, decreases with τ as $l = l_o(1 - h\tau)$ where l_o is the baseline probability of survival (i.e, the probability of surviving if no additional risk from foraging is incurred) and h is a measure of the hazard incurred per unit time foraging. The organism's fitness as a function of τ, is $W(\tau) = ml = (e\tau)(l_o(1 - h\tau))$. So, how much time should the organism spend foraging? Ecologists would like to answer this question by finding the τ which maximizes fitness, $W(\tau)$. This is the optimal foraging time, τ_{opt}, that strikes the best compromise between survival and reproduction and is easily found because $W(\tau)$ is a quadratic in the unknown variable[9]. Ecologists would like to suppose that organisms will come to possess this optimal foraging time as a result of evolution. This approach can be endlessly refined and used for other traits as well. What is the optimal fur thickness in polar bears that compromises warmth and weight? What is the optimal wing span for a wandering albatross that compromises flight distance and ungainlyness on the ground? What is the optimal time for a plant to leave its stomates open, that compromises between gaining CO_2 and losing water vapor? And so forth; we could make models for the optimum phenotype tailored to each of these situations. Is this approach valid? Does evolution actually lead to phenotypes that we can predict by finding the design that maximizes fitness?

Population genetics demonstrates that the license to think about phenotypes as representing an optimal design is limited. At best, the idea that evolution maximizes fitness is an approximation because of possible genetical complications. Beyond this point, geneticists and ecologists part company. Ecologists are not persuaded that genetical complications often prevent evolution from producing near-optimal designs. Do you really mean that a polar bear's fur is too thin because of recombination between two major loci? Are you serious that a plant opens its stomates at the wrong time because of recurrent mutation? The genetic complication that is taken seriously is maladaptation resulting from inbreeding and genetic drift in tiny populations, especially in conservation biology. Otherwise, ecologists are not persuaded that problems with optimization are primarily genetic, although the possibility of genetic complications is acknowledged.

When ecologists are skeptical of optimization, it is usually because either biology or history have been oversimplified. Perhaps the function of the polar bear's fur is not solely warmth, but also protection against ectoparasites. Moreover, perhaps the polar bear evolved its fur in a different place and time, and so that its present-day fur coat was adaptive there and then, but not now. If so, an optimization model to predict fur thickness that omits ectoparasites, and is focused only on conditions in the present-day habitat, is probably incorrect. This problem is not with optimization as such, but with a superficial application of optimization that doesn't take into account all the relevant natural history.

The most important difficulty in applying optimization to organismal design is the assumption, required by the fundamental theorem of natural selection, that the relative fitnesses are independent of the gene-pool composition. If the trait being considered is involved in social interactions, then the relative fitnesses, w_{ij}, depend on the gene pool state, p_t, and the fundamental theorem simply does not apply, even with a benign genetic system. An optimization approach works for socially innocent traits, and not for traits involved in courtship, territorial defense, warning calls, and so forth. When social traits are considered,

[9] The peak of $W(\tau)$ is at $\tau_{\text{opt}} = h/2$. In MATLAB, type `w = '(e*tau)*(lo*(1-h*tau))';` followed by `tau_opt = solve(diff(w,'tau'),'tau')`.

the simultaneous aims of two or more interacting parties are analyzed, in a subject called "game theory." As natural selection propels a socially successful trait to spread throughout a population, the average relative fitness in the population, $\bar{w}$, may progressively decline.

In a similar vein, ecologists want evolutionary biology to license evolutionary functionalism, which is an approach to thinking about why organisms have the traits they do. Evolutionary functionalism is similar to optimization, but less formal—it is used more in discussion and interpretation than in mathematical models. The goal of evolutionary functionalism is to understand a trait's function in terms of how it improves an organism's fitness, and in doing so, to explain why the trait exists. Evolutionary functionalism shares with optimization the requirement of a benign genetic system. Evolutionary functionalism invites one to seek out a use for everything an organism does, for all its traits. If one watches a bird beat its wings, a fish turn somersaults, a kitten clean its fur, or if one sees a red hump on a monkey, a long nose on a seal, or even a spider consume its mate, one should, according to evolutionary functionalism, ascertain why the trait enables the organism who uses it to increase its fitness. This outlook argues against imagining that traits exist for no reason at all, that they just happen by chance, or are accidental by-products of other more prominent functions.

The danger to evolutionary functionalism is the risk of telling "just-so" stories dreamed up to make some trait seem relevant to fitness. Do you mean that my ear-twitch is an adaptation for dislodging the flies that plagued my cave-dwelling ancestors? Claims of adaptive function for traits may be far-fetched. Genetic drift offers a conceptual safeguard against just-so stories. If a claim of adaptive function does not lead to a selection strength higher by a factor of ten than the reciprocal of the population size, then it won't wash, and should be discarded.

As an ecologist, you'll have to take a personal position on where you stand concerning optimization and organismal design. You'll need to make your own personal judgement on whether real traits are close enough to optimal for optimality approaches to be worthwhile. Similarly, you'll have to take a personal stand on whether to use evolutionary functionalism in your thinking. Does this perspective lead you to understand why organisms do what they do, or does it lead you to fantasize function where there isn't any?

5.6 Natural Selection and Population Size

Darwin imagined that natural selection involved resource competition brought about by overpopulation. As we've seen, natural selection is a far more general concept, and takes place regardless of whether there's competition for resources. Natural selection is the differential net production of offspring by different genotypes, and takes place in expanding populations as well as in those whose abundance has become limited by resources. Nonetheless, density-dependent limitation to population expansion does often occur, and in these conditions natural selection upon a trait whose fitness depends on population size is called "density-dependent selection." Fitnesses that depend on population size bring about evolution that is not automatically covered by the fundamental theorem of natural selection, but the fundamental theorem can be easily extended to deal with this situation.

Recall that when we derived the equation for A_1's fraction of the gene pool at time $t+1$ as a result of natural selection we also derived an equation for the population size at time $t+1$. The equations are

$$\begin{aligned} p_{t+1} &= \frac{p_t W_{11}(N_t) + (1-p_t)W_{12}(N_t)}{\bar{W}(N_t, p_t)} p_t \\ N_{t+1} &= \bar{W}(N_t, p_t) N_t \end{aligned}$$

where the average absolute fitness, or factor of geometric increase over one generation, is

$$\bar{W}(N_t, p_t) = p_t^2 W_{11}(N_t) + 2p_t(1-p_t)W_{12}(N_t) + (1-p_t)^2 W_{22}(N_t)$$

Most of population genetics ignores the equation for N_{t+1}; it focuses only on the equation for p_{t+1}, and uses relative selective values, w_{ij}, instead of the absolute selective values, W_{ij}. Here though, we'll redirect the focus back to the ecological origins of natural selection, and focus simultaneously on p_{t+1} and N_{t+1}. When there is density-dependent selection, the W_{ij} each depend on the population size, N. Typically, each W_{ij} is a decreasing function of N; this decrease indicates the presence of a density dependence that lowers fecundity and/or survivorship.

The extension of the fundamental theorem to density-dependent selection is based on the idea of the equilibrium-population size as a function of the gene-pool composition. Imagine that the gene-pool composition is held constant at some value, say p, and the population size is allowed to come to equilibrium. The population size for that p is $\hat{N}(p)$. We can plot the equilibrium population size as a function of p and we get a curve. The high points on it represents a gene pool composition that culminates in a high size of an equilibrium population, whereas low points on it represent a gene-pool composition that leads to a low equilibrium-population size. The way to find the $\hat{N}(p)$ curve is to set the average fitness equal to one, and to solve for N. That is, $\hat{N}(p)$, is the root, N, of

$$1 = \bar{W}(N, p) = p^2 W_{11}(N) + 2p(1-p)W_{12}(N) + (1-p)^2 W_{22}(N)$$

The fundamental theorem of density-dependent natural selection is that the gene pool evolves to a composition that maximizes the equilibrium population size.

It's fun to develop an animation of how the fundamental theorem of density-dependent selection works. We'll adapt the function, `climb_ns.m`, which illustrated a population climbing $\bar{w}$, to apply to density-dependent selection. With a text editor, we'll create the following function to be saved as file called `climb_ds.m`. In this illustration, we'll suppose that the fitnesses each decrease linearly with N, as

$$W_{ij}(N) = 1 + r_{ij} - (r_{ij}/K_{ij})N$$

where r_{ij} and K_{ij} is the intrinsic rate of increase (in discrete time) and carrying capacity of genotype A_iA_j. The interpretation of the coefficients in terms of the familiar r and K of the logistic equation arises because if the population hypothetically consisted entirely of the genotype produced by genotype A_iA_j then its growth would be logistic with parameters r_{ij}

and K_{ij}. Of course, the population does not actually grow logistically, because there are three genotypes in it, so the population is a mixture of three logistically growing subpopulations, each with different r's and K's, and the population's overall growth represents an ever-changing average of the r's and K's of its individuals. The function `climb_ds` is called with the r's and K's of the three genotypes, and the number of iterations.

```
function climb_ds(r11,k11,r12,k12,r22,k22,runlen)
```

Then the graph of $\hat{N}(p)$ is drawn with the following commands,

```
figure
hold on
w11 = '1 + r11 - (r11/k11)*n';
w12 = '1 + r12 - (r12/k12)*n';
w22 = '1 + r22 - (r22/k22)*n';
wbar = symop('p^2','*',w11,'+','2*p*(1-p)','*',w12,'+','(1-p)^2','*',w22);
nhat = solve(symop(wbar,'-','1'),'n');
axis([0 1 0 max([k11 k12 k22])])
p = 0:.01:1;
plot(p,eval(sym2ara(nhat)),'y')
```

This is a little complicated, and here's how it works. The three fitnesses are defined symbolically as `w11`, `w12`, and `w22`. They are then combined symbolically into `wbar`. Then the formula for `nhat` is found symbolically. When the graph is set up, the axis is specified as 0 to 1 on the horizontal, and 0 to the highest of the K_{ij}'s on the vertical. Then a vector of p's from 0 to 1 in steps of 0.01 is defined. Lastly, the curve of `nhat` is plotted in yellow. Now on to the iteration of the equations for p_{t+1} and N_{t+1}. We'll use two trials, and use the mouse to input the initial condition for each:

```
for trial = 1:2
  [p n] = ginput(1);
```

Next, we'll need to compute the fitnesses anew for each generation, because `n` changes in each generation:

```
  for t=1:runlen
   w11 = 1 + r11 - (r11/k11)*n;
   w12 = 1 + r12 - (r12/k12)*n;
   w22 = 1 + r22 - (r22/k22)*n;
   wbar = p^2*w11 + 2*p*(1-p)*w12 + (1-p)^2*w22;
```

Here `w11`, `w12`, `w22`, and `wbar` are now used to represent numerical values, rather than symbolic expressions. Next, `p` and `n` are plotted together as a green asterisk, their new values computed as `pprime` and `nprime`, and `p` and `n` are reassigned for the beginning of the next generation.

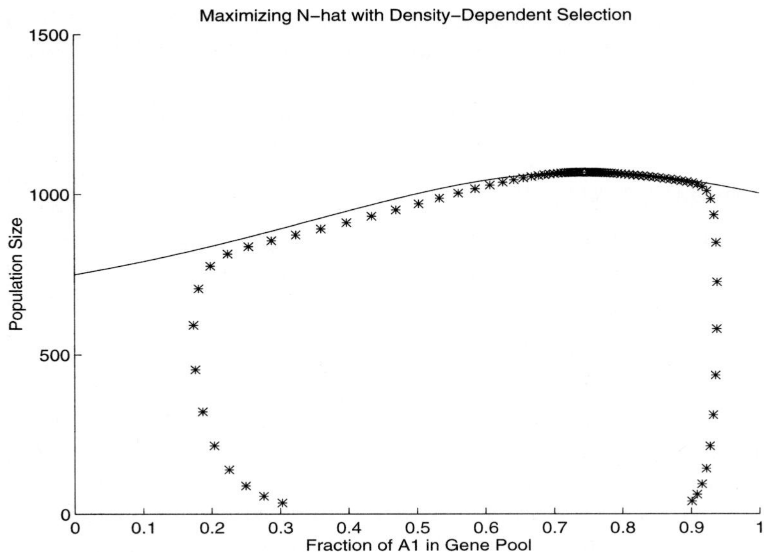

Figure 5.22: Illustration of a density-dependent natural selection causing a population to converge to a peak on the $\hat{N}(p)$ curve.

```
    plot(p,n,'g*')
    pprime = (p*w11 + (1-p)*w12)*p/wbar;
    nprime = wbar*n;
    p = pprime;
    n = nprime;
```

Finally, the loops for the iteration of generations and for the iteration of the two trials are ended with

```
  end
end
```

When these statements have been saved as `climb_ds.m`, it's ready to use.

5.6.1 r- and K-selection

Perhaps the most interesting numerical values to use in the illustration assume a trade-off between r and K. A genotype that is better in both parameters will evolve regardless. If genetic variation remains, it probably involves the genotypes that face a tradeoff between these parameters. Such genetic variation would represent individuals who allocate the energy and nutrients at their disposal into either the highest possible reproduction rate attainable in uncrowded conditions (high r), or into a relatively better, but lower than maximal, reproduction rate in crowded conditions (high K). The evolutionary fate of such genotypes is then illustrated by our typing

```
>> climb_ds(0.6,1000,0.4,1500,0.8,750,50)
```

as shown in Figure 5.22. Here we've assumed that the heterozygote has the best K and the worst r, with the homozygotes intermediate. To make the figure, we clicked the mouse to start the first trial off at a low N and a p near 0.3, and again clicked the mouse to start the second trial off at a low N and a p near 0.9. At first the trajectories bend toward the vertical axes representing $p = 0$ and $p = 1$, because the heterozygote has the lowest r, and selection at low population sizes favors either homozygote. As the population size grows, the heterozygote comes to have the highest fitness, and the trajectories bend back toward the interior of the graph, and come to a polymorphic equilibrium. The equilibrium is at the peak of the $\hat{N}(p)$ curve, and this result shows that density-dependent natural selection produces the gene-pool composition that yields the highest size of equilibrium population.

When there are uncrowded circumstances that favor genotypes having a high r to those having a high K, the selection in these circumstances is often called "r-selection." Selection in crowded conditions favoring genotypes having a high K over those with a high r, is called "K-selection." Reference to r and K is by analogy to the logistic equation. Of course, the fundamental theorem of density-dependent natural selection is not based on the logistic equation as such, and works for any population-dynamic model that leads to a stable equilibrium-population size.

Density-dependent natural selection, including r- and K-selection, are part of what ecologists call "life-history" theory. An organism's life history is the story of what it does throughout its life cycle. Of particular interest are how long an organism lives, when it reproduces, how many offspring it has, and how long it takes care of its offspring. Density-dependent selection is mostly about adaptation to crowding, whereas when an organism does what, or adaptive timing, is the province of age-dependent selection, which comes next.

5.7 Natural Selection and Age

Sure'd be nice to live forever, but don't count on making it much past 100 years. It would also be nice to have children at any age, but don't postpone it much beyond 50 years. Our body has something to say about how long we live and when we have children—it's not just up to us. Natural selection has a hand in determining the average life span and time table of reproduction for every species. Here we delve into how natural selection interacts with age. This section is important, but it's a bit complicated, so please be patient.

To work age into the equations that predict the outcome of natural selection we have to start from scratch. How do we describe a life history in a way useful for modeling? The answer is to suppose that the life history is embodied in the Leslie matrix introduced in Chapter 3. Recall that a Leslie matrix has the age-specific fertilities entered along the top row, and that the probabilities of the organism's surviving from one age class to the next are entered on the subdiagonal of the matrix, as in

$$\begin{matrix} F_1 & F_2 & F_3 \\ P_1 & 0 & 0 \\ 0 & P_2 & 0 \end{matrix}$$

where F_i is the number of offspring to a parent of age-i that survive to the next census, and P_i is the probability of an organism's surviving through age class-i to enter age class-$i+1$ at time $t+1$. The Leslie matrix can represent an organism's life history because the F_i's describe when an organism is having children and how many, and the P_i's describe how long an organism can live—high P_i's mean a long life, and low P_i's mean a short lifespan. Now we'll assume that each genotype at one locus with two alleles, has a Leslie matrix of its own. For example, tell MATLAB about three Leslie matrices, one for each genotype, A_1A_1, A_1A_2, A_2A_2:

```
>> l11 = [1/2 1 3/4; 3/6 0 0; 0 3/6 0];
>> l12 = [1/2 1 3/4; 4/6 0 0; 0 2/6 0];
>> l22 = [1/2 1 3/4; 2/6 0 0; 0 4/6 0];
```

These Leslie matrices all involve three age classes. All three happen to have the same fertility schedule but differ in their survivorship schedule. The A_1A_1 genotype has a 0.5 probability of survival at each age. The A_1A_2 heterozygote survives better at the earlier age but more poorly at the later age, and vice versa for the A_2A_2 homozygote. Which of these would lead the population to grow the fastest? To find out we need to calculate, for each of the Leslie matrices, the geometric growth factor, which is its biggest eigenvalue. The biggest eigenvalue of the first of these matrices is found by our typing

```
>> r11_all=eig(l11); r11 = r11_all(1)
```

The invocation of `eig` returns a vector of all the eigenvalues, and we want the biggest, which is the first one. MATLAB replies with

```
r11 = 1.1056
```

Similarly, the biggest eigenvalues of the other matrices are

```
r12 = 1.1828
r22 = 1.0000
```

Thus, the heterozygote has the Leslie matrix with the highest R.

R_{ij} is the geometric growth factor of a population in the stable age distribution, if *all* the members of the population have the phenotype corresponding to the A_iA_j genotype. Of course, if there is more than one genotype in the population, the population's actual geometric growth factor will be some mixture of the growth potentials of all the genotypes. Our task therefore, is to discover how the gene pool changes if each genotype has a different Leslie matrix, and how the geometric growth rate of the population as a whole is related to the Leslie matrices of the three genotypes in it.

To project the growth of a population with three genotypes in it, we need to carry out the following matrix multiplication

$$\begin{pmatrix} n_{11,1} \\ n_{12,1} \\ n_{22,1} \\ n_{11,2} \\ n_{12,2} \\ n_{22,2} \\ n_{11,3} \\ n_{12,3} \\ n_{22,3} \end{pmatrix}_{t+1} = \begin{pmatrix} p_{t+1}^2 & 0 & 0 & 0 & 0 & 0 & 0 & 0 & 0 \\ 2p_{t+1}(1-p_{t+1}) & 0 & 0 & 0 & 0 & 0 & 0 & 0 & 0 \\ (1-p_{t+1})^2 & 0 & 0 & 0 & 0 & 0 & 0 & 0 & 0 \\ 0 & 0 & 0 & 1 & 0 & 0 & 0 & 0 & 0 \\ 0 & 0 & 0 & 0 & 1 & 0 & 0 & 0 & 0 \\ 0 & 0 & 0 & 0 & 0 & 1 & 0 & 0 & 0 \\ 0 & 0 & 0 & 0 & 0 & 0 & 1 & 0 & 0 \\ 0 & 0 & 0 & 0 & 0 & 0 & 0 & 1 & 0 \\ 0 & 0 & 0 & 0 & 0 & 0 & 0 & 0 & 1 \end{pmatrix} \times$$

$$\begin{pmatrix} F_{11,1} & F_{12,1} & F_{22,1} & F_{11,2} & F_{12,2} & F_{22,2} & F_{11,3} & F_{12,3} & F_{22,3} \\ 0 & 0 & 0 & 0 & 0 & 0 & 0 & 0 & 0 \\ 0 & 0 & 0 & 0 & 0 & 0 & 0 & 0 & 0 \\ P_{11,1} & 0 & 0 & 0 & 0 & 0 & 0 & 0 & 0 \\ 0 & P_{12,1} & 0 & 0 & 0 & 0 & 0 & 0 & 0 \\ 0 & 0 & P_{22,1} & 0 & 0 & 0 & 0 & 0 & 0 \\ 0 & 0 & 0 & P_{11,2} & 0 & 0 & 0 & 0 & 0 \\ 0 & 0 & 0 & 0 & P_{12,2} & 0 & 0 & 0 & 0 \\ 0 & 0 & 0 & 0 & 0 & P_{22,2} & 0 & 0 & 0 \end{pmatrix} \begin{pmatrix} n_{11,1} \\ n_{12,1} \\ n_{22,1} \\ n_{11,2} \\ n_{12,2} \\ n_{22,2} \\ n_{11,3} \\ n_{12,3} \\ n_{22,3} \end{pmatrix}_{t}$$

I know your reaction to this is, Ugh! Still, it's easier than it looks. The basic state variable is a column vector for the number of each genotype in each age class. The Leslie matrices for the three genotypes are combined into one large matrix (to be denoted as $\mathbf{g}$ later on). The survivorship data are entered along a subdiagonal and simply indicate that $n_{ij,a+1,t+1} = P_{ij,a}n_{ij,a,t}$ for each genotype, A_iA_j, and age, a—that is, each individual of a given genotype at time t who lives through an age class is promoted into the next age class for that genotype at time $t+1$. The more tricky part is how the births are handled. The fertility data from the three Leslie matrices are entered along the top row of the large matrix. Then the total number of newborn that will be present at time $t+1$ is

$$b_{t+1} = \Sigma_{ij,a} F_{ij,a} n_{ij,t}$$

which is simply the sum of the offspring produced by parents of all genotypes and age classes. The gene pool fraction among these newborn, p_{t+1}, is

$$p_{t+1} = \frac{\Sigma_a F_{11,a} n_{11,t} + (1/2)\Sigma_a F_{12,a} n_{12,t}}{b_{t+1}}$$

according to Mendel's laws, whereby all the gametes from A_1A_1 parents carry the A_1 allele and half the gametes from A_1A_2 parents carry the A_1 allele. Next, the mating system is assumed to be random union of gametes from parents of all ages, so that the Hardy-Weinberg ratios based on p_{t+1} describe the genotypic ratios among the newborn. These Hardy-Weinberg ratios are placed into a matrix, called the mating matrix, in the first column of the first three rows, with the rest of the entries being zero except on the main diagonal. The product of the mating matrix with the matrix containing the data from the three Leslie matrices, times the state vector at time t, yields the state vector for time $t+1$.

The fundamental theorem of natural selection can be extended to age-dependent selection when the model above is used. In this model, each genotype has its own Leslie matrix and the gametes from parents of all ages combine at random. The fundamental theorem of age-dependent selection is the key needed for an understanding of how natural selection shapes life histories. The extension of the fundamental theorem to age-dependent selection is based on the idea of the asymptotic geometric-growth factor as a function of the gene-pool composition. Imagine that the gene-pool composition is held constant at some value, say p, and that the numbers in the age classes of all genotypes come to their stable age distribution for that value of p. The geometric-growth factor for the population in which the genetically mixed stable age-distribution is attained, is labeled $\hat{R}(p)$. We can plot the asymptotic geometric growth factor as a function of p and we get a curve. The high points on it represent a gene-pool composition that culminates in a high geometric growth factor, whereas the low points on it represent a gene-pool composition that leads to a low geometric-growth factor.

The way to find the $\hat{R}(p)$ curve is to find the largest root of a particular polynomial. The eigenvalues of each Leslie matrix come from a polynomial called the "characteristic polynomial" of the matrix; this polynomial is built up from the elements of the matrix. To calculate the characteristic polynomial of a matrix, MATLAB uses the `poly` command. The R of the genetically mixed population is *not* an average of the eigenvalues of the Leslie matrices from each genotype. Instead, we take the average of the characteristic polynomials and find the biggest root of *that.* So, if `l11`, `l12`, and `l22` are the Leslie matrices for the three genotypes, then the characteristic polynomials for each of these matrices are `poly(l11)`, `poly(l12)`, and `poly(l22)`. The genotypic average of these polynomials is

```
poly_ave = p^2*poly(l11)+2*p*(1-p)*poly(l12)+(1-p)^2*poly(l22)
```

To find the roots of a polynomial, MATLAB uses the `roots` command, and we want the biggest of the roots, which is the first one in the vector of roots. When we plot the biggest root of `poly_ave` found in this way against `p`, we obtain the $\hat{R}(p)$ curve, which is the asymptotic geometric growth factor as a function of p. The fundamental theorem of age-dependent natural selection is that the gene pool evolves to a composition that maximizes the asymptotic geometric growth factor.

To illustrate how age-dependent selection maximizes $\hat{R}(p)$, let's modify the function we previously used to illustrate a climb of the adaptive topography. We'll now develop a function named `climb_ag` as follows. The function will take as its input the three Leslie matrices, and the initial numbers of the three genotypes. The Leslie matrices can be any size, although the examples will use three age classes. The function will carry out two runs, first with the initial n_{11} equal to `na`, the initial n_{12} equal to `nb`, and the initial n_{22} equal to `nc`. The second run will interchange the initial conditions for the homozygotes, so that its initial n_{11} equals `nc`, the initial n_{12} equals `nb`, and the initial n_{22} equals `na`. This setup provides two trajectories from quite different initial conditions, and will yield an informative illustration. The function is also called with the number of iterations to be carried out in each trajectory.

```
function climb_ag(l11,l12,l22,na,nb,nc,runlen)
```

The function's first job is to plot the $\hat{R}(p)$ curve, which is done with the following instructions

```
figure
hold on
poly11 = poly(l11); poly12 = poly(l12); poly22 = poly(l22);
r_hat = [];
for p=0:.1:1
 roots_of_poly=roots(p^2*poly11+2*p*(1-p)*poly12+(1-p)^2*poly22);
 r_hat = [r_hat roots_of_poly(1)];
end
plot(0:.1:1,r_hat)
```

The loop for the two trials and their initial conditions are assigned with

```
for trials=1:2
 if trials==1
  n11 = na; n12 = nb; n22 = nc;
 else
  n11 = nc; n12 = nb; n22 = na;
 end
```

The initial number of newborn and the initial gene-pool fraction of A_1 among the newborn are calculated with

```
 b = n11(1) + n12(1) + n22(1);
 p = (n11(1) + (1/2)*n12(1))/b;
```

Now the matrix combining the information from the three Leslie matrices is synthesized with

```
 a = length(n11);
 n = zeros(3*a,1);
 for i=1:a
  n((i-1)*a + 1) = n11(i);
  n((i-1)*a + 2) = n12(i);
  n((i-1)*a + 3) = n22(i);
 end
 g = zeros(3*a,3*a);
 for i=1:a
  g(1,(i-1)*a + 1) = l11(1,i);
  g(1,(i-1)*a + 2) = l12(1,i);
  g(1,(i-1)*a + 3) = l22(1,i);
```

```
end
for i=1:a-1
 g(i*a+1,(i-1)*a + 1) = l11(i+1,i);
 g(i*a+2,(i-1)*a + 2) = l12(i+1,i);
 g(i*a+3,(i-1)*a + 3) = l22(i+1,i);
end
```

Next, a matrix called `mendel` is made. This matrix aids in calculating the gene pool fraction—it has 1, 1/2, and 0 down its main diagonal for each age class:

```
mendel = [];
for i=1:a
 mendel = [mendel 1 1/2 0];
end
mendel = diag(mendel);
```

The mating matrix, which will contain the Hardy-Weinberg proportions, is initialized with

```
mate = eye(3*a); mate(2,2) = 0; mate(3,3) = 0;
```

The main loop for the iteration is initialized with

```
for t=1:runlen
```

The number of newborn at time $t+1$ is calculated with

```
  bprime = g(1,:)*n;
```

The gene-pool fraction at time $t+1$ is then

```
  pprime = g(1,:)*mendel*n/bprime;
```

Now that p_{t+1} is known, the mating matrix can be filled in with Hardy-Weinberg proportions:

```
  mate(1,1) = pprime^2;
  mate(2,1) = 2*pprime*(1-pprime);
  mate(3,1) = (1-pprime)^2;
```

Finally, we are in a position to predict the full state vector at $t+1$. We do the matrix multiplication of `mate` times `g` times the state vector at t,

```
  nprime = mate*g*n;
```

Now is a good time to plot the actual factor of geometric increase among the newborn, b_{t+1}/b_t, against p_t; a green asterisk will appear.

```
  plot(p,bprime/b,'g*')
```

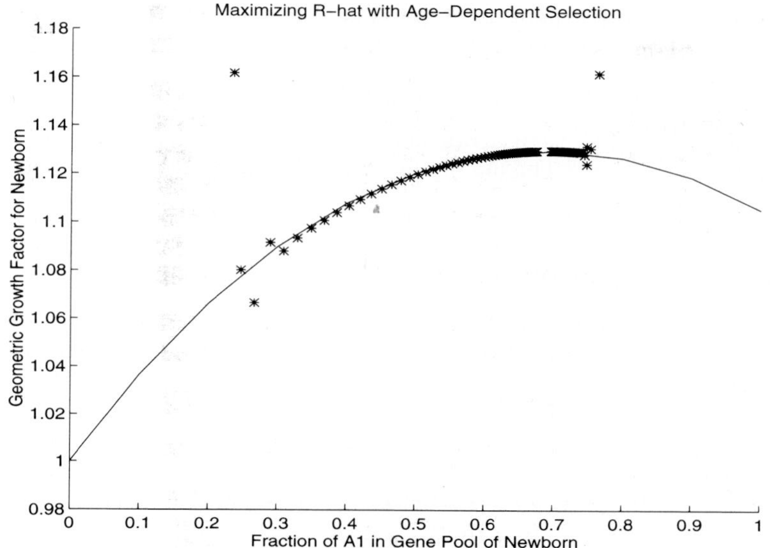

Figure 5.23: Illustration of an age-dependent natural selection causing a population to converge to a peak on the $\hat{R}(p)$ curve.

To conclude the iteration we reassign `b`, `p`, and `n`, and end the loops for the iteration and trials:

```
    b = bprime; p = pprime; n = nprime;
  end
end
```

Once this function has been saved in a file called `climb_ag.m`, it is ready to be used from MATLAB's command line.

To illustrate `climb_ag.m` we'll use the `l11`, `l12`, and `l22` defined previously, and take as our initial numbers the following vectors:

```
>> na = [1; 1; 0];
>> nb = [6; 3; 1];
>> nc = [10; 5; 2];
```

Now typing

```
>> climb_ag(l11,l12,l22,na,nb,nc,60)
```

yields Figure 5.23. One of the trajectories starts out with a p of about 0.25, the other trajectory with a p of about 0.75. After 60 iterations, both trajectories have come close to an equilibrium p at around 0.7, where the population's geometric-growth factor is maximized. This figure illustrates the fundamental theorem of age-dependent natural selection.

The fundamental theorem of age-dependent selection is neat. Because of this theorem, if we want to predict how long animals should live and when they should have their children, all we have to do is figure out what arrangement of F_i's and P_i's yields the highest R in the given ecological circumstance. The life history that maximizes R in a Leslie matrix is called an "optimal life history."

5.7.1 The True Meaning of Death

Let's start by figuring out when to die. We hope never, of course, but let's see what natural selection has to say about this. The issue revolves around tradeoffs. Consider a gene that improves vitality at one age and lowers vitality at another. The net effect on fitness, as measured by the R of a Leslie matrix, is the sum of two contributions—a positive contribution to fitness from the improvement at the age when vitality is increased, plus the negative contribution from the detriment to fitness at the age when vitality is decreased. What we should now do is calculate the effect on fitness of a unit change in survivorship for any given age—you can think of this as a price computed in the currency of fitness. If a gene that raises vitality at one age and lowers it at another shows a net profit in fitness, it will spread within the gene pool. Of the genes affecting vitality at different ages, when there are no more available that show a net profit in fitness, then the survivorship pattern in the species will stop evolving.

Using MATLAB we can directly calculate the fitness value of surviving a little bit better at any given age. Let's walk through the calculation and then bundle what we've done into a MATLAB function. Start with a representative Leslie matrix,

```
>> leslie = [1 1 1 1; .5 0 0 0; 0 .5 0 0; 0 0 .5 0]
leslie =     1.0000    1.0000    1.0000    1.0000
             0.5000         0         0         0
                  0    0.5000         0         0
                  0         0    0.5000         0
```

To find the R for this matrix, type

```
>> e_leslie = eig(leslie);
>> r_leslie = e_leslie(1)
r_leslie = 1.4872
```

Now we want to tinker with the P_i's on the subdiagonal. For example, let's try boosting P_2, which is at position $(3, 2)$, a little bit and see how R changes. By convention[10], the boosts are not given in units of P itself, but in units of the log of P. So let's define another matrix, which is the one we'll modify:

```
>> les = leslie;
```

[10] If you go further with this topic, you'll see that the formulas are simpler with this convention.

To change the element at position (3, 2), we'll take the log, boost it by 0.1, and then convert back to the original units. So type

```
>> les(3,2) = exp(log(les(3,2)) + .1);
>> les
les =      1.0000    1.0000    1.0000    1.0000
           0.5000         0         0         0
                0    0.5526         0         0
                0         0    0.5000         0
```

P_2 has been boosted from 0.5000 to 0.5526. What is the effect of this boost on R? To find the answer, type

```
>> e_les = eig(les);
>> r_les = e_les(1)
r_les = 1.4980
```

To express the effect on R per unit change in the log of P, type

```
>> d_r_over_d_ln_p = (r_les - r_leslie)/.1
d_r_over_d_ln_p = 0.1078
```

This is the fitness value; i.e., the change in R, of a unit change in the log of the organism's survivorship at age 2.

By itself, knowing the fitness value of a change in survivorship at age 2 isn't particularly useful—we need to know the fitness values for changes in survivorship at the other ages too. So, let's check out the fitness value of a boost to the survivorship at age 3. Type

```
>> les = leslie;
>> les(4,3) = exp(log(les(4,3)) + .1);
>> e_les = eig(les);
>> r_les = e_les(1);
>> d_r_over_d_ln_p = (r_les - r_leslie)/.1
d_r_over_d_ln_p = 0.0273
```

This is the fitness value of a unit change in the log of the survivorship at age 3. Now it is obvious that the fitness value of survivorship at age 2 is about five times higher than that at age 3.

So we come to the true meaning of death. Death, or at least its timing, is an adaptation to promote fitness. Consider a gene that increases vitality at age 2, but leads to lower vitality at age 3. Imagine, for example, a novel mutation that raises an organism's metabolic rate, and that the higher rate helps it escape predators but causes wear and tear on the muscles. Compared to the organisms presently in the population, such an organism might survive better at age 2, but more poorly at age 3. Will this new phenotype evolve? It depends on the tradeoff. If the novel mutation gives up 0.1 log survivorship at age 3 to gain 0.1 in log survivorship at age 2, then yes. The gain exceeds the loss. If the novel mutation gives up at age 3 more than about five times what it gains at age 2, then no, because the overall gain in fitness does not exceed the loss. Thus, an overall "selection pressure" in favor of better survival at earlier ages exists and is coupled to poorer survival at later ages. The evolutionary fate of a particular trait that enhances early survival while reducing later survival depends on the details of the tradeoff involved.

Natural selection that improves survival at an earlier age while lowering survival at a later age is bringing about the evolution of "senescence." Senescence is represented as a set of age-specific survivorships, P_i's, that decline monotonically with age. According to evolutionary biology, senescence is an inevitable consequence of natural selection. The genetic basis for senescence should be extremely varied. Because of the ubiquity of tradeoffs between effects at different ages, there should be no single mechanism of senescence whose cure could supply a medical fountain of youth. Alas, all organisms are programmed to die at a rate set in part by natural selection.

But we can sure have a good time when we're alive, so let's move on to happier matters, such as when to have children.

5.7.2 When to Have Children

How natural selection affects the timing of children can be determined the same way that we determined how natural selection affects the timing of death. We change an F_i in the Leslie matrix by a little amount and see how the R changes. This will provide the price, in terms of fitness, of a small change in the organism's fecundity at a given age. So, let's walk through the calculation. Begin by defining the matrix whose elements we'll modify by typing

```
>> les = leslie;
```

Then, to add a bit to the fecundity at say, age 2, type

```
>> les(1,2) = les(1,2) + .1;
>> les
les = 1.0000    1.1000    1.0000    1.0000
      0.5000         0         0         0
           0    0.5000         0         0
           0         0    0.5000         0
```

Notice that the element for F_2 has been increased by 0.1. Now, to get the biggest eigenvalue, R, type

```
>> e_les = eig(les);
>> r_les = e_les(1)
r_les = 1.5101
```

Finally, the change in R per unit change in F_2 is computed by our typing

```
>> d_r_over_d_f = (r_les - r_leslie)/.1
d_r_over_d_f = 0.2290
```

This is the fitness value of a unit change in fecundity at age 2.

To obtain the fitness value of a unit change in fecundity at age 3, just repeat the procedure and type

```
>> les = leslie;
>> les(1,3) = les(1,3) + .1;
>> e_les = eig(les);
>> r_les = e_les(1);
>> d_r_over_d_f = (r_les - r_leslie)/.1
d_r_over_d_f = 0.0771
```

Thus, the fitness value of a unit change in fecundity at age 2 is about three times as high as that at age 3. Therefore, there is also a selection pressure in favor of early reproduction at the cost of lowering success in later reproduction. Whether any particular trait evolves that accomplishes this task—sacrifices reproductive ability in later years to promote reproduction in earlier years—depends on the magnitude of the tradeoffs involved. In this numerical example, as long as the gain in fecundity at age 2 is about 1/3 or more times the loss of fecundity at age 3, a trait producing this tradeoff will evolve because it has a higher net fitness than the trait presently in the population.

So, natural selection's answer to when to have children is: as soon as possible; and if you do postpone, make sure you can do a whole lot better later on than now.

5.7.3 Fitting Life History to Environment

Life-history theory offers to explain why animals from different environments have different life histories. Patterns emerge for some groups of animals when closely related species from different habitats are compared. The black-headed gull, *Larus ridibundus*, feeds near shore where it is susceptible to a higher adult mortality (18% annually) than that of the kittiwake, *Rissa tridactlya*, which feeds offshore (12% annual mortality). The set of eggs a bird lays in its nest and that are raised together is called a "clutch," and the "clutch size" is the number of eggs in the clutch. The nearshore black-headed gull lays a clutch of 3 eggs, the eggs incubate in 23 days, the time to fledgling is 30 days, and the age at first breeding is 2 years. The offshore kittiwake lays a clutch of 2 eggs, incubates for 27 days, fledges in 33 days, and first breeds at 3 or 4 years. The basic picture is that birds from high-mortality environments tend to have bigger clutch sizes, faster egg development, faster juvenile-maturation rates, and earlier dates of first reproduction than those of comparable birds from low-mortality environments.

To see how this pattern might be explained, let's bundle up the MATLAB commands we've just used into a function. The function will accept a Leslie matrix as input, and output two vectors. One vector will show the effect on the biggest eigenvalue, R, of a unit change in the log of P_i for each age, i. The other vector will show the effect on R of a unit change in F_i for each i. Here's the function called `life_his`, which as you know, should be typed with a text editor and saved in a file called `life_his.m`.

```
function [d_r_over_d_ln_p, d_r_over_d_f] = life_his(leslie)
```

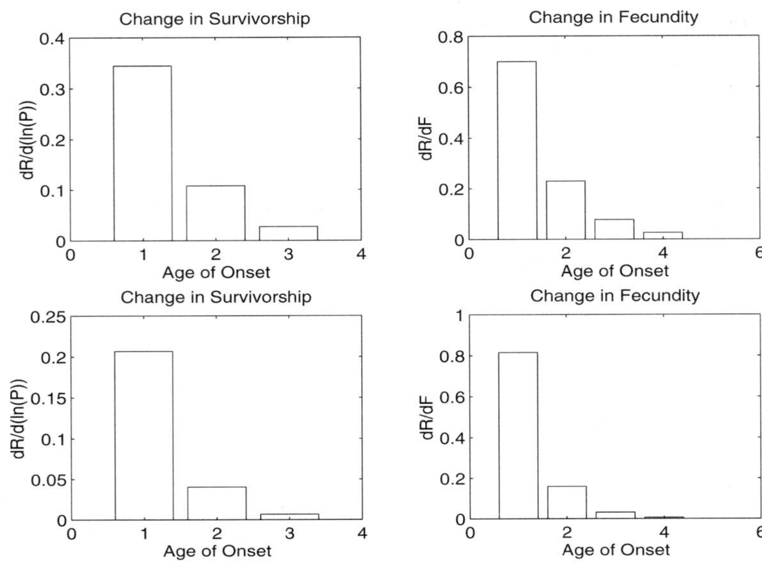

Figure 5.24: The change in the biggest eigenvalue of a Leslie matrix, R, in response to changes in the elements of the Leslie matrix. The top two graphs pertain to a low-risk environment, and the bottom two graphs to a high-risk environment. The graphs on the left describe the value in fitness units of a change to the log of the survivorship as a function of age, and on the right describe the value of changes to the fecundity as a function of age.

```
e_leslie = eig(leslie);
r_leslie = e_leslie(1);
d_ln_p = .1;
d_r_for_d_ln_p = [];
for a = 1:length(leslie)-1
 les = leslie;
 les(a+1,a) = exp(log(les(a+1,a)) + d_ln_p);
 e_les = eig(les);
 r_les = e_les(1);
 d_r_for_d_ln_p = [d_r_for_d_ln_p (r_les - r_leslie)];
end
d_r_over_d_ln_p = d_r_for_d_ln_p/d_ln_p;
d_f = .1;
d_r_for_d_f = [];
for a = 1:length(leslie)
 les = leslie;
 les(1,a) = les(1,a) + d_f;
```

```
 e_les = eig(les);
 r_les = e_les(1);
 d_r_for_d_f = [d_r_for_d_f (r_les - r_leslie)];
end
d_r_over_d_f = d_r_for_d_f/d_f;
```

Once this function is saved in your directory, it's ready to be used.

Let's now find out the fitness values produced when we change all the entries of the Leslie matrix we've been working with so far. Type

```
>> [d_r_over_d_ln_p d_r_over_d_f] = life_his(leslie)
```

and MATLAB replies with

```
d_r_over_d_ln_p = 0.3445     0.1078     0.0273
d_r_over_d_f = 0.7004     0.2290     0.0771     0.0260
```

You'll recognize some of these numbers as those we already determined, during our walk-through of the calculations several paragraphs ago. Let's plot these in green in Figure 5.24. We type

```
>> figure
>> subplot(2,2,1)
>> bar(d_r_over_d_ln_p,'g')
>> subplot(2,2,2)
>> bar(d_r_over_d_f,'g')
```

Next, let's define a new Leslie matrix that represents a more hazardous environment, in which the survival probability drops from 1/2 to 1/4 for each time step:

```
leslie = [1 1 1 1; .25 0 0 0; 0 .25 0 0; 0 0 .25 0];
```

The fitness values produced when we change the entries in this matrix are found by our typing

```
>> [d_r_over_d_ln_p d_r_over_d_f] = life_his(leslie)
```

yielding

```
d_r_over_d_ln_p = 0.2066     0.0404     0.0068
d_r_over_d_f = 0.8156     0.1594     0.0321     0.0064
```

and these are added in red to Figure 5.24 by our typing

```
>> subplot(2,2,3)
>> bar(d_r_over_d_ln_p,'r')
>> subplot(2,2,4)
>> bar(d_r_over_d_f,'r')
```

The big difference between the fitness values from the original low-risk environment and the new high-risk environment reveals the importance of youth relative to old age. In Figure 5.24, notice how the bars for the fitness values of survival and fecundity become smaller as age is increased more quickly in the high-risk environment compared to the low-risk environment. This result means that a high-risk environment has a stronger selection pressure in favor of early reproduction and fast senescence, than does a low-risk environment. Not only that, consider trade-offs between survival and reproduction. How much hazard should an organism expose itself to in order to increase its fecundity by a particular amount? Look at the price in fitness units of survival and fecundity at the same age. Take age 2 for example. In the low-risk environment the value of changing P_2 (in log units) is 0.1078 and of changing F_2 is 0.2290, which is a ratio of about 1 to 2. In the high-risk environment, the value of changing P_2 is 0.0404 and of changing F_2 is 0.1594, which is a ratio of about 1 to 4. Therefore, to bring about 1 unit of fecundity, an organism at age 2 should incur about twice as much hazard in the high-risk environment as in the low-risk environment. All of this adds up to an explanation of why organisms from a high-risk environment should show earlier and greater reproduction, and a shorter lifespan, than comparable organisms from a low-risk environment. Such life-history comparisons are often observed, as with the black-headed gull and kittiwake mentioned earlier.

At this point you've gotten a taste of how evolutionary theory can be applied to explanations of why organisms possess the traits they do in the ecological contexts in which they're found. It would be nice to continue in this vein, but other topics in ecology await us too. But before departing evolutionary ecology for the last time, let me offer some links to other evolutionary subjects that ecologists are working on.

5.8 Hot Links to Other Evolutionary Topics

Interaction is hot. Evolutionary theory started with a focus on individually advantageous and socially innocent traits, traits that improve an individual's survival or reproductive rate regardless of who else is in its population, and regardless of what other species are present as well. Today, the myriad of traits that mediate interactions, both within and between species, have come to prominence in evolutionary ecology. Until relatively recently, there was no common approach to an evolutionary study of social traits because these inherently involve frequency dependence, and the fundamental theorem of natural selection does not apply and therefore cannot provide any unity. In 1982, J. Maynard Smith introduced an approach called "game theory" to evolutionary biology. The idea is that two or more "players" are each involved in maximizing some goal, say fitness, but the tactic each selects depends on the state of the other player as well as its own state. In principle, game theory can be a huge subject, because of the many kinds of games that may take place—there may be two or more players, the players may play only once, or many times, and the rules may be different in different circumstances. From a biological standpoint, special interest is drawn by several specific types of social interactions, such as parental care, mate selection and courtship, territorial defense, and social protection such as warning calls, schooling in fish, and bird flocks. Cooperative parental care, and cooperative foraging also occur. A game can, in principle, capture the spirit of each kind of social interaction, and a hot topic today is seeing if a hypothesized setup for a game can explain the social system that's observed.

Sex is hot. Sex has been hot for a long time. Biologists still can't figure out why we have sex. The answer is not simply to reproduce, because reproduction can take place without sex, as when plants and animals reproduce by budding or by regeneration from fragments. Early on, the idea was that a sexually-reproducing population evolved faster than an asexual one, and that this advantage somehow explained the prevalence of sexual reproduction. But the plot thickened when it was discovered that a disadvantage to sexual reproduction is the cost of maintaining males. It's perhaps of some interest that male biologists were the ones to perceive that their own existence was problematic. The basic problem is that a species of females reproducing asexually can grow twice as fast as a species that is split into two sexes, only one of which does the reproducing. So, male scientists are anxiously trying to find out why most species reproduce sexually, and thereby explain their own existence. Incidentally, biologists define male and femaleness solely by gamete size. Males, by definition, make sperm, which by definition, are smaller than eggs, which by definition, are made by females. All that is essential to maleness and femaleness is gamete size, everything else is constructed either genetically or socially from this starting point. A related issue is the ratio of females to males. The explanation for the approximately 50:50 sex ratio seen in most species is that if there is any one sex in short supply, then members of that sex enjoy more mating success than members of the other sex. This, in turn, makes it advantageous to produce more of the limiting sex. The system then equilibrates at a 50:50 ratio. These are fun questions to think about, and the last word on sex has yet to be uttered.

Metapopulation is hot. The distribution of a population into subpopulations affects its long-term success and chance of extinction, as we saw in Chapter 3. The same is true of a gene within any population that is distributed across subpopulations. The differential net reproductive success of genes from subpopulations is called "group selection." Within a group individual selection for "selfishness" takes place, but the success of the group as a whole depends on "altruists," or team players. If the group fails to produce because of the selfish members in it, then the selfish within it lose out too, so there presumably should emerge some balance between selfish and team-player behavior, that represents a balance between group and individual selection. When biologists think of altruism, they usually think of the social insects (the ants, bees, wasps, and termites), and more recently, mammals such as the naked mole rats, in which the most extreme form of altruism is observed—individuals who forego reproduction themselves to help others reproduce. The organisms who do the reproducing form a "reproductive caste" and the nonreproducing animals are called "workers." W. D. Hamilton showed that special genetics within the ants, bees, and wasps predisposed these groups to evolve altruistic behavior, because the close genetic relationship between a worker and a reproducer basically implied that, by helping a close relative to produce rather than itself reproducing, a worker could produce more copies of its own genes for the next generation. Hamilton's theory also explained further details about social insects, such as why only females occur in the reproductive and worker castes. But termites and mammals have regular diploid genetics, and Hamilton's argument explaining altruism on the basis of kin helping other kin, called "kin selection," does not apply so readily. Instead, group selection arising within the metapopulation structure of these species seems to be a key element.

Complexity is hot. Many species have what is called a "complex life history," which means that the young are produced as larvae that live in one habitat, and then metamorphose into very-different looking adults that live in another habitat and have another life style. The caterpillar that turns into a butterfly, the tiny larvae of lobsters, crabs, and shrimp that metamorphose into the adults that are commercially harvested, the alternation of haploid and diploid generations in algae, mosses and ferns are examples. The life-history theory of density-dependent and age-dependent selection is increasingly being extended to explain the length and size of the stages in complex life histories.

Holism is hot. Populations are not lonely, they live with neighbors. Nonetheless, a population may seem to have the universe to itself either because it interacts very weakly with its neighbors or because its neighbors change very slowly or very quickly relative to itself. The generation-time of an aphid is several weeks and of a tree perhaps 100 years, so evolution in the aphid can thought of, for a while at least, as taking place in a world of constant trees. But what of two aphid species inhabiting the same trees? Surely, evolution in one aphid affects the other, and vice-versa. "Coevolution" is the simultaneous evolution of interacting populations. Coevolution has two components. Evolution within one species affects the abundance of other species. Also, evolution within one species changes the selection pressures within other species. A hummingbird might evolve faster detection of nectar-filled flowers, leading to an increase in the hummingbird population size, which in turn causes a decline in the population size of other species using those flowers. Furthermore, this evolution within one hummingbird may then also cause the other hummingbird species to evolve a preference for a different type of flower, now that its originally preferred flower has been made less available. Similarly, the coevolution of predator and prey has been described as an arms race, in which the prey evolve to run faster and the predators evolve to keep up. Infectious diseases promote the evolution of disease resistance, and conversely, species can evolve to help one another in some circumstances, a phenomenon called "mutualism." What these topics share in common is that evolution within one species is considered in the holistic context of the entire ecological community in which the species is embedded. Also, coevolution (as well as group selection) may serve as a check on run-away frequency-dependent selection within a species. If very expensive traits are needed for mating success within a species—think of huge peacock tails and giant red noses on elephant seals—then the cost is ultimately realized as a lower population size. This lower population size in turn allows an increase in the abundance of competitors or vulnerability to predators, perhaps to the point where the species cannot persist in the community.

The idea of ecological holism and an ecological community brings us to the next chapter. How do populations interact? Stay tuned for the answer.

5.9 Further Readings

Bulmer, M. (1994). *Theoretical Evolutionary Ecology.* Sinauer Associates, Sunderland Massachussets.

Christiansen, F. B. and Feldman, M. W. (1986). *Population Genetics.* Blackwell Scientific Publications, Inc.

Colwell, R. (1981). Group selection is implicated in the evolution of female-biased sex ratios. *Nature*, 290:401–403.

Crow, J. F. and Kimura, M. (1970). *An Introduction to Population Genetics Theory.* Harper & Row.

Eshel, I. (1972). On the neighborhood effect and the evolution of altruistic traits. *Theor. Popul. Biol.*, 3:258–277.

Eshel, I. (1983). Evolutionary and continuous stability. *J. Theor. Biol.*, 103:99–111.

Ewens, W. J. (1969). *Population Genetics.* Methuen & Co. Ltd.

Ewens, W. J. (1979). *Mathematical Population Genetics.* Springer-Verlag.

Fisher, R. A. (1918). The correlation between relatives on the supposition of Mendelian inheritance. *Trans. Roy. Soc. (Edinburgh)*, 52:399–433.

Fisher, R. A. (1958). *The Genetical Theory of Natural Selection, Second Revised Edition.* Dover Publications Inc.

Haldane, J. B. S. (1932). *The Causes of Evolution.* Longmans, Green and Co. (Reprinted 1966, Cornell Paperbacks).

Hamilton, W. D. (1964). The genetical theory of social behavior I., II. *J. Theoret. Biol.*, 7:1–52.

Hamilton, W. D. (1966). The moulding of senescence by natural selection. *J. Theoret. Biol.*, 12:12–45.

Havenhand, J. N. (1995). Evolutionary ecology of larval types. In McEdwards, L., editor, *Ecology Of Marine Invertebrate Larvae*, pages 79–121. CRC Press.

Kingman, J. F. C. (1961). A matrix inequality. *Quart. J. Math.*, 12:78–80.

Lack, D. (1968). *Ecological Adaptations for Breeding in Birds.* Methuen, London.

Maynard Smith, J. (1978). *The Evolution of Sex.* Cambridge University Press.

Maynard Smith, J. (1982). *Evolution and the Theory of Games.* Cambridge University Press.

Mueller, L. D. and Ayala, F. J. (1981). Trade-off between r-selection and K-selection in *Drosophila* populations. *Proc. Nat. Acad. Sci. (USA)*, 78:1303–1305.

Norton, H. T. J. (1928). Natural selection and Mendelian variation. *Proceedings of the London Mathematical Society*, 28:1–45.

Rose, M. J. (1991). *Evolutionary Biology of Aging.* Oxford University Press.

Roughgarden, J. (1971). Density-dependent natural selection. *Ecology*, 52:453–468.

Roughgarden, J. (1977). Coevolution in ecological systems. II. Results from "loop analysis" for purely-density dependent coevolution. In Christiansen, F. and Fenchel, T., editors, *Measuring Selection in Natural Populations*, pages 499–517. Springer-Verlag.

Roughgarden, J. (1979). *Theory of Population Genetics and Evolutionary Ecology: An Introduction.* Macmillan (Reprinted 1996, Prentice Hall).

Roughgarden, J. (1983). The theory of coevolution. In Futuyma, D. J. and Slatkin, M., editors, *Coevolution*, pages 33–64. Sinauer Associates.

Roughgarden, J. (1988). The evolution of marine life cycles. In Feldman, M. W., editor, *Mathematical Evolutionary Theory*, pages 270–300. Princeton University Press.

Roughgarden, J. (1991). The evolution of sex. *American Naturalist*, 138:934–953.

Wade, M. J. (1977). An experimental study of group selection. *Evolution*, 31:134–153.

Williams, G. C. (1975). *Sex and Evolution.* Princeton University Press.

Wilson, D. S. (1980). *The Natural Selection of Populations and Communities.* Benjamin/Cummings.

Wilson, E. O. (1971). *The Insect Societies.* Belknap Press of Harvard University Press.

Wright, S. (1931). Evolution in Mendelian populations. *Genetics*, 16:97–159.

Wright, S. (1969). *Evolution and the Genetics of Populations, Volume 2, The Theory of Gene Frequencies.* The University of Chicago Press.

5.10 Application: Slowing Evolution of Pesticide Resistance

Humans have been waging a war against pest insects for thousands of years, with mixed results[11]. Even today, we spend billions of dollars on chemical insecticides and herbicides, and still lose large portions of our annual harvests to pest insects or to weedy competitors. One of the reasons our modern technologies fail us is that pests have evolved biological responses that confer resistance to our chemical weapons. For instance as of 1990, over 500 species of insects were known to be resistant to at least one chemical insecticide.

The newest technology in our war against pests is genetic engineering—particularly the genetic engineering of plants that produce their own insecticide. Plants with such an ability have now been engineered with a gene from the bacteria *Bacillus thuringiensis* (or just Bt),

[11]This section is contributed by Peter Kareiva.

so that the plants produce a bacterial toxin (called Bt toxin) that is poisonous only to plant-feeding insects. Ecologists are worried that this new technology is being hyped beyond its real promise—with little hope that the plants can provide durable protection because pest insects are likely to evolve the ability to detoxify the Bt toxin.

One possibility for deterring pest evolution is to include in all fields sacrificial plants that do not produce the toxin. The idea is that pest insects will find and destroy these plants, and in the process will face reduced selection for the evolution of resistance. In this chapter, you have already seen general models that show how the rate of evolutionary change is proportional to selection pressures. Here, we modify these models so we can consider the value of mixing susceptible plants with genetically engineered plants, as a strategy for slowing the rate of pest evolution.

We need to create a function for tracking the frequency of the resistance gene in our pest population:

```
function [p,w] = five1(po,h,s,bt,cost,runlen)
```

where `po` is the initial frequency of the susceptible gene, `h` is the degree of dominance of the resistant gene, `s` is the selection intensity (e.g. 0.99 means 99% of homozygous susceptible individuals will die when feeding on Bt plants), `bt` is the fraction of Bt plants in the system, `cost` is the cost of carrying the resistant gene on non Bt plants, and `runlen` is how long the simulation should be run. The notation we use for assigning fitness values, like `wss1`, is that the second two letters refer to genotype, and the number is 1 if on Bt plant, 0 if on a non-Bt plant.

```
p=[po]; w=[];
wrr1 = 1.0; wrs1 = 1-h*s; wss1 = 1-s;
wrr0 = 1-cost; wrs0 = 1-h*cost; wss0 = 1.0;
for t=1:runlen
```

The $\bar{w}$ for insects on, and for insects off, the Bt plants are

```
  wbarbt = p(t)^2*wss1 + 2*p(t)*(1-p(t))*wrs1 + (1-p(t))^2*wrr1;
  wbarnon = p(t)^2*wss0 + 2*p(t)*(1-p(t))*wrs0 + (1-p(t))^2*wrr0;
```

Then the `pprime` on and off the Bt plants are

```
  ppBt = (p(t)*wss1 + (1-p(t))*wrs1)*p(t)/wbarbt;
  ppnon = (p(t)*wss0 + (1-p(t))*wrs0)*p(t)/wbarnon;
```

And the overall `pprime` and `wbar` are

```
  pprime = bt*ppBt + (1-bt)*ppnon;
  wbar =    bt*wbarbt + (1-bt)*wbarnon;
```

The function then concludes with

```
  p = [p pprime];
  w = [w wbar];
end
```

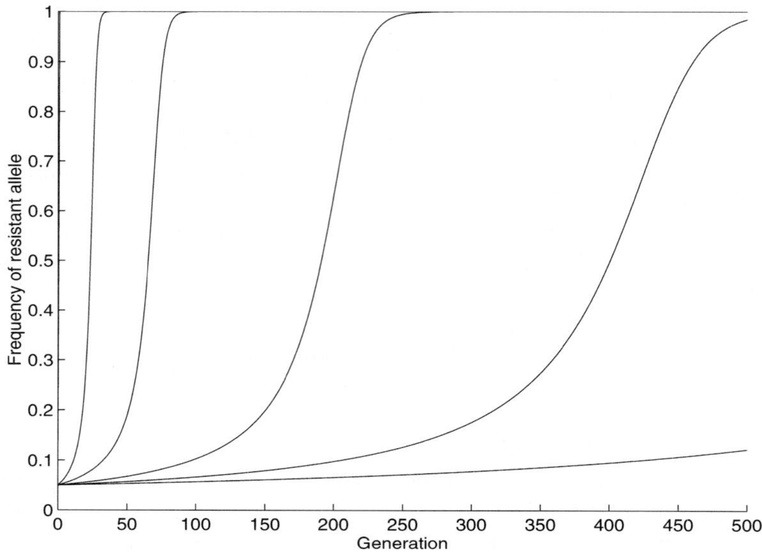

Figure 5.25: The change in the frequency of the resistant allele over time for selection intensities of 1, 0.5, 0.25, 0.1, 0.05, and 0.025 (from the left-most curve near the vertical axis to the right-most curve), with the following parameter values: initial frequency of the susceptible allele is 0.95, dominance is 1, fraction of plants expressing the Bt toxin is 1, and the cost of the resistant allele is 0.

Once this function has been entered with a text editor and saved as `five1.m`, we can use it to investigate the effects of different mixtures of Bt plants and normal susceptible plants, where 100 minus the percentage of Bt plants is the percentage of susceptible plants. At the command line, type

```
>> figure; hold on;
>> for s=[1 .5 .25 .1 .05 .025]
>>    [p w]=five1(.95,1,s,1,0,500);
>>    plot(0:500,(1-p))
>> end;
```

which yields Figure 5.25, showing how a pesticide-resistant allele spreads throughout the pest population through time.

Now we can easily experiment with a different mixture of Bt/non-Bt plants. We can also investigate the consequences of a cost to the pest, of having a resistance gene when the pest is not feeding on plants protected by Bt toxins. The idea is that it may cost an insect something to detoxify the Bt toxins, and this cost makes these resistant insects less fit when they are competing against susceptible genotypes under conditions that do not include any

Bt toxins. Moreover, in making the figure we assumed that the resistant allele is recessive, and we can ask how our results would change if we altered this assumption, and explore cases in which there is no dominance for either allele, or dominance for the resistant allele.

All in all, the interesting lesson from this model is that it is possible to slow pest evolution if fields are planted with a modest proportion of susceptible crops. These are not simply the musings of theoretical population biologists. The first commercial plantings of crops engineered to include Bt toxins have typically involved susceptible seeds lacking any gene producing Bt toxins. Only time will tell whether models such as those explored above are correct in their predictions about the effectiveness of this strategy for slowing the evolution of pesticide resistance.

Chapter 6

Populations at Play

Let's now turn our gaze outwards, away from what happens within a species, and toward interactions among species. Species depend on other species. Animals depend on plants, predators on prey, and parasites upon their hosts. Species also compete with one another. Different bird species eat the same seeds, and different plants species take root in the same ground and grasp for light from the same sun. Species help other species too. Bacteria living in the root nodules of peas fix nitrogen that fertilizes the plants who provide their home. Carrying pollen from flower to flower, butterflies receive meals of sweet nectar in reward. This chapter is about species interactions, how species play around with one another.

Why would you want to know what is in this chapter? Because we, as a species, also depend on other species. We may wish to conserve a valuable species and to eliminate a harmful one. But how? The answer depends on what species do to each other. Conversely, innocent human activity may inadvertently alter the balance between two species, so that one drives the other to extinction. Innocent human activities may provide comfort and sustenance to a disease that may infect us later. But when? Again, the answer depends on how species interact. We'll start with competition between species. There's usually only so much light, water, nutrients, and food to go around, so how can several species competing for limiting resources coexist without one crowding the others out? Then we'll move to various flavors of the predator-prey interaction, wherein one species exploits another. What enables a prey species to support the predators upon it? What enables a host population to support a disease? Will a disease pop up again even if it is eliminated? And how about crop pests? What makes a lady bug or wasp a good biological control for aphids? We'll end with issues such as when a species should contract with another species to carry out a desired function, like removing ectoparasites, or providing shelter. This chapter introduces models for species interactions. But there are many more kinds of species interactions than there are models, and even when a model exists, it often must be greatly tailored to take into account the personalities of the species involved. Still, general themes emerge, so don't go away.

6.1 Two-Species Lotka-Volterra Model of Competition

Take a walk, a walk through a park or woods, or around a college campus. Where I live there are two common species of blue jays. One has a black headcrest and is called a Steller's jay; the other has a smoothly rounded head and is called a scrub jay. Yet only the scrub jay is found at the Stanford campus, and only the Steller's jay is found a few miles to the west, in the hills with coastal redwood trees. Why don't both birds occur together on both the campus and the coastal hills? It's not likely that the climate is sufficiently different between the campus and the nearby hills to make the habitat of either species lethal for the other. Nor is it likely that either species just hates the other's habitat, and simply won't choose to live there under any circumstances. Instead, it's plausible, though not demonstrated, that competition for resources is maintaining the spatial segregation into different habitats. Perhaps the Steller's jay has some advantage at foraging in dark forest, and the scrub jay an advantage at foraging in open grasslands, and these advantages translate into each species crowding out the other in the place where it is best. Competition theory is about this kind of phenomenon. When can two species that compete for resources coexist throughout each other's range, and when must they segregate spatially because they can't coexist together locally?

Competition[1] between species that are very similar to one another, such as two blue jays, can be modeled as a modification of the model for competition within a species. Recall from Chapter 4 that intraspecific competition may be modeled in continuous time with the logistic equation,

$$\frac{dN}{dt} = \frac{rN(K-N)}{K}$$

where N is the population size at time t, r is the intrinsic rate of increase, and K is the carrying capacity. This model may be extended to represent competition between species as well. Let's use a subscript to denote a property of a species, so that N_1 is the population size of species-1, r_1 the intrinsic rate of increase of species-1, K_1 the carrying capacity of the environment for species-1, and similarly for species-2 with N_2, r_2, and K_2. The interaction between these species is described with two new parameters, called competition coefficients. The effect of a member of species-2 upon a member of species-1 *relative* to the effect of another member of species-1 upon that member of species-1 is denoted as α_{12}. The reciprocal competition coefficient, α_{21}, measures the effect of a member of species-1 upon a member of species-2 relative to another member of species-2 upon that member of species-2. Using these α's, a model of competition may be proposed as

$$\begin{aligned}\frac{dN_1}{dt} &= \frac{r_1 N_1(K_1 - N_1 - \alpha_{12}N_2)}{K_1}\\ \frac{dN_2}{dt} &= \frac{r_2 N_2(K_2 - N_2 - \alpha_{21}N_1)}{K_2}\end{aligned}$$

These equations are called the Lotka-Volterra competition equations. If both α's are 0,

[1] Competition between species is called "interspecific competition," and competition within a species is called "intraspecific competition."

there is no interspecific competition, and the two species simply grow according to separate logistic equations, independently of each other.

It's one thing to make up models out of thin air, and another to know that the models can describe reality, at least sometimes. So, let's return to the *Paramecium* populations that were used earlier to illustrate the logistic model, and see if they can illustrate the Lotka-Volterra model too.

6.1.1 Using the Lotka-Volterra Model with Data

G. F. Gause in 1934 provided the data on the growth of *Paramecium* populations growing in the laboratory. The experiments in which *P. aurelia* and *P. caudatum* were grown by themselves have been analyzed in Chapter 4. Now we add data from three replicate experiments in which these two species were grown together so that they competed for the same food supply. The data are entered into MATLAB by typing

```
>> pac1 = [2 2; 4 4; 7 7; 25 14; 43 26; 81 65; 140 72; ...
>>          180 126; 224 136; 240 120; 204 105; 375 65; 370 115; ...
>>          285 60; 400 65; 325 50; 370 60; 320 40; 300 45; 325 60];
>> pac2 = [2 2; 4 5; 6 9; 13 5; 41 41; 74 60; 195 135; ...
>>          164 92; 160 148; 240 90; 230 90; 215 65; 230 50; ...
>>          285 60; 360 65; 275 40; 350 40; 350 60; 405 65; 330 40];
>> pac3 = [2 2; 9 8; 16 13; 25 15; 89 19; 120 26; 270 57; ...
>>          144 88; 280 88; 400 70; 275 45; 320 67; 305 85; 450 45; ...
>>          402 72; 405 65; 370 65; 300 20; 370 35; 280 40];
```

The data for a replicate, say `pac1`, is stored as a matrix whose first column is the abundance of *P. aurelia* and the second column is *P. caudatum*. The abundances of these populations through time are graphed by our typing

```
>> figure
>> hold on
>> plot(0:length(pac1)-1,pac1)
>> plot(0:length(pac2)-1,pac2)
>> plot(0:length(pac3)-1,pac3)
```

This results in Figure 6.1.

In principle, one could estimate all six parameters of the Lotka-Volterra equations from this data set, but doing so invites statistical error in the estimates because the data don't span all combinations of N_1 and N_2 very well. Instead, it makes better use of Gause's overall experimental design to estimate the r's and K's from the experiments in which each species was grown in isolation. We'll hypothesize therefore, that the r's and K's don't change when the species are put together, and that any difference between the population growth occurring when they are in isolation and when they are together is caused by competition, as represented by the magnitude of the competition coefficients. So, what we'll do is to estimate the α's from the data above, and we'll assume that the r's and K's are the same

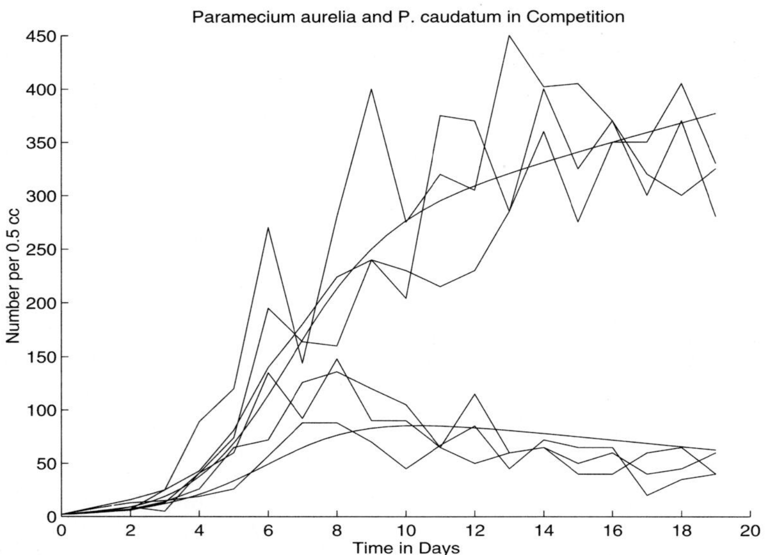

Figure 6.1: The abundance of three replicates of *Paramecium aurelia* at top, and *P. caudatum*, at bottom, grown together in competition. The smooth curves are the predictions of the Lotka-Volterra competition equations.

as we obtained in Chapter 4. If this approach generates α's that result in a Lotka-Volterra model that poorly fits the data, then the model is rejected. Alternatively, the Lotka-Volterra model may turn out to fit very well after all, and we'll be happy with the outcome.

As we did in Chapter 4, we'll use the `fmins` command to find the values of α that minimize the sum of squared deviations between the data and model's predictions—that is, we'll find the best fit of the Lotka-Volterra model to the data, and then judge if that fit is good enough to be acceptable. To find out what the Lotka-Volterra model predicts for any set of r's, K's and α's, we need to integrate the differential equations of the model from an initial condition. This is done in MATLAB with the `ode23` function that we've used before in Chapter 1. `ode23` requires that we define the differential equation to be solved in a separate file. Therefore, with a text editor, create the following file, which is only four lines long, and save it as `lvcomp2.m`.

```
function ndot = lvcomp2(t,n)
 global r k alpha;
 ndot(1,1) = r(1)*n(1)*(k(1)-n(1)-alpha(1,2)*n(2))/k(1);
 ndot(2,1) = r(2)*n(2)*(k(2)-n(2)-alpha(2,1)*n(1))/k(2);
```

This function accepts as input time, `t`, which is not in fact used, and the column vector, `n`, whose first element is N_1 and second element is N_2. The parameters in the model, the

r's, K's and α's, are defined as global. Specifically, `r` and `k` are column vectors whose first elements are r_1 and K_1, and whose second elements are r_2 and K_2, respectively. The competition coefficients are stored in the matrix `alpha`, where `alpha(1,2)` is α_{12} and `alpha(2,1)` is α_{21}. The elements on the main diagonal of `alpha` equal 1 by convention, even though they're not actually used here. Finally, the function returns a column vector, `ndot`, whose first element is dN_1/dt and second element is dN_2/dt.

We're half through. Next, we need to calculate the sum of the squared deviations between what is predicted by the Lotka-Volterra model, and the data. So, create the following file and save it as `fitalpha.m`:

```
function sumdevsq = fitalpha(param,n1,n2)
 global r k alpha
 alpha = [1 param(1); param(2) 1];
 [t_ode n_ode] = ode23('lvcomp2',0,length(n1)-1,[n1(1); n2(1)]);
 n1predict = interp1(t_ode,n_ode(:,1),0:(length(n1)-1));
 n2predict = interp1(t_ode,n_ode(:,2),0:(length(n2)-1));
 sumdevsq = sum((n1predict-n1).^2)+sum((n2predict-n2).^2);
```

Although this function is short too, it could have been even shorter if MATLAB weren't so fussy. `fitalpha` takes as input `param`, which is a vector of the competition coefficients that must be fitted to the data. The first element of `param` is α_{12} and the second element is α_{21}. The next two items of input are column vectors with the abundances of N_1 and N_2 through time, respectively. Now on to the body of the function. First, the global parameters are defined. Next the present estimates of the competition coefficients are copied from `param` to the `alpha` matrix. Then `ode23` is called with the name of the model to be solved, the start time, the stop time, and a column vector of the initial conditions. `ode23` returns a pair of vectors which are a set of times and values of N_1 and N_2 at those times. Unfortunately, the times returned by `ode23` aren't spaced one day apart, so the next two lines in the function interpolate N_1 and N_2 with a one-day spacing based on what `ode23` provided us with; it uses the `interp1` command. Finally, the sum of the squared deviations for both populations is computed in the last line of the function.

Once both `lvcomp2.m` and `fitalpha.m` have been saved on the disk, we can proceed to estimate the competition coefficients. The r's and K's for the *P. aurelia* and *P. caudatum* were found in Chapter 4, and now is a good time to remind MATLAB of this, by our typing

```
>> ra = .7816;     rc = .6283;
>> ka = 559.6860; kc = 202.4931;
```

Continuing the preparation, let's assign the parameters by typing

```
>> global r k alpha
>> r = [ra;rc];
>> k = [ka;kc];
>> alpha = [1 0;0 1];
```

This assignment implies that species-1 is *P. aurelia* and species-2 is *P. caudatum*; i.e., the species are numbered alphabetically. The values assigned to `r` and `k` don't change, but the `alpha` matrix does change during the fitting process. At last then, we get to do it. Upon typing

```
>> best_pac1 = fmins('fitalpha',[0 0],[0 0.01 0.01],[],pac1(:,1),pac1(:,2))
```

MATLAB replies (after a little thinking time) with its best estimate of the competition coefficients from the first replicate:

```
best_pac1 = 2.5033    0.3799
```

This indicates that α_{12} is estimated as 2.5033 and α_{21} as 0.3799. `fmins` is called with this information: the name of the function to be minimized (`'fitalpha'`); the initial guess for the competition coefficients (`[0 0]`); the options, which are not to print out intermediate results, to use 0.01 as the accuracy when the competition coefficients are determined, and 0.01 as the accuracy when the sum of squares (`[0 0.01 0.01]`) are minimized; and the column vectors of data for N_1 (`pac1(:,1)`) and for N_2 (`pac1(:,2)`), respectively. The estimates for the other replicates are generated by our typing

```
>> best_pac2 = fmins('fitalpha',[0 0],[0 0.01 0.01],[],pac2(:,1),pac2(:,2))
>> best_pac3 = fmins('fitalpha',[0 0],[0 0.01 0.01],[],pac3(:,1),pac3(:,2))
```

To summarize the output from the three replicates, and take the average and standard error, we type

```
>> a12 = [best_pac1(1) best_pac2(1) best_pac3(1)]
>> a12_ave = mean(a12)
>> a12_se = std(a12)/sqrt(3)
>> a21 = [best_pac1(2) best_pac2(2) best_pac3(2)]
>> a21_ave = mean(a21)
>> a21_se = std(a21)/sqrt(3)
```

which collectively yields

```
a12 = 2.5033    3.3826    2.0877
a12_ave = 2.6579
a12_se = 0.3817
a21 = 0.3799    0.5190    0.3404
a21_ave = 0.4131
a21_se = 0.0542
```

Well, now that we have the competition coefficients, let's see how well they work. We can use `ode23` together with `lvcomp2` to generate the N_1 and N_2 through time predicted by the Lotka-Volterra model. Typing

```
>> alpha = [1 a12_ave; a21_ave 1]
```

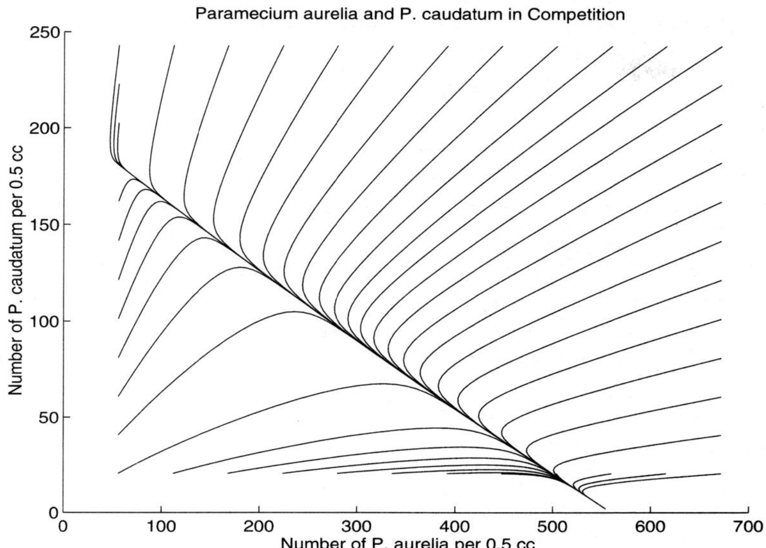

Figure 6.2: A flow of trajectories from different initial conditions for *Paramecium aurelia* and *P. caudatum*.

assigns the `alpha` matrix as

```
alpha = 1.0000    2.6579
        0.4131    1.0000
```

Then, to call `ode23`, we type

```
>> [time n] = ode23('lvcomp2',0,length(pac1)-1,[pac1(1,1);pac1(1,2)]);
```

This yields a set of time points and values of N_1 and N_2 at those times. These are then added to Figure 6.1 by our typing

```
>> plot(time,n,'c')
```

The curve for N_1 and N_2 predicted by the Lotka-Volterra model with these competition coefficients passes nicely through the data points, although the fit for *P. aurelia* is somewhat better than than for *P. caudatum*. Thus, interspecific competition between *P. aurelia* and *P. caudatum* in Gause's laboratory culture appears to be satisfactorily described by the Lotka-Volterra model.

Is the Lotka-Volterra model useful only for reproducing the data used to estimate its coefficients? No. The Lotka-Volterra competition equations also predict what would happen in experiments started with different initial conditions. Gause started his three replicates all with two individuals from both species. Suppose instead that a whole spectrum of initial conditions is used. How could we graph the results of many such experiments? An attractive presentation is to put N_1 on the horizontal axis and N_2 on the vertical axis, and to plot consecutive points of (N_1, N_2) from a given initial condition. The resulting curve, called a trajectory, expresses the history of population sizes through time from that initial condition. The set of all the trajectories from various initial conditions yields an overall pattern called a "flow."

Here's a little program that produces a picture of the flow of the Lotka-Volterra equations. It's called `lvflow2.m`, and should be entered with a text editor and saved in your directory or folder. It's really simple, in principle. All it does is generate initial conditions around a box that runs from $0.1K_1$ to $1.2K_1$ on the horizontal and from $0.1K_2$ to $1.2K_2$ on the vertical. For each initial condition around this box, it calls `ode23` to generate a solution of the Lotka-Volterra model, and then plots the solution. The final picture reveals the flow.

```
function lvflow2(t_stop)
 global r k alpha
 figure
 hold on
 for n1=[0.1*k(1) 1.2*k(1)]
   for n2=0.1*k(2):0.1*k(2):1.2*k(2)
     [time n] = ode23('lvcomp2',0,t_stop,[n1;n2]);
     plot(n(:,1),n(:,2))
     end
   end
 for n2=[0.1*k(2) 1.2*k(2)]
   for n1=0.2*k(1):0.1*k(1):1.1*k(1)
     [time n] = ode23('lvcomp2',0,t_stop,[n1;n2]);
     plot(n(:,1),n(:,2))
     end
   end
```

To use this program, with `r`, `k`, and `alpha` already defined, just type

```
>> lvflow2(25)
```

and Figure 6.2 appears. What this picture is showing is that all the trajectories approach a more-or-less straight line that runs diagonally from K_2 on the vertical axis down to K_1 on the horizontal axis. Once the trajectories get near this diagonal line they follow it down and to the right and wind up at K_1. In the long run, the Lotka-Volterra model predicts that *P. aurelia* crowds out *P. caudatum*, and that these two species will not coexist under the conditions of the experiment.

Is the exclusion of one species by another the inevitable result of competition? To answer this question, we need to discover the range of possible outcomes of the Lotka-Volterra model.

6.1.2 Outcome of Two-Species Competition

The basic approach to determining the outcomes possible with a multispecies model begins the same way as our approach to any other dynamic model we've looked at before—we identify all the equilibria and then check to see if they are stable. Here we can identify four equilibria right off. The equilibria are values of N_1 and N_2 which make both dN_1/dt and dN_2/dt equal to zero and therefore are found from

$$\begin{aligned} 0 &= \frac{r_1 N_1 (K_1 - N_1 - \alpha_{12} N_2)}{K_1} \\ 0 &= \frac{r_2 N_2 (K_2 - N_2 - \alpha_{21} N_1)}{K_2} \end{aligned}$$

Well, either N_1 or N_2 may be zero, or the expression in parentheses may be zero. Putting these possibilities together in all possible combinations yields four pairs of N_1 and N_2 that make both equations equal zero simultaneously. They are,

1. $N_1 = 0$ and $N_2 = 0$.
2. $N_1 = K_1$ and $N_2 = 0$.
3. $N_1 = 0$ and $N_2 = K_2$.
4. N_1 and N_2 satisfy

$$\begin{aligned} 0 &= K_1 - N_1 - \alpha_{12} N_2 \\ 0 &= K_2 - N_2 - \alpha_{21} N_1 \end{aligned}$$

If this equilibrium is positive and stable, it represents species-1 and species-2 coexisting.

Now let's look at the first equilibrium. The total-extinction equilibrium is stable if both r's are negative. Otherwise, at least one of the species can grow in the habitat. It's easy to dismiss this equilibrium as uninteresting, but it's very important in indicating whether the habitat is occupyable at all. If this equilibrium is stable, then either the habitat is lousy to begin with, or it has been destroyed.

The plot thickens when we examine the second and third equilibria. You may suspect that to examine the stability of these, we'll have to calculate something like the λ used in the last two chapters. You're right, but let's postpone this a little while. For the second and third equilibria we can look at an intuitive condition that tells us about stability. Then we'll develop a λ-like criterion for the fourth equilibrium later.

The intuitive criterion for stability at the second and third equilibria is called the "invasion-when-rare" condition. If species-1 is alone in the habitat, and a little bit of species-2 is introduced, we ask: Can species-2 increase? If it can, then the equilibrium with species-1 alone is unstable; specifically, it is unstable to a perturbation in the direction of N_2. If species-2 can't invade when it is rare, then the species-1 alone equilibrium may be stable—you'll also need to check if it is stable with respect to perturbations in N_1 itself (which it is if r_1 is positive). In summary: *The species-1-alone equilibrium is unstable if species-2 can increase when it is rare. The species-1-alone equilibrium is stable if species-2 cannot increase when it is rare, provided species-1 is stable by itself.* And similarly for the species-2-alone equilibrium.

When we are investigating two species, it turns out that we can classify all the possible outcomes of competition according the Lotka-Volterra competition model just by listing all the possible states of the increase-when-rare condition.

The Increase-When-Rare Criterion

The condition for increase when rare can almost be read off directly from the Lotka-Volterra equations. Consider species-1. The dynamics of species-1 are

$$\frac{dN_1}{dt} = \frac{r_1 N_1 (K_1 - N_1 - \alpha_{12} N_2)}{K_1}$$

When N_1 is positive, dN_1/dt is positive if the expression in parentheses is positive. If $N_2 = K_2$ and N_1 are nearly zero, the expression in parentheses is approximately

$$K_1 - \alpha_{12} K_2$$

For species-1 to increase when rare, we want this to be positive. So, rearranging to express a comparison between the α and the K's, the criterion for species-1 to increase when rare is

$$\alpha_{12} < \frac{K_1}{K_2}$$

The invasion-when-rare condition means that if the competition received by an individual of species-1 from an individual of species-2 is low enough and/or if species-1 has a high enough carrying capacity relative to species-2, it can invade. In general, a species' invasibility is promoted by its being little impacted from competition with other species and by its having a relatively high carrying capacity.

The condition for species-2 to invade when rare is derived in a similar way, and works out to be

$$\alpha_{21} < \frac{K_2}{K_1}$$

Now let's use these conditions to classify the possible outcomes of two species in competition according to the Lotka-Volterra model.

Species-1 Can Invade When Rare and Species-2 Can't

Here,

$$\begin{aligned} \alpha_{12} &< \frac{K_1}{K_2} \\ \alpha_{21} &> \frac{K_2}{K_1} \end{aligned}$$

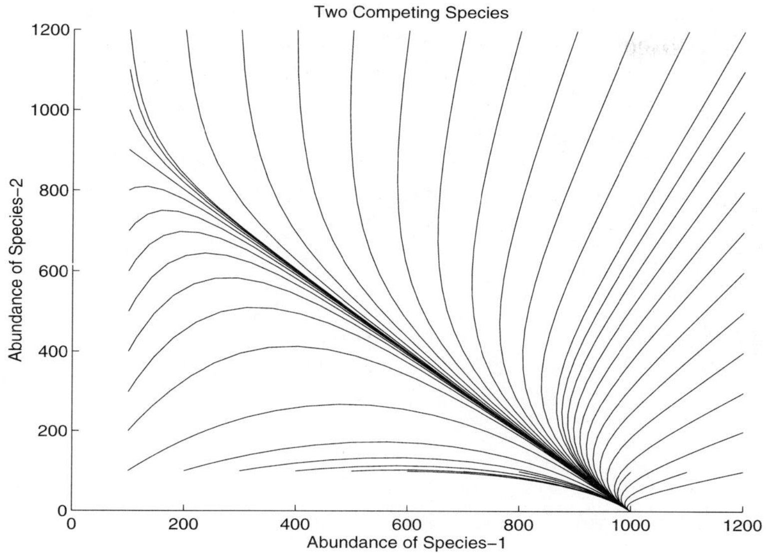

Figure 6.3: The flow of trajectories if species-1 can increase when rare and species-2 can't.

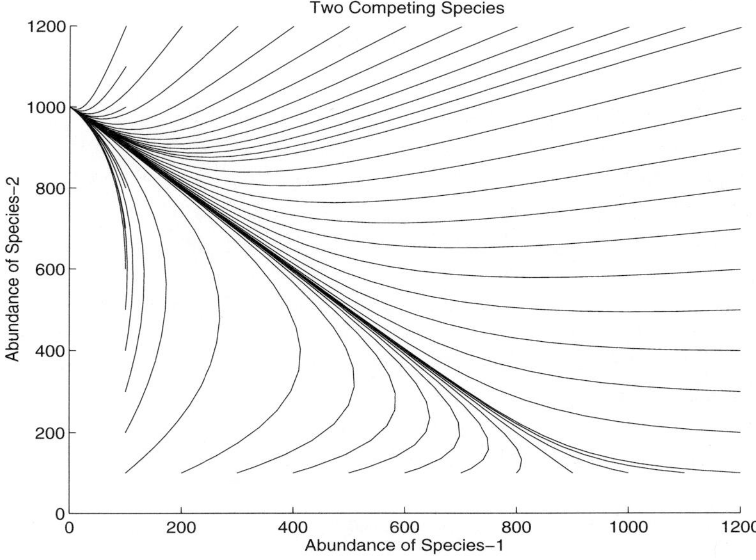

Figure 6.4: The flow of trajectories if species-2 can increase when rare and species-1 can't.

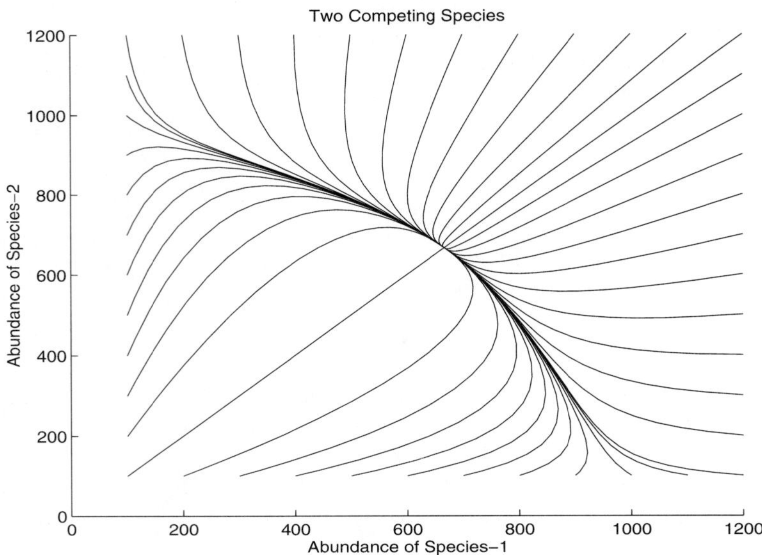

Figure 6.5: The flow of trajectories if both species can increase when rare.

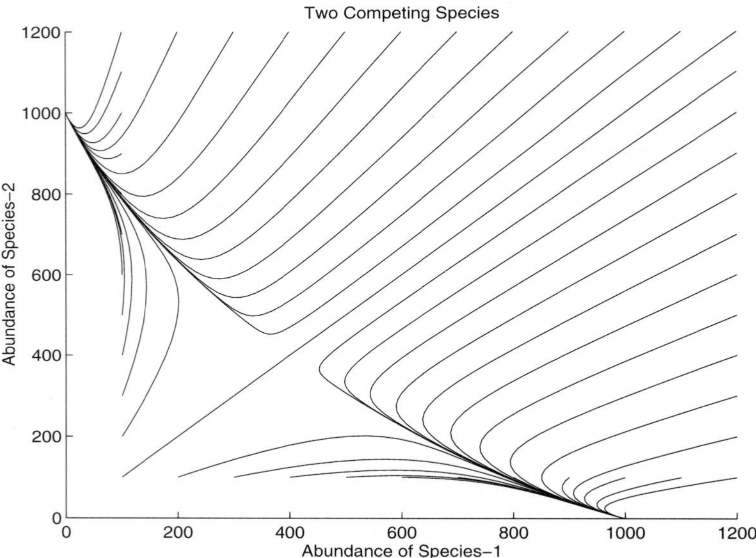

Figure 6.6: The flow of trajectories if neither species can increase when rare.

When this is true, species-1 will inevitably crowd out species-2 from the habitat.

It turns out that the competition experiments between *Paramecium* are an instance of this case. With species-1 being *P. aurelia* and species-2 being *P. caudatum*, the first inequality works out to be a comparison between

```
alpha(1,2) = 2.6579
k(1)/k(2) = 2.7640
```

Clearly, α_{12} is less than K_1/K_2. Therefore, species-1 can invade a habitat containing species-2 at equilibrium. The second equality involves a comparison between

```
alpha(2,1) = 0.4131
k(2)/k(1) = 0.3618
```

In this comparison, α_{21} is greater than K_2/K_1. Therefore, species-2 cannot invade a habitat containing species-1 at equilibrium. The net result is that species-1 (*P. aurelia*) inevitably crowds out species-2 (*P. caudatum*) from the habitat.

Another illustration can be generated by use of the `lvflow2` function we developed earlier, together with some hypothetical values for `r`, `k` and `alpha`. Specifically, type

```
>> global r k alpha
>> r = [1; 1]
>> k = [1000; 1000]
>> alpha = [1 0.5; 1.5 1]
>> lvflow2(25)
```

and Figure 6.3 emerges. All the trajectories end up approaching the horizontal axis at K_1. We'll compare this flow with that in the other possible cases.

Species-1 Can't Invade When Rare and Species-2 Can

This case is the mirror image of the previous. The conditions for this are

$$\begin{aligned} \alpha_{12} &> \frac{K_1}{K_2} \\ \alpha_{21} &< \frac{K_2}{K_1} \end{aligned}$$

Now species-2 inevitably crowds out species-1 from the habitat.

An illustration of the flow of trajectories in this case is generated by our typing

```
>> alpha = [1 1.5; 0.5 1]
>> lvflow2(25)
```

and Figure 6.4 emerges. All the trajectories end up approaching the vertical axis at K_2.

Both Species Can Invade When Rare

If both species can increase when rare, then they can coexist in the long term because neither can crowd out the other. This case arises when

$$\alpha_{12} < \frac{K_1}{K_2}$$
$$\alpha_{21} < \frac{K_2}{K_1}$$

If the interspecific competition in both directions is weak enough, meaning that the α's are low enough, then these conditions are met and coexistence is possible. Thus, for two species, high interspecific competition is generally a barrier to coexistence and to the maintenance of biodiversity.

This case has led to the idea that, in order to coexist, species must occupy different "niches." In ecology, a niche is the name for a species' occupation, and a "habitat" is the name for a species' address. A bird that looks for insects under leaves has the niche of a "foliage gleaner," and its habitat might be the northeastern deciduous forest of North America. This case of two-species competition, wherein both α's are low enough, is often interpreted to mean that coexisting species must occupy different niches in the sense that the competition coefficients between a foliage-gleaning bird species and a seed-eating bird species are low because each is using different resources from the other. While it's clearly true that being in different niches allows coexistence simply because there's then no competition to worry about, it's not necessary for species to be in different niches to coexist, provided more than two species are involved. As we'll see later, with three or more species, special schemes of competition allow coexistence even with large α's.

An illustration of the flow of trajectories in the case with two species is generated by our typing

```
>> alpha = [1 0.5; 0.5 1]
>> lvflow2(25)
```

and Figure 6.5 emerges. The illustration shows that the coexisting species approach an equilibrium point. The pattern of trajectories in this case is called a "stable node." As we'll see later, when there are three species, coexistence does not necessarily imply an equilibrium. Three competing species may coexist in a never-ending oscillation of abundance. But when there are two species, trajectories of coexisting competitors wind up at an equilibrium point.

Neither Species Can Invade When Rare

If neither species can invade when rare, then the first species into the habitat gets it. This is called a "priority effect." It means that there are two stable equilibria, and the system winds up at one or the other, depending on which was closer at the start. This case arises when

$$\alpha_{12} > \frac{K_1}{K_2}$$

$$\alpha_{21} > \frac{K_2}{K_1}$$

To generate an illustration, type

```
>> alpha = [1 1.5; 1.5 1]
>> lvflow2(25)
```

and Figure 6.6 appears. The trajectories end up either on the horizontal axis at K_1 or on the vertical axis at K_2 depending on where they started out. The pattern of trajectories here is called a "saddle."

6.1.3 Ecological Stability With Multiple Species

Although all possible outcomes of two-species competition can be classified by considering all possible pairs of the invasion-when-rare criterion, we're just lucky. Figure 6.5 does show trajectories that approach a stable equilibrium. But if all species can invade when rare, in principle there still may not be a stable equilibrium point. In fact, if all species can invade when rare, and there isn't any stable equilibrium point, then something weird is going on, like an oscillation of some sort—this situation is actually illustrated later with three species. So, in general, we can't avoid learning how to check the stability of a multi-species equilibrium, because we can't rely on the invasion-when-rare criterion alone. In MATLAB it turns out to be quite easy to investigate the stability of a multi-species equilibrium, at least numerically.

Recall from Chapter 4, that an equilibrium is considered to be stable if the system returns to equilibrium when it is pushed slightly away from the equilibrium. With one species, in continuous time, we considered

$$\frac{dN}{dt} = F(N)$$

and calculated

$$\lambda = F'(\hat{N})$$

If λ, which is the derivative of $F(N)$ evaluated at the value of N being checked, is negative, then the equilibrium is stable. When we are considering two or more species, the idea is the same, only more derivatives are involved. With two species we start with two equations,

$$\begin{aligned}\frac{dN_1}{dt} &= F_1(N_1, N_2)\\ \frac{dN_2}{dt} &= F_2(N_1, N_2)\end{aligned}$$

Now we make a matrix of all the possible derivatives. The top row is the derivative of F_1 with respect to N_1 and then of F_1 with respect to N_2. The bottom row is the derivative of F_2 with respect to N_1 and then of F_2 with respect to N_2. As you know, in MATLAB, a derivative of an expression is taken with the `diff` command, so we have to make a matrix that looks like this:

```
diff(F1,'n1')  diff(F1,'n2')
diff(F2,'n1')  diff(F2,'n2')
```

Next, instead of one λ, we have two of them. The two λ's are eigenvalues of this matrix. For stability, we require *both* of the eigenvalues to be negative, or if they are complex numbers, to have negative real parts[2]. Incidentally, the matrix whose eigenvalues determine whether an equilibrium is stable is called the "Jacobian" of the system, evaluated at equilibrium.

Enough talk, let's do it. For the two-species competition model, let's look at two cases that involve an equilibrium where both N_1 and N_2 are positive, and check out the stability. First, to tell MATLAB about the Lotka-Volterra model, we type

```
>> dn1 = 'r1*n1*(k1-n1-a12*n2)/k1'
>> dn2 = 'r2*n2*(k2-n2-a21*n1)/k2'
```

We can obtain a list of all the equilibria. Using the `solve` command, we type

```
>> [n1hat,n2hat] = solve(dn1,dn2,'n1,n2')
```

MATLAB replies with two column vectors, one for $\hat{N}_1$ and the other for $\hat{N}_2$.

```
n1hat = [                         0]
        [                         0]
        [                        k1]
        [-(k1-a12*k2)/(-1+a21*a12)]

n2hat = [                         0]
        [                        k2]
        [                         0]
        [(-k2+a21*k1)/(-1+a21*a12)]
```

The first element from each vector is the first equilibrium, the second element the second equilibrium, and so forth. The fourth equilibrium is the one of interest here—it's the one in which both species coexist. This is the equilibrium obtained when both expressions in parentheses of the Lotka-Volterra model are zero; i.e., we solve for N_1 and N_2 in

$$\begin{aligned} 0 &= K_1 - N_1 - \alpha_{12} N_2 \\ 0 &= K_2 - N_2 - \alpha_{21} N_1 \end{aligned}$$

This equilibrium is called the "interior" equilibrium, as distinguished from the "boundary" equilibria, in which one or more of the species is absent. The interior equilibrium in a Lotka-Volterra model can also be found by our solving a system of linear equations. Rearrange

[2]This requirement for stability is derived for two species just as it was for one species, in Chapter 4. Specifically, we consider small deviations from the equilibrium, and expand both $F_1(N_1, N_2)$ and $F_2(N_1, N_2)$ as Taylor series around the equilibrium point $\hat{N}_1$ and $\hat{N}_2$. Then, keeping only first-order terms in the Taylor series, we obtain a pair of linear differential equations with constant coefficients for the deviations from equilibrium. By solving this system of differential equations, we see that the deviations from equilibrium go to zero as time tends to infinity, when the eigenvalues of the matrix of coefficients are negative or have negative real parts.

the equations above to look as follows,

$$\begin{aligned} N_1 + \alpha_{12} N_2 &= K_1 \\ \alpha_{21} N_1 + N_2 &= K_2 \end{aligned}$$

In matrix notation, this pair of equations then becomes

$$\begin{pmatrix} 1 & \alpha_{12} \\ \alpha_{21} & 1 \end{pmatrix} \begin{pmatrix} N_1 \\ N_2 \end{pmatrix} = \begin{pmatrix} K_1 \\ K_2 \end{pmatrix}$$

MATLAB will obtain formulas to solve a system of linear equations with the `linsolve` command. To use this command, define the matrices

```
>> alpha = sym('[1 a12; a21 1]')
>> k = sym('[k1; k2]')
```

Then typing

```
>> nhat4 = linsolve(alpha,k)
```

yields

```
nhat4 = [-(k1-a12*k2)/(-1+a21*a12)]
        [(-k2+a21*k1)/(-1+a21*a12)]
```

which is the same as the fourth of the equilibria obtained previously with `solve`. Thus, if we know we're specifically interested in the interior equilibrium of a Lotka-Volterra model, we can use `linsolve` to zoom in directly on this one, and not have to bother with all the boundary equilibria that `solve` also finds.

Now that we have the equilibria, the next step is to calculate the matrix of derivatives, the Jacobian. We type

```
>> j11 = simplify(diff(dn1,'n1'))
>> j12 = simplify(diff(dn1,'n2'))
>> j21 = simplify(diff(dn2,'n1'))
>> j22 = simplify(diff(dn2,'n2'))
```

which collectively yield

```
j11 = -r1*(-k1+2*n1+a12*n2)/k1
j12 = -r1*n1*a12/k1
j21 = -r2*n2*a21/k2
j22 = -r2*(-k2+2*n2+a21*n1)/k2
```

Now we have the formulas we need[3], so let's plug in some numbers and see how this all works out.

Figure 6.5, in which the trajectories approached a stable equilibrium point, was made with the following parameter values:

[3] The Jacobian can be calculated in a single step with the `jacobian` command in MATLAB, but here, to illustrate how the matrix is assembled from scratch, I've computed each element explicitly. Later in the chapter we'll use the `jacobian` command.

```
>> r1 = 1; r2 = 1;
>> k1 = 1000; k2 = 1000;
>> a12 = .5; a21 = .5;
```

With these values, the interior equilibrium is found by our typing

```
>> n1 = eval(sym(n1hat,4,1))
>> n2 = eval(sym(n2hat,4,1))
```

yielding

```
n1 = 666.6667
n2 = 666.6667
```

To evaluate the Jacobian, type

```
>> jac = [eval(j11) eval(j12); eval(j21) eval(j22)]
```

yielding

```
jac = -0.6667   -0.3333
      -0.3333   -0.6667
```

The eigenvalues of this matrix are found with the now-familiar `eig` command:

```
>> lambda = eig(jac)
```

yielding

```
lambda = -0.3333
         -1.0000
```

Both these eigenvalues are negative, so the equilibrium is stable. To make a plot of these eigenvalues, we can use the `compass` command. Type

```
>> figure
>> subplot(2,1,1)
>> compass(lambda)
```

and the top panel of Figure 6.7 appears. The pattern of trajectories in this case, as previously illustrated, is a stable node. An equilibrium is a node when all the eigenvalues are real and of the same sign. The node is stable if they are all negative, and is unstable if they are all positive.

Figure 6.6 illustrated the case wherein either species excluded the other, the exclusion depending on who had the initial numerical advantage, leading to a priority effect. This case used the same r's and K's as before, but different α's:

```
>> a12 = 1.5; a21 = 1.5;
```

There is a positive interior equilibrium, exhibited by our typing

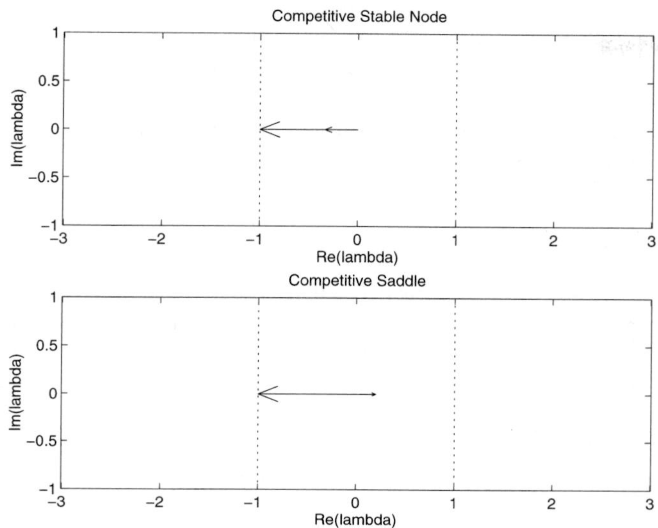

Figure 6.7: Eigenvalues of the Jacobian matrix for the two-species Lotka-Volterra competition model evaluated at the interior equilibrium. Top: The stable equilibrium from Figure 6.5. Bottom: The unstable equilibrium from Figure 6.6.

```
>> n1 = eval(sym(n1hat,4,1))
>> n2 = eval(sym(n2hat,4,1))
```

yielding

```
n1 = 400
n2 = 400
```

To find the Jacobian and its eigenvalues, type

```
>> jac = [eval(j11) eval(j12); eval(j21) eval(j22)]
>> lambda = eig(jac)
```

yielding

```
jac = -0.4000   -0.6000
      -0.6000   -0.4000


lambda = 0.2000
        -1.0000
```

One of the eigenvalues is positive, so the equilibrium is unstable. But the other is negative. This means that trajectories approach the equilibrium from one direction and leave it from another direction. Notice therefore in Figure 6.6 that in approaching the vicinity of the equilibrium, the trajectories move to the northeast if they started from below and to the southwest if they started from above; they then end up at the closest equilibrium. A graph of these eigenvalues is added as the bottom panel of Figure 6.7 by our typing

```
>> subplot(2,1,2)
>> compass(lambda)
```

The pattern of trajectories in this case, as previously illustrated, is a saddle. An equilibrium is a saddle if all but one of the eigenvalues are real and negative, and one of the eigenvalues is real and positive. The one positive eigenvalue makes the equilibrium unstable, but the remaining negative eigenvalues imply that the trajectories close in on the equilibrium point from many directions, and then depart along a special direction.

Now that you know how to analyze a multispecies population-dynamic model, let's poke into some biological scenarios generated by competition between species.

6.1.4 Altitudinal Zonation and Species Borders

Remember the scrub jay and the Steller's jay? The Steller's jay is a high-elevation blue jay, and the scrub jay a low-elevation blue jay. How will these species zone altitudinally? If we start at sea level and hike up a mountain, at what elevation will we see our first Steller's jay and at what elevation will we see our last scrub jay?

We can use a competition model to predict the location of species borders, and the pattern of zonation along an environmental gradient. The idea is to apply the competition equations point by point along the gradient. For example, let x denote elevation, where x goes from sea level, 0, to 1000 m. Suppose the high-elevation species has a K that starts at 100 and increases to 200 at 1000 m, and the low-elevation species has a K of 200 at sea level that decreases to 100 at 1000 m. To tell MATLAB these assumptions, we type

```
>> kh = '100*(1+x/1000)';
>> kl = '200*(1-x/2000)';
```

and to graph these assumptions, we type

```
>> x = 0:50:1000;
>> figure
>> hold on
>> plot(x,eval(sym2ara(kl)),'g');
>> plot(x,eval(sym2ara(kh)),'c');
```

as shown in Figure 6.8. If only one of the blue jays were present in the region, then it would be found throughout the entire elevation gradient. But with two species present, we can ask for each spot, x, whether both species can coexist there. This assumes that a "spot" is really a piece of habitat large enough to contain a population of birds, say 10 km^2. We'll

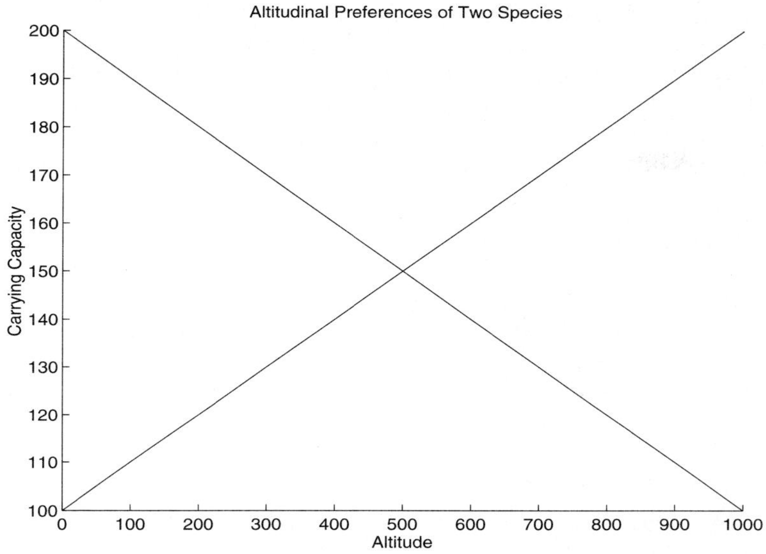

Figure 6.8: The carrying capacity as a function of elevation for two species.

imagine that the overall transect is really a sequence of squares, like bathroom tiles, running from sea level to 1000 m elevation, and we'll be asking if the bird populations can coexist within each of these adjacent tiles.

Well, the K's are only half the story, so let's suppose, rather arbitrarily, that the competition coefficients of the low-elevation species against the high elevation species is 0.6 and the reciprocal coefficient is 0.8, and type

```
>> ahl = '0.6';
>> alh = '0.8';
```

OK, now the lowland species can increase when rare at spot x if α_{lh} is less than K_l/K_h. So let's plot this condition at each x along the transect. Type

```
>> figure
>> hold on
>> plot(x,eval(sym2ara(symop(kl,'/',kh))),'g');
>> plot([0 1000],[eval(alh) eval(alh)],'g');
```

and Figure 6.9 appears. The lowland species can increase when rare, to the left of the intersection of these curves, and the intersection point is the location of the species border itself—the highest elevation of the lowland species. The numerical value of this border can be found by our typing

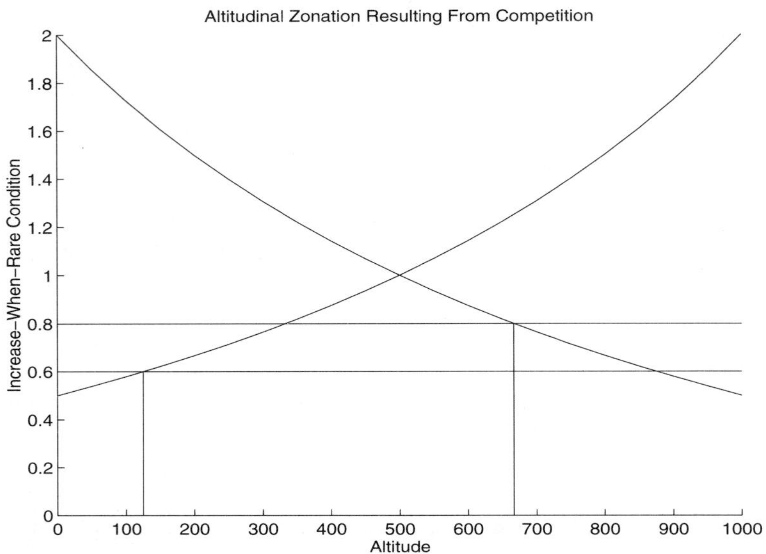

Figure 6.9: The pattern of zonation along an environmental gradient for two species, predicted by the Lotka-Volterra competition model. The low-elevation species occurs alone in the left interval, both species co-occur in the middle interval, and the high-elevation species occurs alone in the right interval.

```
>> xl = solve(symop(alh,'-',kl,'/',kh))
```

yielding

```
xl = 666.6666666666667
```

To add to the graph this location, as a vertical line from the intersection point down to the horizontal axis, type

```
>> plot([eval(xl) eval(xl)],[0, eval(alh)],'g');
```

The high-elevation species is treated in the same way. Its increase-when-rare condition is added to the graph with

```
>> plot(x,eval(sym2ara(symop(kh,'/',kl))),'c');
>> plot([0 1000],[eval(ahl) eval(ahl)],'c');
```

Its species border, the lowest elevation of the high-altitude species, is computed by our typing

```
>> xh = solve(symop(ahl,'-',kh,'/',kl))
```

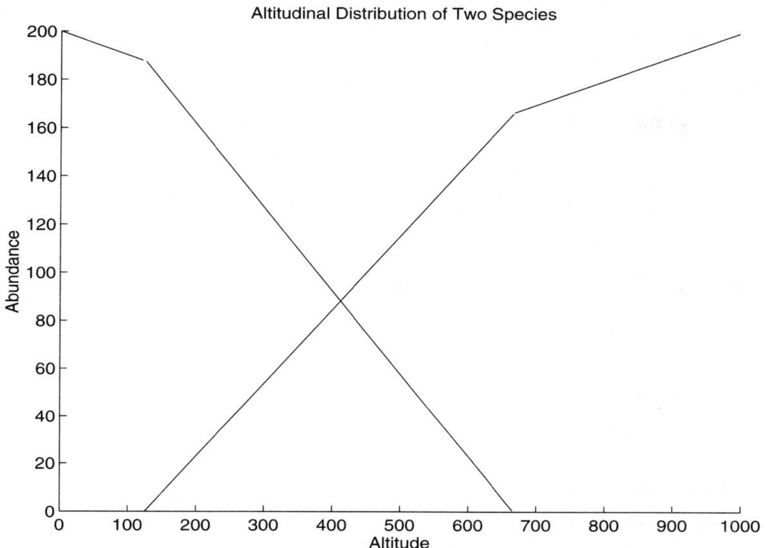

Figure 6.10: The distribution of abundance of the two species whose zonation pattern is predicted in the preceding figure.

yielding

```
xh = 125.
```

To draw a vertical line to demark this boundary, type

```
>> plot([eval(xh) eval(xh)],[0, eval(ahl)],'c');
```

Figure 6.9 now illustrates the complete zonation pattern. The elevations from sea level to x_h contain only the scrub jay, the elevations from x_h to x_l contain both blue jay species, and the elevations from x_l to 1000 m contain only the Steller's jay. What would happen if the competition coefficients were closer to 1? The zone of overlap would shrink. In fact, if both α's equal 1, there would be an abrupt transition from the scrub jay to the Steller's jay and practically no zone of overlap.

The distribution of abundance of both species along the gradient can also be predicted. Where the scrub jay is alone, its abundance is K_l; where the Steller's jay is alone, its abundance is K_h; where they co-occur, the abundances are given by the interior equilibrium solution of the Lotka-Volterra model, as presented in the previous section. MATLAB already knows about `kh` and `kl`, and telling it about the interior equilibrium requires our typing

```
>> nl = '(kl-alh*kh)/(1-alh*ahl)';
>> nl = subs(nl,kh,'kh');
```

```
>> nl = subs(nl,kl,'kl');
>> nl = subs(nl,ahl,'ahl');
>> nl = subs(nl,alh,'alh');
>> nh = '(kh-ahl*kl)/(1-alh*ahl)';
>> nh = subs(nh,kh,'kh');
>> nh = subs(nh,kl,'kl');
>> nh = subs(nh,ahl,'ahl');
>> nh = subs(nh,alh,'alh');
```

To plot the abundance in each of the three zones, type

```
>> figure
>> hold on
>> x = 0:10:eval(xh);
>> plot(x,eval(sym2ara(kl)),'g')
>> x = eval(xh):10:eval(xl);
>> plot(x,eval(sym2ara(nl)),'g')
>> plot(x,eval(sym2ara(nh)),'c')
>> x = eval(xl):10:1000;
>> plot(x,eval(sym2ara(kh)),'c')
```

which results in Figure 6.10. So, the last thing to do is actually take the hike from sea level to 1000 m and see if the birds do occur at the elevations predicted in this figure.

6.1.5 Intermediate Disturbance Principle

Space is often a limiting resource for plants and sessile animals. On land, grasses compete with dandelions for space on a lawn. At sea, barnacles compete with mussels and sea weeds for space on rocks. Sessile organisms are vulnerable to a mortality that is called a "disturbance" by ecologists—this is just a synonym for a density-independent mortality that opens up space previously covered by living organisms. Hurricanes and fires are disturbances to a forest. Logs and chunks of ice that crash against a rocky shore are disturbances to the animals attached there. Disturbance is interesting ecologically because it interacts with competition in a way that promotes species diversity. Here's how the idea works. First, it's very commonly observed that species of sessile organisms compete with one another in a "hierarchy." This means that if there are say, three species, A, B, and C, then A can exclude either B or C or both, B can exclude C, and C is at the bottom of the totem pole. Second, A doesn't get to be the top competitor free of charge—it is exposed to extra risk from disturbance. Typically, to become top competitor, the top competitor has to overgrow a subordinate competitor. But by being on top, it is also the first species to be impacted by disturbances that rain down upon it. So the situation arises in which disturbance, which is just a form of mortality and might be thought of as bad, can prevent the top competitor from completely dominating the subordinates, so that species diversity is higher than what would occur in the absence of disturbance. Of course, enough disturbance would render the habitat unlivable. So, the idea is widespread in ecology that an intermediate degree of disturbance is a good thing, because it promotes species diversity, which is considered good to begin with. This idea is called the "intermediate disturbance principle" and we can adapt the Lotka-Volterra model to show how this principle comes about.

To add disturbance to the Lotka-Volterra competition model, we include loss terms $-d_1 N_1$ and $-d_2 N_2$ into the equations, as follows

$$\begin{aligned} \frac{dN_1}{dt} &= \frac{r_1 N_1 (K_1 - N_1 - \alpha_{12} N_2)}{K_1} - d_1 N_1 \\ \frac{dN_2}{dt} &= \frac{r_2 N_2 (K_2 - N_2 - \alpha_{21} N_1)}{K_2} - d_2 N_2 \end{aligned}$$

Next, to bring the loss terms into the parentheses, we rearrange as follows:

$$\begin{aligned} \frac{dN_1}{dt} &= \frac{r_1 N_1 (K_1(1 - \frac{d_1}{r_1}) - N_1 - \alpha_{12} N_2)}{K_1} \\ \frac{dN_2}{dt} &= \frac{r_2 N_2 (K_2(1 - \frac{d_2}{r_2}) - N_2 - \alpha_{21} N_1)}{K_2} \end{aligned}$$

Therefore, the equilibria with disturbance are the same as the equilibria without disturbance, provided that we replace each K_i with $K_i(1 - d_i/r_i)$. In particular, the species-alone equilibria are

$$\begin{aligned} \hat{N}_{1\text{alone}} &= K_1(1 - d_1/r_1) \\ \hat{N}_{2\text{alone}} &= K_2(1 - d_2/r_2) \end{aligned}$$

The coexistence equilibria are

$$\begin{aligned} \hat{N}_{1\text{coexist}} &= \frac{K_1(1 - d_1/r_1) - \alpha_{12} K_2(1 - d_2/r_2)}{1 - \alpha_{12}\alpha_{21}} \\ \hat{N}_{2\text{coexist}} &= \frac{K_2(1 - d_2/r_2) - \alpha_{21} K_1(1 - d_1/r_1)}{1 - \alpha_{12}\alpha_{21}} \end{aligned}$$

The increase-when-rare conditions for species-1 and species-2 are

$$\begin{aligned} K_1(1 - d_1/r_1) - \alpha_{12} K_2(1 - d_2/r_2) &> 0 \\ K_2(1 - d_2/r_2) - \alpha_{21} K_1(1 - d_1/r_1) &> 0 \end{aligned}$$

Using these formulas we can illustrate the intermediate disturbance principle with a numerical example. It's convenient to replace d_1 with dh_1 and d_2 with dh_2, where d is a measure of overall disturbance, and h_1 measures the exposure to hazard from this disturbance in species-1 and h_2 the exposure to hazard in species-2. So, to tell MATLAB about all this, type

```
>> n1_alone = 'k1*(r1-h1*d)/r1';
>> n2_alone = 'k2*(r2-h2*d)/r2';
```

```
>> n1_coexist = '(k1*(r1-h1*d)/r1-a12*k2*(r2-h2*d)/r2)/(1-a12*a21)';
>> n2_coexist = '(k2*(r2-h2*d)/r2-a21*k1*(r1-h1*d)/r1)/(1-a12*a21)';
>> dn1_rare = 'k1*(r1-h1*d)/r1-a12*k2*(r2-h2*d)/r2';
>> dn2_rare = 'k2*(r2-h2*d)/r2-a21*k1*(r1-h1*d)/r1';
```

Now, by convention, let's label species-1 as the dominant competitor, the one that crowds out the other in the absence of disturbance. Our question then is how much disturbance is needed to allow species-2 to increase when rare? The answer is to solve for the d that makes the increase-when-rare criterion for species-2 equal to zero. A higher d than this then will allow species-2 to increase when rare. MATLAB finds this minimum level of disturbance necessary for species-2 to invade, when we type

```
>> d_min = solve(dn2_rare,'d')
```

yielding

```
d_min = -(k2-a21*k1)/(-k2/r2*h2+a21*k1/r1*h1)
```

As the disturbance rises, however, it continues to impact species-1 more heavily, and eventually species-1 can no longer remain in the system. This maximum disturbance level that allows species-1 to remain, is found by our typing

```
>> d_max = solve(dn1_rare,'d')
```

yielding

```
d_max = -(k1-a12*k2)/(-k1/r1*h1+a12*k2/r2*h2)
```

Finally, as the disturbance is raised still more, eventually even species-2 is driven from the system. This upper bound to the allowable disturbance occurs at the d for which the species-2 alone equilibrium is zero, and is found by our typing

```
>> d_ub = solve(n2_alone,'d')
```

yielding

```
d_ub = r2/h2
```

In summary, only species-1 occurs if the disturbance is between 0 and `d_min`, both species occur if the disturbance is between `d_min` and `d_max`, only species-2 occurs if the disturbance is between `d_max` and `d_ub`, and nothing occurs if the disturbance exceeds `d_ub`.

To generate a numerical example, suppose that species-1 is the competitive dominant in that it is unaffected by species-2 ($\alpha_{12} = 0$), but ungenerously clobbers species-2 in return ($\alpha_{21} = 2$). Species-1's r and K are lower because of the effort it puts into competition, and its exposure to hazard is higher. These assumptions are recorded with MATLAB by our typing

```
>> r1=.75; k1=750; a12=0; h1=1;
>> r2=1; k2=1000; a21=2; h2=.75;
```

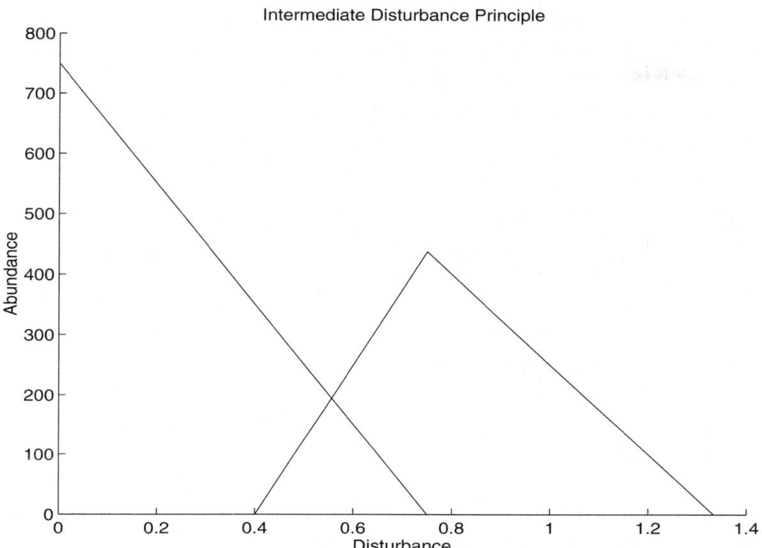

Figure 6.11: The abundance of two species as a function of the level of disturbance. Both species coexist at an intermediate degree of disturbance, and thus illustrate the intermediate-disturbance principle.

The critical disturbance levels are then evaluated as

```
>> eval(d_min)
ans = 0.4000
>> eval(d_max)
ans = 0.7500
>> eval(d_ub)
ans = 1.3333
```

To illustrate the numerical results, plot the abundance of both species against the level of disturbance. Type

```
>> figure
>> hold on
>> d = 0:eval(d_min)/10:eval(d_min);
>> plot(d,eval(n1_alone),'g')
>> d = eval(d_min):(eval(d_max)-eval(d_min))/10:eval(d_max);
>> plot(d,eval(n1_coexist),'g')
>> plot(d,eval(n2_coexist),'r')
>> d = eval(d_max):(eval(d_ub)-eval(d_max))/10:eval(d_ub);
>> plot(d,eval(n2_alone),'r')
```

This yields Figure 6.11. Although both species coexist in the interval of intermediate disturbance, between `d_min` and `d_max`, the abundance of both species combined generally declines with the level of disturbance. Thus it is diversity, not necessarily total biomass, that is maximized at an intermediate level of disturbance.

The intermediate-disturbance principle has been both useful and controversial in conservation biology and park management. If natural fires are prevented in a park, then the vegetation becomes dominated by the superior competitors who crowd out other plant species, including many with beautiful flowers. To reinstate the natural disturbance regime, park personnel allow so-called controlled burning. This practice is controversial because, of course, the burning may escape control and do a lot of damage. Still, unless controlled burning is allowed, a park is destined to become a less diverse system.

The intermediate-disturbance principle was originally developed to explain diversity among the sessile invertebrates of the marine rocky intertidal zone. It's now clear, though, that a complicating factor there is the rate of return of larvae from the ocean, which is called the "larval settlement rate," or the "larval recruitment rate." If the rate of larvae arriving from the ocean to a stretch of intertidal coast is very high, then the intermediate-disturbance principle works just fine. But if the larval arrival rate is low, the species never build up enough abundance to crowd one another out anyway. In these circumstances, which may be true of most rocky intertidal habitat worldwide, a superior competitor can't exclude subordinate species even in the absence of disturbance, and so the intermediate-disturbance principle becomes irrelevant. Probably the most important application of the intermediate-disturbance principle now is to the maintenance of diversity in terrestrial vegetation.

6.2 Three-Species Lotka-Volterra Competition

As we move to what happens when three species compete, a floodgate of new possibilities opens. It's neither practical nor interesting to analyze exhaustively all the possible outcomes of three-species competition, or for that matter, of competition involving even more species. Instead, let's look into several scenarios of special biological or mathematical interest.

6.2.1 Limiting Similarity

In 1959, G. E. Hutchinson raised the question of how different two species had to be from each other, for them to coexist. The idea is that if species are different from one another, then they use different resources, and therefore don't compete very much, and therefore can coexist. Hutchinson surveyed the bill lengths of coexisting bird species, and the skull lengths of coexisting mammal species and found that the ratio of the larger to the smaller was typically about 1.3. This led Hutchinson to surmise, as an empirical generalization, that a length ratio of 1.3 between species led to competition coefficients generally low enough to permit their coexistence. The assumption in this statistic is that bill length in birds, and skull length in mammals, somehow correlates with the kind of food being eaten. In insectivorous birds and lizards, for example, bill length and skull length correlate with the size of their insect prey; this correlation implies that large predators consume different species of insects than small predators. Almost ten years later, MacArthur and Levins used the three-species Lotka-Volterra equations to offer a theoretical explanation of Hutchinson's empirical generalization.

The setup involves what might be called "nearest-neighbor competition." The idea is that there is an axis, called a "niche axis," such that the closer two species are to one another along this axis, the higher the competition between them. Body size provides such an axis for many species. The closer two *Anolis* lizards are in body size, the stronger the competition between them because they take more of the same prey. Similar arguments have been made for body size in birds, mammals, butterflies, and so forth. This assumption states that each species competes most strongly with the species most similar to it in body size, and compete less so with any other species. Suppose we label the species by the order of their body sizes, so species-1 is the smallest, species-2 has a medium body size, and species-3 the largest. The competition then is strongest between species-1 and species-2, and less so between species-1 and species-3. In this sense, species compete mostly with their nearest neighbors on the niche axis. This situation may be ideally represented with a matrix of competition coefficients in which interspecific competition is indicated only along the diagonals above and below the main diagonal, with zeros elsewhere, as in the following MATLAB definition,

```
>> alpha = sym('[1 a 0; a 1 a; 0 a 1]')
```

Here, the parameter `a` indicates how close the species are to each other along the axis. If `a` is near 1, the species have nearly the same body size, whereas if `a` is near 0, they are very different in body size. In *Anolis* lizards, for example, if one species is about twice the body size of the other, the competition coefficient is too small to measure, and is effectively zero. If one species is only 10% larger than the other, then the competition coefficient between them is effectively one. So, we'll use `a` as a measure of how similar two species are to each other, where the detailed calibration of `a` in terms of physical measurements such as body size in millimeters, depends on the kind of organism being discussed. If we assume the three species all have the same K's, the vector of carrying capacities is defined as

```
>> k = sym('[k; k; k]')
```

Once an `alpha` matrix and `k` vector are defined, the equilibrium-population sizes at the interior equilibrium are generated by our typing

```
>> nhat = linsolve(alpha,k)
```

which yields

```
nhat = [   k*(a-1)/(2*a^2-1)]
       [k*(-1+2*a)/(2*a^2-1)]
       [   k*(a-1)/(2*a^2-1)]
```

Now imagine that we're going to take species-1 and species-3, and move their body sizes closer to one another, squeezing species-2 in the middle. Thus `a` starts out near zero, and increases as we make species-1 and species-3 more similar to each other. As a increases, the abundance of the middle species drops, and it drops all the way to zero when a equals 1/2. Thus, the three species can coexist if their difference in body sizes is large enough that a < 1/2. Conversely, only two species can coexist if the body sizes are so similar that a > 1/2. The value of a = 1/2 is called the "limiting similarity."

In the 30 years since the idea of limiting similarity was proposed, it has been extensively modeled, with a great many variants on the theme above. You may easily check out some of these variants now, if you wish. For example, look at coexistence among more than three species with nearest-neighbor competition. I've tried up to seven species and MATLAB always finds a simple answer. With four species, there's no limiting similarity. With five species, the second and third can get squeezed out, depending on the degree of similarity. And so forth. One may also vary the assumption of equal K's. I've been especially interested in how the similarity of an invading species to the species already present affects the invader's chance of success. The role of coevolution in possibly enhancing initial differences between competing species has also been extensively studied. Generally speaking, nearest-neighbor competition one way or another causes species to be spaced out along a niche axis.

6.2.2 Diffuse Competition

Suppose there's no single niche axis. Just the reverse: Suppose there are as many axes as there are species, and that all the species are separated from one another by the same distance in a multidimensional space. With three species we have an equilateral triangle in niche space. Interspecific competition occurs among *all* the species living together just because they're there in the same place. This is called "diffuse competition." It may be modeled when an alpha matrix is defined as

```
>> alpha = sym('[1 a a; a 1 a; a a 1]')
```

Using the same k vector as before, the interior equilibrium is

```
>> nhat = linsolve(alpha,k)
nhat = [1/(2*a+1)*k]
       [1/(2*a+1)*k]
       [1/(2*a+1)*k]
```

Here there's no limit to a that prevents this equilibrium from being positive. To show that this equilibrium is stable if a < 1, you can compute the Jacobian and its eigenvalues symbolically with the jacobian and eigensys commands. Thus, all the species may be arbitrarily similar to one another and still coexist. Furthermore, this idea can be extended to four, five, or any number of species. You may wish to check this out; I've tried it with four species, which MATLAB handles just fine. The bottom line is that competition itself does not prevent species from coexisting—it's whether a species has a big enough competitive *advantage* that is crucial. Here, there's lots of competition among all the species, but no one has a big advantage over the others, so they all coexist.

6.2.3 Hierarchical Competition

If one species does have a big competitive advantage over the others it may still coexist with them if it is disadvantaged in some other way. The idea of hierarchical competition came up before when we looked into the intermediate disturbance principle. Among competitors for space, one species often can overgrow the others, and the first subordinate can outgrow the others below it, and so forth, leading to a hierarchy. We'll see later that models of competition by plants for soil resources such as nitrogen leads to a perfect hierarchy as well. So, let's see what's needed for coexistence in spite of strong hierarchical competition. To define a `k` vector, type

```
>> k = sym('[k1; k2; k3]')
```

An `alpha` matrix to indicate complete competitive domination and perfect hierarchical ordering is defined with

```
>> alpha = sym('[1 0 0; 1 1 0; 1 1 1]')
```

The interior equilibrium for this system is then found as

```
>> nhat = linsolve(alpha,k)
nhat = [     k1]
       [-k1+k2]
       [-k2+k3]
```

So, all we need for coexistence is that $K_1 < K_2 < K_3$; that is, the carrying capacities have to line up in a hierarchy that is the reverse of the competitive hierarchy. This too can be generalized to four, five, or any number of species. So the bottom line is that competing species can coexist if none has a big advantage that is not matched with some compensating disadvantage.

6.2.4 Competitive Oscillations

When three species compete, another new feature of the Lotka-Volterra equations appears: the possibility of sustained oscillations. We'll look at two examples. To exhibit the oscillations, we need to call on one of MATLAB's numerical differential-equation-solvers. Up to now we've used the `ode23` command, but now we'll use the `ode45` command, which is more accurate but slower. Recall that `ode23` requires the model to be defined in a separate function, and `ode45` does too. So, with a text editor, type the following function, and save it as `lvcomp3.m`:

```
function ndot = lvcomp3(t,n)
 global r k alpha;
 ndot(1,1) = r(1)*n(1)*(k(1)-n(1)-alpha(1,2)*n(2)-alpha(1,3)*n(3))/k(1);
 ndot(2,1) = r(2)*n(2)*(k(2)-alpha(2,1)*n(1)-n(2)-alpha(2,3)*n(3))/k(2);
 ndot(3,1) = r(3)*n(3)*(k(3)-alpha(3,1)*n(1)-alpha(3,2)*n(2)-n(3))/k(3);
```

This is the three-species counterpart to `lvcomp2.m`, which was used earlier for two competing species.

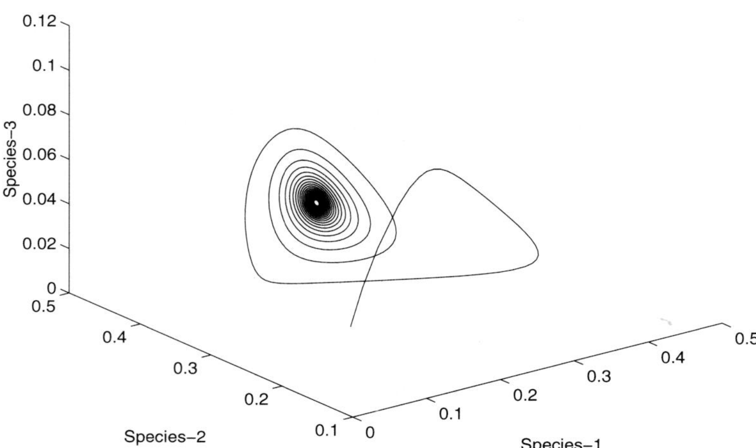

Figure 6.12: Three competing species coming to a stable equilibrium with a relatively low r.

Stability Depends on r

The first example of a three-species oscillation traces to a paper by Strobeck, who pointed out for the first time that the stability of the interior equilibrium in a Lotka-Volterra model depends on the r's. Strobeck's observation showed that the classical approach to teaching about competition models in ecology was incorrect once more than two species were involved. Textbooks traditionally teach the two-species Lotka-Volterra model based on what are called "isoclines." For two species there are two lines—along one dN_1/dt is zero and along the other dN_2/dt is zero. If these lines intersect at a positive N_1 and N_2, then an equilibrium exists. By looking at the sign of dN_1/dt and dN_2/dt in all combinations around the equilibrium, one can deduce whether the equilibrium is stable. The problem is that the isoclines are independent of the r's. Well, when there are two species, the r's don't turn out to affect whether an equilibrium is stable or not, they only affect how fast trajectories move toward or away from the equilibrium. But with three species, the r's do affect whether an equilibrium is stable, and therefore the traditional way of teaching about the Lotka-Volterra model with two species cannot be generalized to three species or more. Strobeck provided a numerical example in which the r's affected the stability. Here is that example.

Define the `k` and `alpha` parameters as

```
>> global r k alpha
>> k = [1; 9/19; 4/19]
```

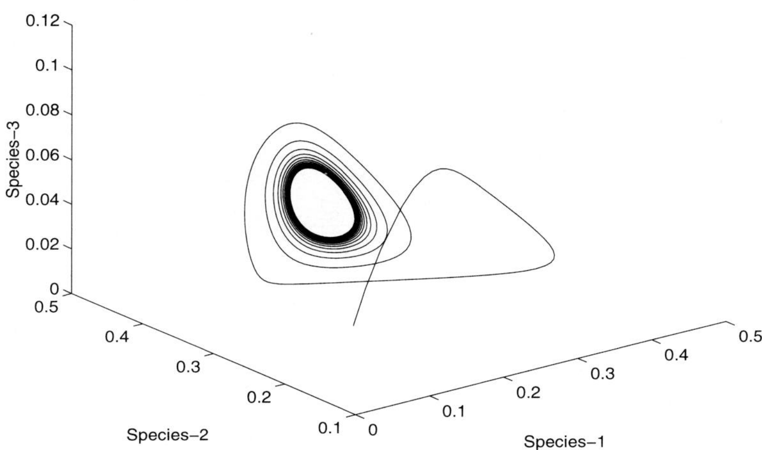

Figure 6.13: Three competing species entering a permanent oscillation, called a limit cycle, with a relatively high r.

```
>> alpha = [1 2 4; 1/3 1 2; 1/3 1/3 1]
```

For the first set of r's define

```
>> r = [6.255; 1; 6.255]
```

Now, to generate a numerical solution, call `ode45` and specify a time run from 0 to 1000, and an initial condition of `[1/19;3/19;1/36]`.

```
>> [time n] = ode45('lvcomp3',0,1000,[1/19;3/19;1/36]);
```

`ode45` returns a set of times, `time`, and a matrix of population sizes at those times, `n`. The first column of `n` is N_1, the second, N_2, and the third column is N_3. So, let's plot these in a 3D graph. We type

```
>> figure
>> plot3(n(:,1),n(:,2),n(:,3))
```

and Figure 6.12 appears. This shows the three species approach a stable equilibrium. The trajectory winds into the equilibrium point. A pattern of trajectories that spirals into an equilibrium is called a "stable focus."

Next, define the r's in species-1 and species-3 to be bigger:

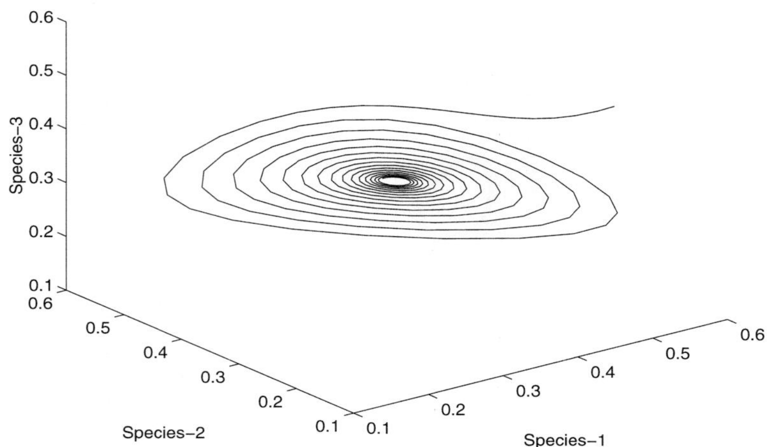

Figure 6.14: Three competing species approaching an equilibrium point.

```
>> r = [6.655; 1; 6.655]
```

Strobeck showed that the equilibrium point was unstable with these r's and yet each of the three species still can increase when rare. Therefore there are no stable equilibria. To see what does happen, generate another numerical solution by typing

```
>> [time n] = ode45('lvcomp3',0,1000,[1/19;3/19;1/36]);
```

To plot the result, type

```
>> figure
>> plot3(n(:,1),n(:,2),n(:,3))
```

Figure 6.13 shows the resulting graph. The three species end up oscillating forever around a loop in the graph. This permanent oscillatory pattern is called a "limit cycle." Thus, three coexisting competing species do not necessarily come to equilibrium, but may coexist in a limit cycle. This transition as r is increased from a stable equilibrium to a limit cycle surrounding an unstable equilibrium is called a "Hopf bifurcation" in mathematics.

Intransitive Competition

Another kind of a three-species oscillation was offered by May and Leonard; it involves what is called "intransitive competition." The idea is that species-1 can exclude species-2,

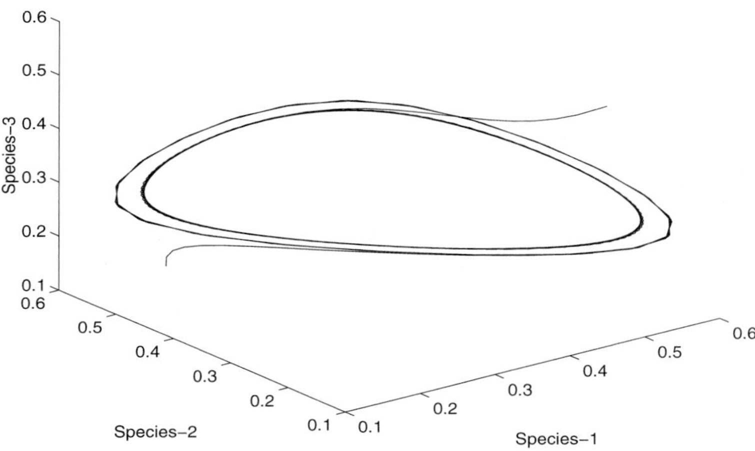

Figure 6.15: Three competing species approaching different cycles from two different initial conditions.

species-2 can exclude species-3, and species-3 can exclude species-1. Put them all together, and a cycle can result. Specifically, define the parameters as

```
>> global r k alpha
>> r = [1; 1; 1]
>> k = [1; 1; 1]
>> a = 0.5; b = 1.45;
>> alpha = [1 a b; b 1 a; a b 1]
```

Generating a numerical solution and plotting the trajectory with

```
[time n] = ode45('lvcomp3',0,350,[3/6;1/6;3/6]);
figure
plot3(n(:,1),n(:,2),n(:,3))
```

produces Figure 6.14, showing all three species converging with damped oscillation to a stable equilibrium point.

Next, increase the parameter b from 1.45 to 1.50. Type

```
>> b = 1.50;
>> alpha = [1 a b; b 1 a; a b 1]
```

We'll need two numerical solutions to illustrate what happens now, so type

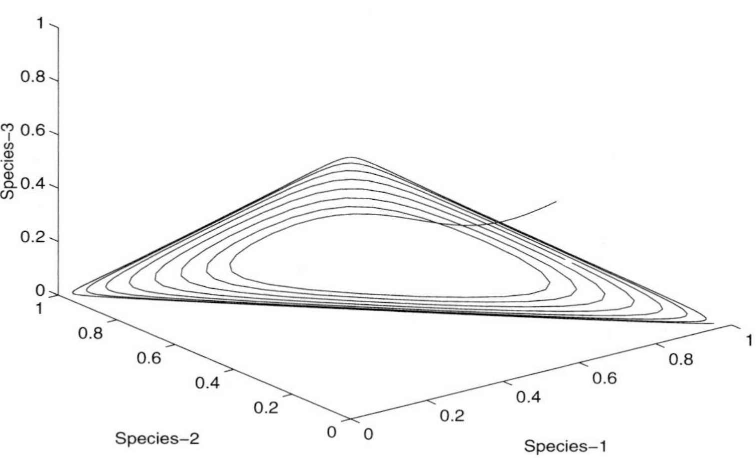

Figure 6.16: Three competing species oscillating in an ever-increasing period.

```
>> [timea na] = ode45('lvcomp3',0,150,[3/6;1/6;3/6]);
>> [timeb nb] = ode45('lvcomp3',0,150,[1/6;3/6;1/6]);
```

and plot them in yellow and red with

```
>> figure
>> plot3(na(:,1),na(:,2),na(:,3),'y',nb(:,1),nb(:,2),nb(:,3),'r')
```

This results in Figure 6.15. The three species now converge to various cycles depending on their initial conditions.

Finally, increase b still more, from 1.50 to 1.55. Type

```
>> b = 1.55;
>> alpha = [1 a b; b 1 a; a b 1]
```

and generate a numerical solution and graph with

```
>> [time n] = ode45('lvcomp3',0,250,[4/6;1/6;3/6]);
>> figure
>> plot3(n(:,1),n(:,2),n(:,3))
```

This yields Figure 6.16. This case is particularly interesting because the trajectories wind out; that is, they take ever more time to complete a go-around as they move closer to the axes.

The two examples of three-species oscillations in the Lotka-Volterra model are probably not empirically important. But they're interesting conceptually because they show that competition does not necessarily stabilize an ecological system. The *arrangement* of the competition among the species, together with any compensatory tradeoffs between competitive ability and the r's and K's, determine whether coexistence occurs, even in the presence of strong competition, and determine whether coexistence leads to a stable equilibrium point or to a limit cycle. None of these insights emerge in the the two-species Lotka-Volterra model. One needs to work with three or more species for these new possibilities to arise.

6.3 Resource-Based Competition Models

Resource-based competition models are built up from assumptions about how specific resources are consumed by competing organisms. These models may be easier to use than the Lotka-Volterra model because their parameters are tied more directly to the mechanism of competition than are the parameters of the Lotka-Volterra model. Here is the major resource-based competition model used for plant populations.

The idea is that the growth of a plant population depends on the uptake of nutrients from the soil. The biomass of species-1, B_1 in grams, is assumed to change according to

$$\frac{dB_1}{dt} = B_1(u_1 R - d_1)$$

where u_1 is the growth coefficient resulting from the uptake of the soil resource, R, and d_1 is the loss rate per unit biomass in species-1. The uptake kinetics are taken to be linear here, but usually Michaelis-Menton uptake kinetics are used in the literature. I'm not persuaded that Michaelis-Menton kinetics offer sufficiently improved accuracy to be worth the messiness they bring to the formulas. The kinetics of biomass change in species-2 follows the same type of formula:

$$\frac{dB_2}{dt} = B_2(u_2 R - d_2)$$

Meanwhile, as the plants are growing, nutrients are being withdrawn from the soil by the plants. Also, nutrients are being added back to the soil by decomposition at rate S, which is taken as a fixed source term for the nutrients. Nutrients are being leached out of the soil too, as a result of water runoff from rain, at rate aR. This all leads to

$$\frac{dR}{dt} = S - aR - w_1 u_1 B_1 R - w_2 u_2 B_2 R$$

where w_1 and w_2 are the amounts of nutrient per unit biomass. These three equations comprise a model for two competing plant populations whose growth depends on a common limiting nutrient. Well, what happens?

6.3.1 The R^* Principle

The main conclusion of the resource-based models of competition is that one species always ends up the winner, and all the others are crowded out. Moreover, all the species can be ranked in a competitive hierarchy, such that the highest can exclude everyone else, the second highest can exclude every species below it, and so forth. Tilman has termed this conclusion the R^* principle. Here's how this principle works. To define the model to MATLAB, type

```
>> db1 = 'b1*(u1*r-d1)'
>> db2 = 'b2*(u2*r-d2)'
>> dr = 's-a*r-w1*u1*b1*r-w2*u2*b2*r'
```

Then, to solve for all the possible equilibria, type

```
>> [b1hat,b2hat,rhat] = solve(db1,db2,dr,'b1,b2,r')
```

yielding

```
b1hat = [-(-s*u1+a*d1)/w1/d1/u1]
        [                     0]
        [                     0]

b2hat = [                     0]
        [                     0]
        [-(-s*u2+a*d2)/w2/d2/u2]

rhat = [d1/u1]
       [  s/a]
       [d2/u2]
```

The first and third of the equilibria that happen to be returned by MATLAB are for species-1 and species-2 alone, respectively. The second equilibrium happens to be that for both species being extinct. There are no other equilibria, and in particular, there are no equilibria with both species coexisting. Now the R^* principle is that the winner is identified when the equilibrium with the lowest `rhat`, which is also denoted as R^* in Tilman's notation, is found. For example, if `d1/u1` is lower than `d2/u2` then species-1 is the winner, and if the reverse is true, then species-2 is the winner. The basis for the R^* principle is an increase-when-rare argument. When a species is rare, it cannot invade against the species who can survive at the lowest resource concentration.

Let's look into a numerical example. To define some parameters for MATLAB, type

```
>> global u d s a w;
>> u1 = .1; u2 = .05;
>> u = [u1; u2];
>> d1 = .1; d2 = .1;
>> d = [d1; d2];
>> w1 = .1; w2 = .1;
>> w = [w1; w2];
>> s = 10; a = .5;
```

Let's see who the winner is predicted to be by typing

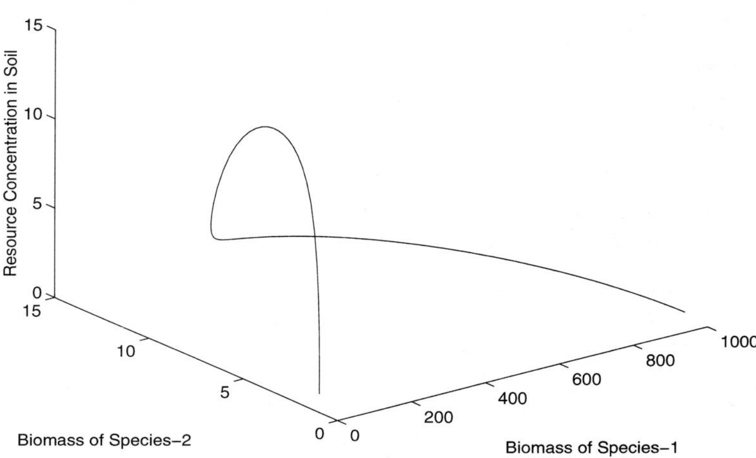

Figure 6.17: Two plant populations competing for a resource.

```
>> for i=1:3
>>  eval(sym( rhat,i,1))
>>  eval(sym(b1hat,i,1))
>>  eval(sym(b2hat,i,1))
>> end
```

which yields

```
ans = 1
ans = 950.0000
ans = 0

ans = 20
ans = 0
ans = 0

ans = 2
ans = 0
ans = 900
```

The first of these equilibria has an **rhat**, i.e., an R^*, equal to 1, which is the lowest. So species-1 should win.

To illustrate the trajectories of these competing populations, we can once again use MATLAB's numerical integration command, `ode45`. So, use a text editor to define the model for MATLAB and save it as `rcomp2.m`

```
function xdot = rcomp2(t,x)
 global u d s a w;
 xdot(1,1) = x(1)*(u(1)*x(3)-d(1));
 xdot(2,1) = x(2)*(u(2)*x(3)-d(2));
 xdot(3,1) = s-a*x(3)-w(1)*u(1)*x(1)*x(3)-w(2)*u(2)*x(2)*x(3);
```

In this function, the state variable is a column vector `x` whose first element is B_1, second element is B_2, and third element is R. Now typing

```
>> [time x] = ode45('rcomp2',0,100,[1;1;1]);
>> figure
>> plot3(x(:,1),x(:,2),x(:,3))
```

generates Figure 6.17. The trajectory ends up at 950 on the B_1 axis, as predicted.

6.3.2 Lotka-Volterra Weeds

The resource-based model of plant competition is perhaps not very interesting in itself, because it predicts a monotonous monoculture of the most efficient resource user. When embedded in a metapopulation context, however, the party gets lively. Recall the Logistic Weed from Chapter 4, where P habitat patches out of a total of H are occupied by plants, m is the probability that an occupied patch sends a colonist to an unoccupied patch, and μ is the probability that the plants in an occupied patch die out:

$$\frac{dP}{dt} = mP(H - P) - \mu P$$

Recall also that this model is equivalent to a logistic model, as seen when the right-hand side is expanded, as

$$\frac{dP}{dt} = (mH - \mu)P - mP^2$$

In this equation, we can equate r with $(mH - \mu)$ and r/K with m. Solving these two expressions for r and K implies that we can identify r with $(mH - \mu)$ and K with $H - \mu/m$.

Now let's suppose that there are more than one species of plants and that they compete for soil resources according to the resource-based model of competition discussed in the previous section. If so, the plant species can be ranked in a perfect competitive hierarchy based on their R^*'s. Let's label the species according to their R^*'s, with the dominant species (the one with the lowest R^*) being species-1. Metapopulations of three species can then be modeled as

$$\frac{dP_1}{dt} = m_1 P_1(H - P_1) - \mu_1 P_1$$

$$\begin{aligned}\frac{dP_2}{dt} &= m_2P_2(H - P_1 - P_2) - \mu_2P_2 - m_1P_2P_1\\ \frac{dP_3}{dt} &= m_3P_3(H - P_1 - P_2 - P_3) - \mu_3P_3 - m_1P_3P_1 - m_2P_3P_2\end{aligned}$$

In this scheme, species-1 acts as if it has the whole place to itself, just as it does in a single-species model. Species-2 gains a patch when it disperses to a place in which neither it nor species-1 is already there; it loses a patch by extinction at rate μ_2; and it also loses a patch if one that it currently occupies is colonized by species-1. And similarly for species-3. Just as the single-species metapopulation model was equivalent to a logistic equation, this model turns out to be equivalent to special case of the Lotka-Volterra competition equations. For example, take the second species, and expand it to read

$$\frac{dP_2}{dt} = (m_2H - \mu_2)P_2 - m_2P_2^2 - (m_2 + m_1)P_2P_1$$

Now r_2 must equal $(m_2H - \mu_2)$, r_2/K_2 must equal m_2 and $r_2\alpha_{21}/K_2$ must equal $(m_2 + m_1)$, and similarly for the other species. Thus, the Lotka-Volterra version for competing metapopulations uses the following parameters for species-i:

$$\begin{aligned} r_i &= m_iH - \mu_i \\ K_i &= H - \mu_i/m_i \\ \alpha_{ij} &= 1 + m_j/m_i && (j < i) \\ &= 1 && (j = i) \\ &= 0 && (j > i) \end{aligned}$$

The equilibrium at which these metapopulations coexist throughout the overall region is found by our loading an `alpha` matrix and `k` vector with these parameters and using the `linsolve` command to ask MATLAB for the answer. Simple expressions emerge if all the μ's are the same. In this case the solution shows that a positive equilibrium exists if the competitive dominance hierarchy is compensated by a reverse hierarchy in dispersal ability, m_i. Alternatively, if all the m's are the same, a positive equilibrium exists if the competitive dominance hierarchy is compensated by a reverse hierarchy in extinction probability, μ_i. Regardless of details, the conclusion is the same—for there to be coexistence, the competitive-dominance hierarchy has to be compensated by a sufficiently strong reverse hierarchy in the K's, which are themselves constructed from both the extinction and dispersal rates, μ_i and m_i.

Surprisingly, this model foretells of sorrow and woe for the competitive dominant at the hand of human development. One might think that the species to survive habitat loss from human development would be the best competitor—why not? This would seem to be the least fragile. But because the plant species are coexisting, and because there is dominance-hierarchy competition, the competitive dominant has to have the lowest K. But this also makes it the most vulnerable. Look at what goes into K_1—it's H, the total number of patches, minus μ_1/m_1. Well, if K_1 is the lowest of them all, then either species-1 is suffering a high extinction rate, μ_1, or is a lousy disperser with a low m_1, or both. Either way, there can't be much difference between H and μ_1/m_1. Hence, as H is lowered, the competitive dominant has its K hit zero first, and is the first species to become extinct. After its extinction, the remaining species may undergo some additional extinctions. Once the shakedown is over, the most dominant of the remaining species will then also be the most vulnerable to further habitat loss. And so on. Habitat loss will knock off, one by one, the competitive dominants.

6.4 Predator-Prey Interaction

Animals prey on other animals, and many charismatic creatures such as lions, tigers, bears, eagles, snakes, and sharks are all predators. Of course, most predation occurs with less dramatic actors, and in the arthropod world predators include wasps and spiders; among marine invertebrates are starfish, snails and sea anemonies; and among terrestrial vertebrates are robins, blue jays, fly catchers, small lizards, and frogs—the list is endless. The common denominator faced by all predators is dependency on their prey, and should the prey disappear, so will the predators depending on them. So, the most basic questions of any predator-prey interaction are how many predators the prey population can support, and how much predation would lead to overexploitation of the prey and the ultimate extinction of the predator. Reciprocally, the prey are affected by the predator, and a secondary issue is whether the predator levies a serious toll of mortality on the prey. If the predators disappeared, would the prey increase much, or only slightly? Curiously, we tend to like predators and dislike prey. The prey in a predator-prey interaction are often potential pests to humans, so if the predators are removed and the prey population therefore explodes, the price is paid by people. A third recurring theme in the ecology of the predator-prey interaction is the presence of natural cycles. There are many examples in nature of predator populations exploding for a while, decimating their prey, then dying of starvation, until the prey population rebounds. The most famous of these is the recurring cycle of lynx (a wild cat) and hares in Canada, but other examples include the crown-of-thorns starfish on Pacific coral reefs, cyclic outbreaks of insect populations, and recurring epidemics of diseases. So, we'll look into whether a predator can depend on its prey, whether prey can be held in check by predators, and whether predator and prey populations tend to oscillate with one another. We explore some rather generic models first, and then check out models tailored to insects and to pathogens because these have great practical importance for people.

6.4.1 Volterra Predator-Prey Model

The first predator-prey model in ecology was introduced by the mathematician Vito Volterra in the 1920's. It's a neat model because it isolates the predator-prey interaction. The idea is that the prey can grow exponentially in the absence of the predator, and that the predators die off exponentially in the absence of prey. So, there's no density dependence within either predator or prey. But the big news is predation. The predators are assumed to contact prey as though by random collision. The rate of collision is proportional to the product $N_1 N_2$ where N_1 is the number of prey and N_2 is the number of predators. Of course, both predators and prey may be very clever at finding and avoiding one another, but it's assumed that statistically the rate of predation works out to be proportional to the product of abundances. This is called a "mass-action" assumption. It doesn't imply that the animals are just big stupid particles bumping into one another in the dark. Automobile traffic flowing across bridges can be modeled with mass-action assumptions because of the large number of individuals involved, even though the driver of each car might have a sophisticated behavioral repertoire. So here is Volterra's model:

$$\begin{aligned}\frac{dN_1}{dt} &= rN_1 - aN_1N_2\\ \frac{dN_2}{dt} &= baN_1N_2 - dN_2\end{aligned}$$

r is the intrinsic rate of increase in the prey, d is the death rate in the predators, a is a coefficient relating the prey-capture rate to the rate of predator-prey collisions, and b is a coefficient describing the number of prey captures needed to produce the birth of a predator.

Let's find out what this model has to tell us. To acquaint MATLAB with it, type

```
>> dn1 = 'r*n1-a*n1*n2'
>> dn2 = 'b*a*n1*n2-d*n2'
```

The equilibria are then found by our typing

```
>> [n1hat,n2hat] = solve(dn1,dn2,'n1,n2')
```

yielding

```
n1hat = [     0]
        [d/b/a]

n2hat = [  0]
        [r/a]
```

There are two equilibria, one with both species extinct, the other with both predator and prey coexisting. Wait a second though. These equilibria do not turn out to be stable. So, let's take a look at a trajectory of N_1 and N_2 through time, as predicted by this model. We're going to need one of MATLAB's differential-equation-solvers. With a text editor, type in and save the function `volterra.m`:

```
function ndot = volterra(t,n)
 global r a b d;
 ndot(1,1) = r*n(1)-a*n(1)*n(2);
 ndot(2,1) = b*a*n(1)*n(2)-d*n(2);
```

Now type

```
>> global r a b d;
>> r = 0.5; a = 0.01; b = 0.02; d = 0.1;
>> n1hat_n = eval(sym(n1hat,2,1));
>> n2hat_n = eval(sym(n2hat,2,1));
```

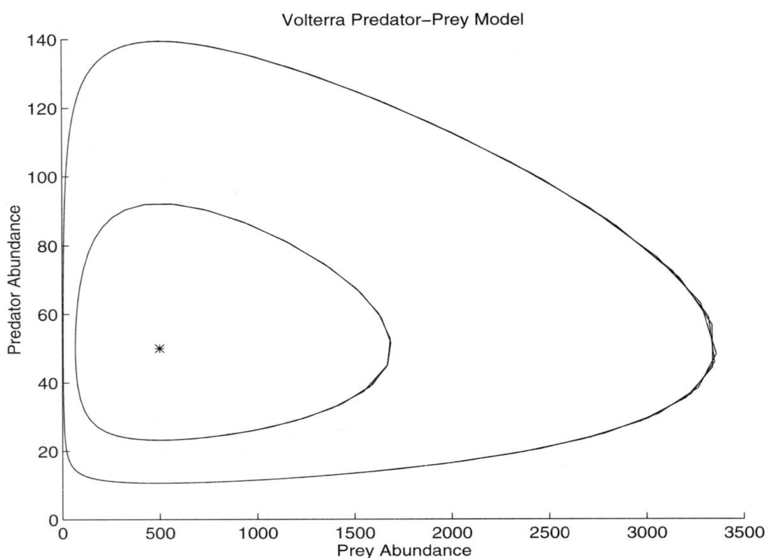

Figure 6.18: The flow of trajectories of the Volterra predator-prey model.

That will produce a numerical example of the interior equilibrium and yield

```
n1hat_n = 500
n2hat_n = 50
```

Then, to illustrate two trajectories, one started at the equilibrium values divided by two, and another at the equilibrium divided by four, type

```
>> figure
>> hold on
>> plot(n1hat_n,n2hat_n,'*')
>> [time n] = ode45('volterra',0,100,[n1hat_n/2; n2hat_n/2]);
>> plot(n(:,1),n(:,2))
>> [time n] = ode45('volterra',0,100,[n1hat_n/4; n2hat_n/4]);
>> plot(n(:,1),n(:,2))
```

and Figure 6.18 appears. The figure shows two examples of the fact that the trajectory from *any* initial condition is a closed loop—the trajectory moves counter-clockwise and after some time returns to the initial condition. The set of all the trajectories comprises a family of closed loops, an infinity of them—one for each initial condition. This means that predator and prey always coexist in this model; neither drives the other to extinction. Instead, they oscillate with one another, and the amplitude of the oscillation depends only on the initial condition. That is, if the predator and prey are far away from the equilibrium

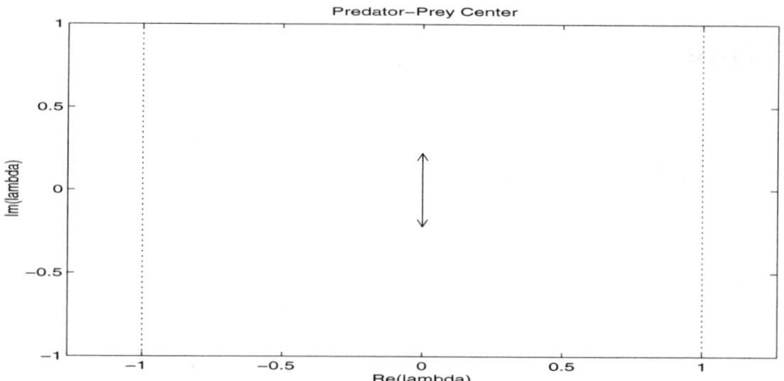

Figure 6.19: The eigenvalues of the Jacobian matrix for the Volterra predator-prey model, evaluated at the interior equilibrium.

point to begin with, they exhibit great fluctuation through time and together remain far away from equilibrium, whereas if they start out close to equilibrium, they show small fluctuations through time and remain near the equilibrium. Thus, the equilibrium point is neutrally stable. To confirm the equilibrium's neutral stability, check out the eigenvalues of the model's Jacobian, evaluated at the equilibrium. Type

```
>> n1 = n1hat_n;
>> n2 = n2hat_n;
>> jac = [eval(diff(dn1,'n1')) eval(diff(dn1,'n2')); ...
>>        eval(diff(dn2,'n1')) eval(diff(dn2,'n2'))]
```

yielding

```
jac =        0   -5.0000
        0.0100         0
```

The main diagonal elements of `jac` are zero; we'll discuss that condition later. To find the eigenvalues, type

```
>> lambda = eig(jac)
```

yielding

```
lambda = 0 + 0.2236i
         0 - 0.2236i
```

The eigenvalues have a zero real part, that is, they are purely imaginary, as graphed in Figure 6.19. This figure was made by our typing

```
>> figure
>> compass(lambda)
```

The fact that the real parts of `lambda` are zero is *not* enough to enable us to say for sure that the equilibrium is neutrally stable. But this is necessary for neutral stability, and is a good clue to what is illustrated in the Figure 6.18[4]. Mathematicians call the flow of trajectories consisting of an infinite number of concentric loops around the equilibrium a "center."

I'll bet you're more interested in the biological lessons to be drawn from the Volterra model than in its mathematical features. Well, there are four main lessons to learn from it. First, predators have the power to check the exponential growth of a prey population. In this model, if the predators are removed, the prey numbers happily proceed to infinity. Second, any prey population can support some predator population. That is, predator and prey always coexist for any combination of the parameters, r, d, a and b. Third, we can use the formulas for the equilibrium abundances as approximate predictors of the *average* abundances through time. If the trajectories were circles, the equilibrium point would be exactly the average through time, and because the trajectories are not quite circular, the equilibrium point is only approximately the average through time. But let's look at these formulas anyway. The equilibrium abundance of the predator is r/a. Now think about this. It does not depend on d. That is, for a given a, the abundance of predators is set by the productivity of the prey population, r. Conversely, the equilibrium abundance of the prey is $d/(ba)$. This is independent of r, and means the abundance of the prey, for a given ab, is set by the death rate of the predators. Fourth, the predator-prey interaction has an inherent tendency to oscillate.

We'll now begin to add more features to this model, piece by piece. Some of the lessons will change, but not very much. One prediction though, does change. The flow consisting of concentric loops is not "structurally stable." This means that as we add features, the flow becomes qualitatively different than that consisting of a set of concentric loops.

It's a final tribute to Volterra, as well as of practical importance, to mention the "Volterra principle," which describes the impact of a general pesticide on a predator-controlled prey population. Many prey of insect predators are pests, like scale insects whose predators include wasps. If the abundance of scale insects is controlled by their predators, that abundance is approximately $d/(ba)$. Now suppose a general pesticide is introduced, such as DDT. The immediate effect is to lower r in the prey and to increase d in the predators. But because the abundance of the pests is controlled by the predator, increasing d leads to a net increase in the number of pests, and to a net decline in the number of predators, whose abundance is approximately r/a. Although the pesticide does hurt the prey, the prey are kept under control by the predators, and hurting the predators leads to a net increase in the number of pests. The Volterra principle shows that trying to intervene in a predator-prey system is tricky if natural predators are controlling the prey's abundance.

---

[4]Neutral stability to the equilibrium in the Volterra model is proved by a demonstration that a quantity is conserved along any trajectory. The system is said to have an "integral of motion."

6.4.2 Adding Density Dependence to the Prey

Among the possibly important omissions in the Volterra predator-prey model, is density dependence in the prey. This is added by our assuming that the prey grow logistically, rather than exponentially, in the absence of the predator. The predator-prey model now becomes

$$\begin{aligned} \frac{dN_1}{dt} &= rN_1(K - N_1)/K - aN_1N_2 \\ \frac{dN_2}{dt} &= baN_1N_2 - dN_2 \end{aligned}$$

To tell MATLAB about this new model, type

```
>> dn1 = 'r*n1*(k-n1)/k-a*n1*n2'
>> dn2 = 'b*a*n1*n2-d*n2'
```

and to find the equilibria, type

```
>> [n1hat,n2hat] = solve(dn1,dn2,'n1,n2');
>> n1hat = simple(n1hat)
>> n2hat = simple(n2hat)
```

which collectively yields

```
n1hat = [    0]
        [    k]
        [d/b/a]

n2hat = [                 0]
        [                 0]
        [r/a-r/b/a^2/k*d]
```

Now there are three equilibria: one for total extinction; one for the prey alone and the predator extinct; and one for coexistence of predator and prey. The third equilibrium, which is the interior equilibrium where both coexist, is perhaps most pleasingly formated as

$$\begin{aligned} \hat{N_1} &= \frac{d}{ba} \\ \hat{N_2} &= \frac{r}{a}(1 - \frac{d}{baK}) \end{aligned}$$

Unlike those portrayed in the Volterra model, these prey are not automatically able to support the predators. To do that, the prey's carrying capacity, K, must be big enough. The minimum K that will support the predator population is found by our typing (or by inspection) as

```
>> k_min = solve(sym(n2hat,3,1),'k')
```

which yields

```
k_min = d/b/a
```

So, if $K > d/(ba)$, the equilibrium at which prey and predator coexist is positive. $d/(ba)$ is identical to $\hat{N}_1$ at the interior equilibrium. So the requirement for predator-prey coexistence boils down to saying that the prey's K must be higher than its equilibrium abundance when the predator is present. Otherwise the prey population has not even the potential to be large enough to support the predators dependent on it.

To illustrate trajectories with this model, use a text editor to create and save the file `ppden.m`:

```
function ndot = ppden(t,n)
 global r a b d k;
 ndot(1,1) = r*n(1)*(k-n(1))/k-a*n(1)*n(2);
 ndot(2,1) = b*a*n(1)*n(2)-d*n(2);
```

To use this function, define some parameters as

```
>> global r a b d k;
>> r = 0.5; a = 0.01; b = 0.02; d = 0.1;
```

Let's start with a rather low K and evaluate the interior equilibrium with

```
>> k = 750;
>> n1hat_n = eval(sym(n1hat,3,1))
>> n2hat_n = eval(sym(n2hat,3,1))
```

yielding

```
n1hat_n = 500
n2hat_n = 16.6667
```

Then a figure with this equilibrium, together with six trajectories from a variety of initial conditions, is generated by our typing

```
>> figure
>> hold on
>> plot(n1hat_n,n2hat_n,'*')
>> [time n] = ode45('ppden',0,100,[n1hat_n/2; n2hat_n/2]);
>> plot(n(:,1),n(:,2))
>> [time n] = ode45('ppden',0,100,[n1hat_n/4; n2hat_n/4]);
>> plot(n(:,1),n(:,2))
>> [time n] = ode45('ppden',0,100,[2*n1hat_n; 2*n2hat_n]);
>> plot(n(:,1),n(:,2))
>> [time n] = ode45('ppden',0,100,[4*n1hat_n; 4*n2hat_n]);
>> plot(n(:,1),n(:,2))
```

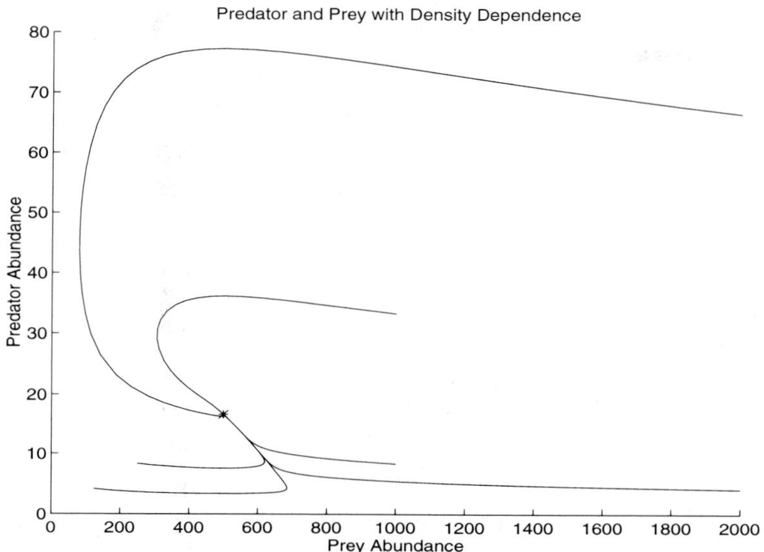

Figure 6.20: The flow of trajectories of a predator-prey model with density dependence in the prey. The prey's carrying capacity is relatively low.

```
>> [time n] = ode45('ppden',0,100,[2*n1hat_n; n2hat_n/2]);
>> plot(n(:,1),n(:,2))
>> [time n] = ode45('ppden',0,100,[4*n1hat_n; n2hat_n/4]);
>> plot(n(:,1),n(:,2))
```

and Figure 6.20 appears. The predator and prey clearly approach a stable equilibrium, and they do so with nary a hint of oscillation. To confirm this, compute the eigenvalues of the Jacobian at the equilibrium. Type

```
>> n1 = n1hat_n;
>> n2 = n2hat_n;
>> jac = [eval(diff(dn1,'n1')) eval(diff(dn1,'n2')); ...
>>        eval(diff(dn2,'n1')) eval(diff(dn2,'n2'))]
>> lambda = eig(jac)
```

This yields collectively

```
jac = -0.3333   -5.0000
       0.0033         0

lambda = -0.2721
         -0.0613
```

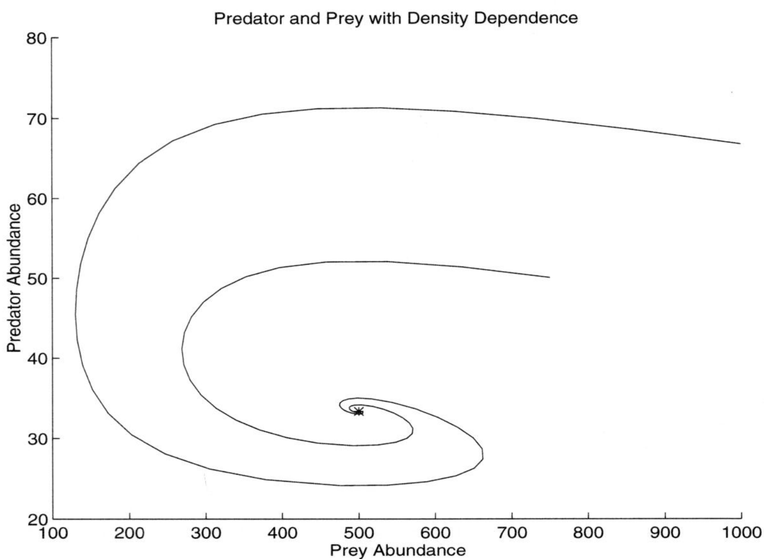

Figure 6.21: The flow of trajectories of a predator-prey model with density dependence in the prey. The prey's carrying capacity is relatively high.

These eigenvalues are both real and negative, so they indicate a stable node. They may be plotted in a figure, which we will add more to later, by our typing

```
>> fig_lambda = figure;
>> subplot(2,1,1)
>> compass(lambda)
```

Although Figure 6.20 shows no oscillations, perhaps if we increase K, oscillations will appear. After all, if K goes to infinity, then we recover the Volterra model of the previous section in which oscillations were evident. So, if we repeat the preceding steps but use a higher K, we type

```
>> k = 1500;
>> n1hat_n = eval(sym(n1hat,3,1))
n1hat_n = 500
>> n2hat_n = eval(sym(n2hat,3,1))
n2hat_n = 33.3333
```

A plot with two trajectories is then made by our typing

```
>> figure
>> hold on
```

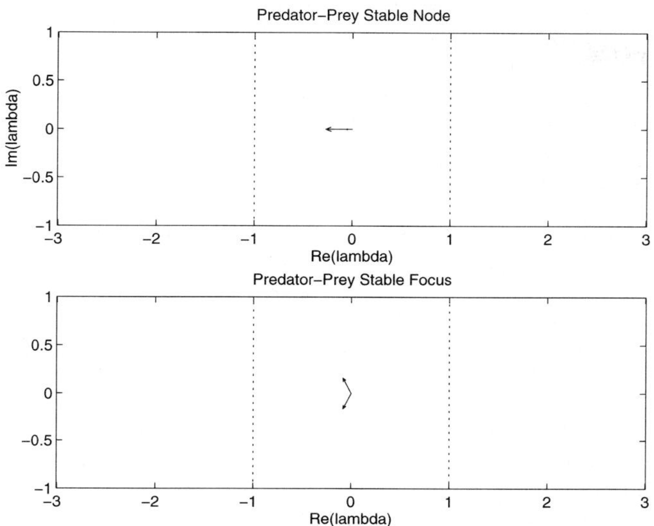

Figure 6.22: The eigenvalues of the Jacobian matrix for a predator-prey model with density dependence in the prey evaluated at the interior equilibrium. Top, a stable node with a relatively low carrying-capacity in the prey. Bottom, a stable focus with a relatively high carrying-capacity in the prey.

```
>> plot(n1hat_n,n2hat_n,'*')
>> [time n] = ode45('ppden',0,500,[1.5*n1hat_n; 1.5*n2hat_n]);
>> plot(n(:,1),n(:,2),'y')
>> [time n] = ode45('ppden',0,500,[2*n1hat_n; 2*n2hat_n]);
>> plot(n(:,1),n(:,2),'r')
```

and Figure 6.21 pops up. It shows a stable equilibrium, but the trajectories approach the equilibrium in a spiral. Mathematicians call this flow of trajectories a "stable focus." To see the eigenvalues of the Jacobian at this equilibrium, type

```
>> n1 = n1hat_n;
>> n2 = n2hat_n;
>> jac = [eval(diff(dn1,'n1')) eval(diff(dn1,'n2')); ...
>>        eval(diff(dn2,'n1')) eval(diff(dn2,'n2'))]
>> lambda = eig(jac)
```

which yields

```
jac = -0.1667   -5.0000
       0.0067         0
```

```
lambda = -0.0833 + 0.1624i
         -0.0833 - 0.1624i
```

The eigenvalues are now complex numbers, whose real part is negative. A focus results when a pair of eigenvalues are complex, and the focus is stable when the real part of the eigenvalues is negative. The focus is unstable if the real part of the eigenvalues is positive. These are added to make Figure 6.22 by our typing

```
>> figure(fig_lambda)
>> subplot(2,1,2)
>> compass(lambda)
```

All in all, adding density dependence in the prey has a large effect on a predator-prey system. If the prey's K is low enough, then its productivity is too low to support the predator population, and as K is raised, coexistence becomes possible. If K is raised high enough, then the inherent tendency of predator and prey to oscillate is revealed as an oscillatory approach to the equilibrium.

The Volterra principle, that a general insecticide clobbers the prey more than the predator, is still more or less true even though density dependence in the prey has been added. The prey's abundance strongly depends on d, the predator's death rate, and the predator's abundance strongly depends on r, the prey's intrinsic rate of increase. So a pesticide that simultaneously increases d and lowers r will cause a net increase in the prey and a net decline in the predator. And if the prey is a crop pest, then all that good money put into buying pesticide is wasted.

6.4.3 Tricks of Stability Analysis with Two Variables

The interaction between predators and prey can be foretold by an examination of the eigenvalues of the Jacobian matrix evaluated at the equilibrium point, as you now know only too well. As we've just seen, the tendency for predator and prey populations to oscillate is expressed when the eigenvalues are complex, and is not expressed when the eigenvalues are real. And the equilibrium is stable if both eigenvalues are negative or have negative real parts. Because so much can be learned from the eigenvalues, it would be nice if there were a quick and dirty way to determine, just by looking at the Jacobian matrix, if the eigenvalues have negative real parts and if they are complex. Well, when two species are being considered, there are some tricks that make matters easier, and here they are.

To begin, we can get the Jacobian matrix symbolically rather quickly: instead of computing it element by element as we have until now, we can use the `jacobian` command. `jacobian` must be called with a symbolic column vector for the model's differential equations, and a symbolic row vector for the model's variables. So, define `dn1`, `dn2`, `n1hat`, and `n2hat`—we'll just continue to use the definitions of these from the previous section. Next, type

```
>> jacob = jacobian(sym([dn1 '; ' dn2]), sym('[n1,n2]'))
```

which yields

```
jacob = [r*(k-n1)/k-r*n1/k-a*n2,     -a*n1]
        [                 b*a*n2, b*a*n1-d]
```

The only tricky part here is that the column matrix of formulas for the model's differential equations was assembled with the fact that a symbolic expression is a string[5]. `[dn1 '; ' dn2]` manufactures a large string built up from three smaller strings, `dn1`, `'; '`, and `dn2`; the large string is then passed to `sym`, which makes the desired symbolic column vector. The other argument, `sym('[n1,n2]')`, is the usual way to generate a symbolic row vector. Now that we have the symbolic Jacobian, we have to substitute the equilibrium values for `n1` and `n2`, which is done with

```
>> jacob = subs(jacob,sym(n1hat,3,1),'n1');
>> jacob = subs(jacob,sym(n2hat,3,1),'n2');
>> jacob = simple(jacob)
```

yielding

```
jacob = [   -d*r/b/a/k, -d/b]
        [b*r-1/a*r/k*d,    0]
```

So this is the symbolic matrix whose eigenvalues we need to examine.

Next, let's see how the eigenvalues of a general 2×2 matrix are related to its elements. Define a general matrix as

```
>> matrix = sym('[a b; c d]');
```

and ask MATLAB to give us its eigenvalues symbolically. We use the `eigensys` command and type

```
>> lambdas = eigensys(matrix)
```

yielding

```
lambdas = [1/2*a+1/2*d+1/2*(a^2-2*a*d+d^2+4*b*c)^(1/2)]
          [1/2*a+1/2*d-1/2*(a^2-2*a*d+d^2+4*b*c)^(1/2)]
```

By itself, this looks pretty messy. But matters are about to improve. The "trace" of a matrix is defined as the sum of its diagonal elements. The trace is computed symbolically as

```
>> trace = symop(sym(matrix,1,1),'+',sym(matrix,2,2))
```

yielding

```
trace = a+d
```

[5]The differential equations *must* be assembled exactly as illustrated, or otherwise the string won't make sense to MATLAB's symbolic toolkit.

The "determinant" of a matrix is a formula concocted to be zero if any two or more of the rows or columns are the same except for a multiplicative constant. To give the formula for the determinant of a matrix, MATLAB uses the `determ` command, as in

```
>> determinant = determ(matrix)
```

which yields the formula for the determinant of a 2×2 matrix as

```
determinant = a*d-b*c
```

So by inspecting a 2×2 matrix, we see that the trace is just the sum of elements on the main diagonal, and the determinant is the product of the elements going from top left to lower right minus the product of the elements going from lower left to top right. Now return to the formula for the eigenvalues. We can write these as

$$\begin{aligned} \lambda_1 &= (\text{trace} + \sqrt{(\text{trace})^2 - 4(\text{determinant})})/2 \\ \lambda_2 &= (\text{trace} - \sqrt{(\text{trace})^2 - 4(\text{determinant})})/2 \end{aligned}$$

Check it out. You'll see that this is the same as the formula returned by the `eigensys` command. Whether the eigenvalues are complex is controlled by the sign of the expression whose square root is being taken. This expression is called the "discriminant" and is defined in MATLAB by

```
>> discriminant = symop(trace,'^','2','-','4','*',determinant)
```

yielding

```
discriminant = (a+d)^2-4*a*d+4*b*c
```

Now look at the various possible combinations of the trace and determinant, and see how the eigenvalues are affected by each combination. Here are the possibilities, arranged in order of an increasing determinant:

(determinant) < 0. Here the discriminant is positive, so the eigenvalues are real. Furthermore, one must be positive and the other negative, regardless of the sign of the trace. Therefore, this case necessarily describes a saddle, and is unstable.

$0 <$ (determinant) $<$ (trace)$^2/4$. Here the discriminant is again positive, so the eigenvalues are again real. Both eigenvalues have the same sign, and the pattern is therefore a node. The two possibilities are

- (trace) > 0. Both eigenvalues are positive, so the equilibrium is an unstable node.
- (trace) < 0. Both eigenvalues are negative, so the equilibrium is a stable node.

(trace)$^2/4 <$ (determinant). Now the discriminant is negative, so the eigenvalues are complex. The pattern is therefore a focus. The real part is the trace of the determinant. The two possibilities are (trace) > 0. The real part is positive, so the equilibrium is an unstable focus.

(trace) < 0. The real part is negative, so the equilibrium is a stable focus.

All in all, *if either the trace is positive or the determinant is negative, then the equilibrium is unstable.* Conversely, *if both the trace is negative and the determinant is positive, then the equilibrium is stable.*

Let's apply these criteria to the predator-prey model with density dependence in the prey, and see what happens. Type

```
>> trace = simple(symop(sym(jacob,1,1),'+',sym(jacob,2,2)))
```

yielding

```
trace = -d*r/b/a/k
```

which is clearly negative. Next the determinant is found by our typing

```
>> determinant = simple(determ(jacob))
```

yielding

```
determinant = d*r-d^2/b/a*r/k
```

which is positive if $K > d/(ba)$. Recall that K must be greater than $d/(ba)$ for the interior equilibrium to be positive. Therefore, if the interior equilibrium is positive, it is stable.

The value of K that marks the transition from a stable node to a stable focus is the K that makes the discriminant equal to zero. MATLAB finds the discriminant upon our typing

```
>> discriminant = simple(symop(trace,'^','2','-','4','*',determinant))
```

which yields

```
discriminant = d^2*r^2/b^2/a^2/k^2-4*d*r+4*d^2/b/a*r/k
```

To find the value(s) of K that make this zero, type

```
>> k_nodefocus = solve(discriminant,'k')
```

yielding

```
k_nodefocus = [1/8/b^2/a^2*(4*d*b*a+4*d^(1/2)*b*a*(d+r)^(1/2))]
              [1/8/b^2/a^2*(4*d*b*a-4*d^(1/2)*b*a*(d+r)^(1/2))]
```

The first of these roots turns out to be the relevant one. So, to evaluate it numerically, type

```
>> r = 0.5; a = 0.01; b = 0.02; d = 0.1;
>> k_nodefocus_n = eval(sym(k_nodefocus,1,1))
```

yielding

```
k_nodefocus_n = 862.3724
```

Recall that Figure 6.20 was made with $K = 750$ and Figure 6.21 with $K = 1500$. This analysis shows why these values work, because 750 is less than 862 and therefore leads to a stable node, whereas 1500 is greater than 862 and therefore leads to a stable focus.

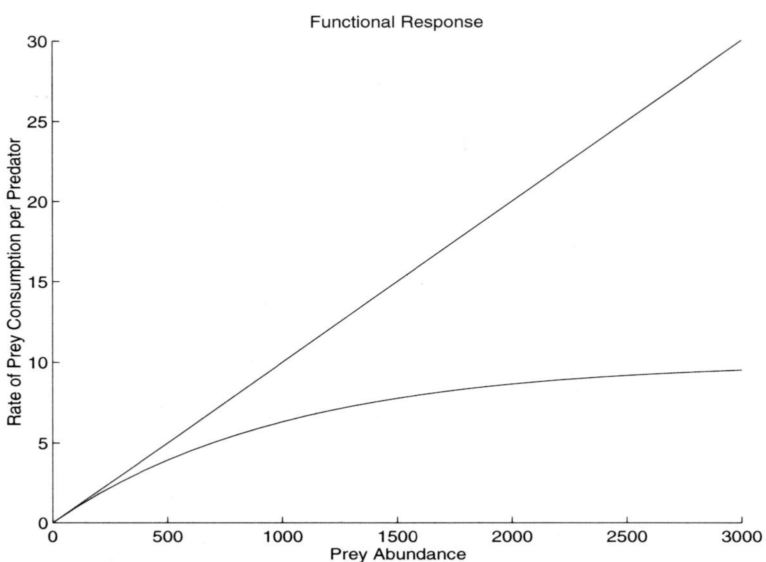

Figure 6.23: A linear (Type-1) and asymptotic (Type-2) functional response.

6.4.4 Adding Satiation to Predators

Predators can gorge themselves on prey only up to a limit. Predators cease to control the abundance of prey when they are satiated, and this fact may greatly affect predator-prey population dynamics. Predator satiation is added to the predator-prey model by a change in the predation term.

The predation term in the predator-prey models we've looked at so far is aN_1N_2. We can view this term as aN_1 times N_2, which may be written with parentheses, as $(aN_1)N_2$. The expression in parentheses is called the "functional response" of a predator and describes the rate at which an average predator consumes prey as a function of the prey abundance. Clearly (aN_1) is a straight line with slope a when plotted against N_1, as illustrated in Figure 6.23. This linear functional response is also called a Type-1 functional response.

Satiation is added to a predator-prey model with the assumption that the predator's consumption rate levels off at some maximum as the prey abundance increases. A functional response with this property is

$$c(1 - e^{-aN_1/c})$$

and is called a Type-2 functional response. It starts out like the linear functional response with a slope of a, but then levels off at an asymptote of c. These two functional responses are graphed in Figure 6.23; to create it, type

```
>> type1 = 'a*n1'
```

```
>> type2 = 'c*(1-exp(-a*n1/c))'
>> a = 0.01; c = 10;
>> n1 = 0:100:3000;
>> figure
>> hold on
>> plot(n1,eval(type1))
>> plot(n1,eval(type2))
```

The predator birth term we've looked at so far is baN_1N_2. This can be viewed as $(b(aN_1))N_2$. Here $(b(aN_1))$ is called the "numerical response" of the predator. It describes the birth rate of predators as a function of prey abundance. Typically, the numerical response is assumed to be a constant, b, times the functional response where b describes how many prey are needed to produce a predator. So, if an asymptotic functional response is used instead of a linear functional response, the numerical response is just

$$bc(1 - e^{-aN_1/c})$$

Predator satiation therefore leads to the following predator-prey model:

$$\begin{aligned}\frac{dN_1}{dt} &= rN_1(K - N_1)/K - c(1 - e^{-aN_1/c})N_2 \\ \frac{dN_2}{dt} &= bc(1 - e^{-aN_1/c})N_2 - dN_2\end{aligned}$$

To see what the addition of predator satiation does to predator-prey population dynamics, alert MATLAB to the new model. Type

```
>> dn1 = 'r*n1*(k-n1)/k-c*(1-exp(-a*n1/c))*n2'
>> dn2 = 'b*c*(1-exp(-a*n1/c))*n2-d*n2'
```

The equilibria are found, as usual, by our typing

```
>> [n1hat,n2hat] = solve(dn1,dn2,'n1,n2')
```

yielding

```
n1hat = [                     0]
        [                     k]
        [-log((b*c-d)/b/c)/a*c]

n2hat = [                                                        0]
        [                                                        0]
        [-r*log((b*c-d)/b/c)*c*b*(k*a+log((b*c-d)/b/c)*c)/k/a^2/d]
```

We again find that the prey may not support the predator. To find out how large the prey's carrying capacity must be, type

```
>> k_min = solve(sym(n2hat,3,1),'k')
```

yielding

```
k_min = -log((b*c-d)/b/c)/a*c
```

This minimum K is identical to $\hat{N}_1$ at the interior equilibrium. So again the requirement for predator-prey coexistence boils down to saying that the prey's K must be higher than its equilibrium abundance when the predator is present. Otherwise the prey population is not potentially large enough to support the predators.

To find out if the equilibrium is stable, we can apply what we've learned from the previous two sections. When K is a little bit higher than `k_min`, the equilibrium is a stable node, with two real and negative eigenvalues to the Jacobian. As K is increases, the discriminant of the 2×2 Jacobian goes from positive to negative, and is zero at the K that marks the transition from a stable node to a stable focus. As K is increased still further, the trace changes from negative to positive, and is zero at the K that marks a transition from a stable focus to an unstable focus. As we'll see, when the equilibrium is an unstable focus, it is surrounded by a closed loop, called a limit cycle, that represents a predator-prey oscillation. To derive all this, first compute formulas for the trace, determinant, and discriminant of the Jacobian. Type

```
>> jacob = jacobian(sym([dn1 ';  ' dn2]), sym('[n1,n2]'));
>> jacob = subs(jacob,sym(n1hat,3,1),'n1');
>> jacob = subs(jacob,sym(n2hat,3,1),'n2');
>> jacob = simple(jacob);
>> trace = simple(symop(sym(jacob,1,1),'+',sym(jacob,2,2)));
>> determinant = simple(determ(jacob));
>> discriminant = simple(symop(trace,'^','2','-','4','*',determinant));
```

So, to find the K for the transition from node to focus, type

```
>> k_nodefocus = solve(discriminant,'k');
```

This yields two roots, the second of which turns out to be the relevant one. Next, the K for the transition from a stable focus to an unstable focus, combined with a limit cycle, is found by our typing

```
>>k_hopf = solve(trace,'k');
```

When a parameter such as K is tuned, the formation of a permanent oscillation arising out of a previously stable focus is called a "Hopf bifurcation[6]." I haven't bothered to print out the explicit formulas for the transition K's because they're too messy.

To generate illustrations of the predator-prey populations for the cases of a stable node, stable focus, and unstable focus/limit cycle, let's evaluate the transition K, and then pick some K's accordingly. For example, type

```
>> global r a b d c k;
>> r = 0.5; a = 0.01; b = 0.02; d = 0.1; c = 10;
```

[6] Recall another example of this with three competing species in Figures 6.12-13.

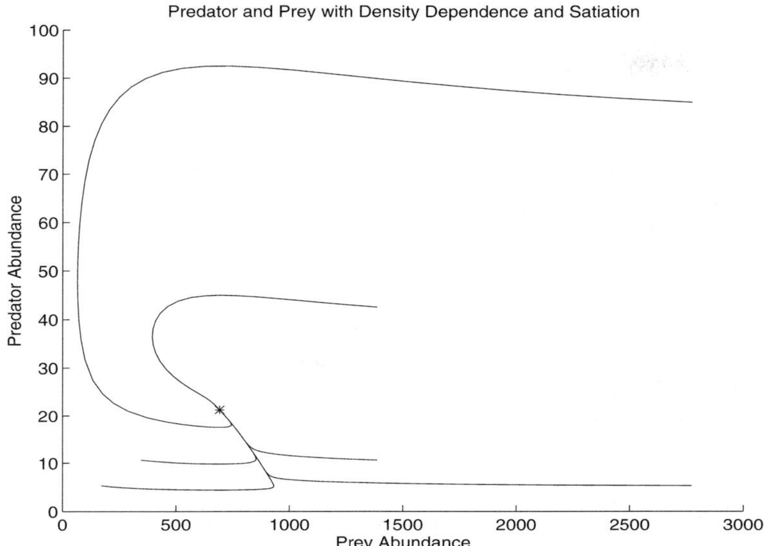

Figure 6.24: The flow of trajectories in a predator-prey model with density dependence in the prey and satiation in the predators. K in the prey is 1000, and the pattern is a stable node.

Now type

```
>> k_min_n = eval(k_min)
>> k_nodefocus_n = eval(sym(k_nodefocus,2,1))
>> k_hopf_n = eval(k_hopf)
```

yielding collectively

```
k_min_n = 693.1472
k_nodefocus_n = 1.1617e+03
k_hopf_n = 2.9520e+03
```

So let's select K's at 1000, 2000, and 4000, and we should see examples of a stable node, stable focus and unstable focus/limit cycle, respectively.

To generate numerical trajectories, use a text editor to create and save the following file as `ppdensat.m`:

```
function ndot = ppdensat(t,n)
 global r a b d c k;
 ndot(1,1) = r*n(1)*(k-n(1))/k-c*(1-exp(-a*n(1)/c))*n(2);
 ndot(2,1) = b*c*(1-exp(-a*n(1)/c))*n(2)-d*n(2);
```

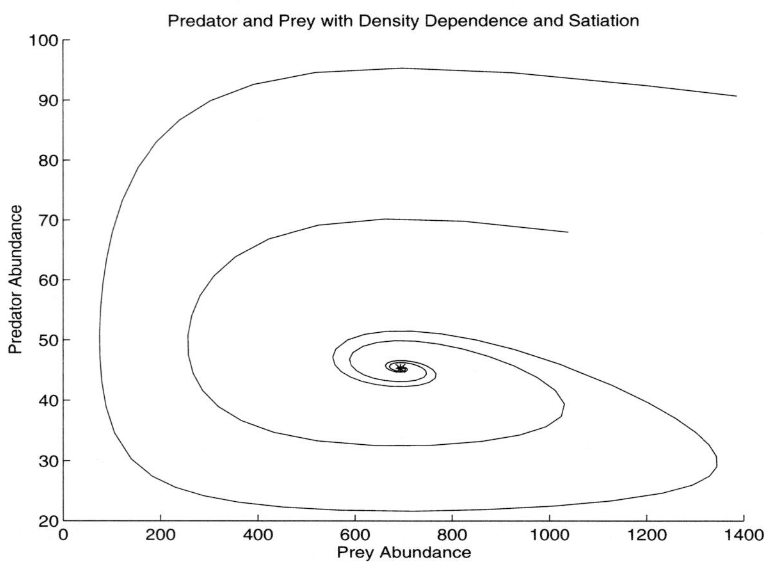

Figure 6.25: The flow of trajectories in a predator-prey model with density dependence in the prey and satiation in the predators. K in the prey is 2000, and the pattern is a stable focus.

Now let's use this function, together with one of MATLAB's differential-equation-solvers, and a K of 1000. Figure 6.24 then pops up upon our typing

```
>> k = 1000;
>> n1hat_n = eval(sym(n1hat,3,1));
>> n2hat_n = eval(sym(n2hat,3,1));
>> figure
>> hold on
>> plot(n1hat_n,n2hat_n,'*')
>> [time n] = ode45('ppdensat',0,100,[n1hat_n/2; n2hat_n/2]);
>> plot(n(:,1),n(:,2))
>> [time n] = ode45('ppdensat',0,100,[n1hat_n/4; n2hat_n/4]);
>> plot(n(:,1),n(:,2))
>> [time n] = ode45('ppdensat',0,100,[2*n1hat_n; 2*n2hat_n]);
>> plot(n(:,1),n(:,2))
>> [time n] = ode45('ppdensat',0,100,[4*n1hat_n; 4*n2hat_n]);
>> plot(n(:,1),n(:,2))
>> [time n] = ode45('ppdensat',0,100,[2*n1hat_n; n2hat_n/2]);
>> plot(n(:,1),n(:,2))
>> [time n] = ode45('ppdensat',0,100,[4*n1hat_n; n2hat_n/4]);
```

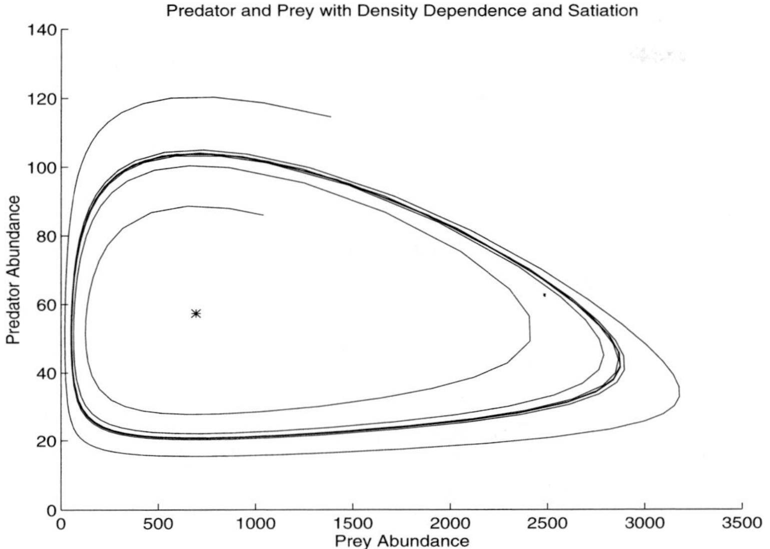

Figure 6.26: The flow of trajectories in a predator-prey model with density dependence in the prey and satiation in the predators. K in the prey is 4000, and the equilibrium is an unstable stable focus surrounded by a limit cycle. The transition from the previous figure that results as K is increased, is a Hopf bifurcation.

```
>> plot(n(:,1),n(:,2))
```

The numerical values of the equilibrium, the Jacobian, and its eigenvalues, are obtained by our typing

```
>> n1 = n1hat_n
>> n2 = n2hat_n
>> jac = [eval(diff(dn1,'n1')) eval(diff(dn1,'n2')); ...
>>        eval(diff(dn2,'n1')) eval(diff(dn2,'n2'))]
>> lambda = eig(jac)
```

yielding collectively

```
n1 = 693.1472
n2 = 21.2694
jac = -0.2995   -5.0000
       0.0021         0
lambda = -0.2583
         -0.0412
```

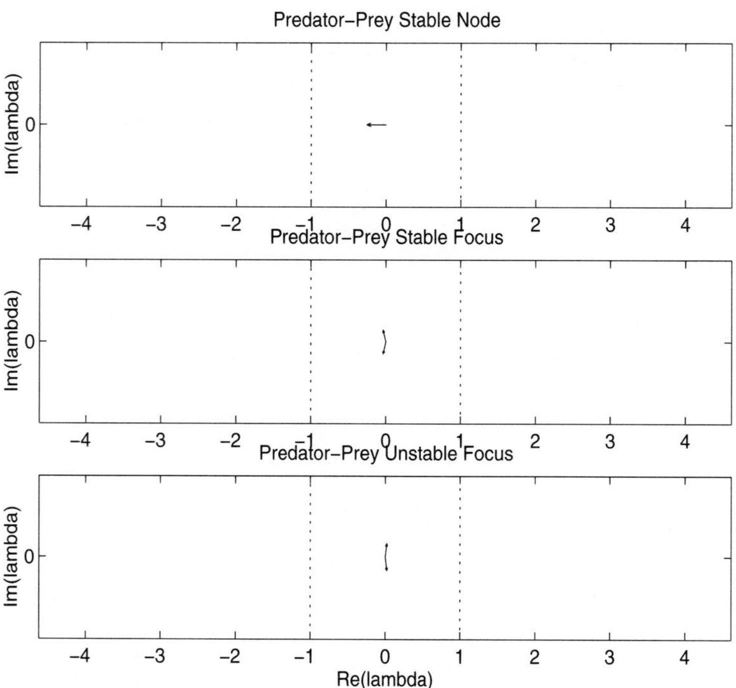

Figure 6.27: The eigenvalues of the Jacobian for a predator-prey model with density dependence in the prey and satiation in the predators. Top, a stable node with a relatively low K in the prey. Middle, a stable focus with an intermediate K in the prey. Bottom, an unstable focus with a relatively high K in the prey.

The eigenvalues are clearly negative, and Figure 6.24 shows a stable node, as predicted. A compass plot of the eigenvalues is made with

```
>> fig_lambda = figure;
>> subplot(3,1,1)
>> compass(lambda)
```

Next, let's try a K of 2000. Type

```
>> k = 2000;
>> n1hat_n = eval(sym(n1hat,3,1));
>> n2hat_n = eval(sym(n2hat,3,1));
>> figure
```

```
>> hold on
>> plot(n1hat_n,n2hat_n,'*')
>> [time n] = ode45('ppdensat',0,500,[1.5*n1hat_n; 1.5*n2hat_n]);
>> plot(n(:,1),n(:,2),'y')
>> [time n] = ode45('ppdensat',0,500,[2*n1hat_n; 2*n2hat_n]);
>> plot(n(:,1),n(:,2),'r')
```

and Figure 6.25 pops up. The numerical values of the equilibrium, Jacobian, and its eigenvalues are again generated when we type

```
>> n1 = n1hat_n
>> n2 = n2hat_n
>> jac = [eval(diff(dn1,'n1')) eval(diff(dn1,'n2')); ...
>>        eval(diff(dn2,'n1')) eval(diff(dn2,'n2'))]
>> lambda = eig(jac)
```

yielding collectively

```
n1 = 693.1472
n2 = 45.2921
jac = -0.0730   -5.0000
       0.0045         0
lambda = -0.0365 + 0.1460i
         -0.0365 - 0.1460i
```

The eigenvalues are now complex, and have negative real parts. So the pattern of trajectories should be a stable focus, and Figure 6.25 in fact reveals this focus. These eigenvalues are plotted with

```
>> figure(fig_lambda)
>> subplot(3,1,2)
>> compass(lambda)
```

Both Figure 6.24 and 6.25 reveal the same patterns as Figures 6.20 and 6.21. Now we come to the new feature that results from adding the assumption of predator satiation. With a K of 4000, the equilibrium is predicted to be an unstable focus. So to see what happens, type

```
>> k = 4000;
>> n1hat_n = eval(sym(n1hat,3,1));
>> n2hat_n = eval(sym(n2hat,3,1));
>> figure
>> hold on
>> plot(n1hat_n,n2hat_n,'*')
>> [time n] = ode45('ppdensat',0,500,[1.5*n1hat_n; 1.5*n2hat_n]);
>> plot(n(:,1),n(:,2),'y')
>> [time n] = ode45('ppdensat',0,500,[2*n1hat_n; 2*n2hat_n]);
>> plot(n(:,1),n(:,2),'r')
```

The numerical values of the equilibria, Jacobian, and its eigenvalues are again produced by our typing

```
>> n1 = n1hat_n
>> n2 = n2hat_n
>> jac = [eval(diff(dn1,'n1')) eval(diff(dn1,'n2')); ...
>>        eval(diff(dn2,'n1')) eval(diff(dn2,'n2'))]
>> lambda = eig(jac)
```

This collectively yields

```
n1 = 693.1472
n2 = 57.3034
jac = 0.0402   -5.0000
      0.0057         0
lambda = 0.0201 + 0.1681i
         0.0201 - 0.1681i
```

The eigenvalues are complex with positive real parts, so the equilibrium is an unstable focus, and trajectories started near the equilibrium wind outwards and away from it, as shown in Figure 6.26. Where do the trajectories end up? At a particular closed loop. It is called a "limit cycle" because this particular closed loop is attained by all trajectories in the limit as time becomes very large. That there is only one particular loop to which all trajectories converge differs fundamentally from the pattern called a "center" found in the Volterra model, which consists of an infinity of loops.

The eigenvalues for this case are graphed with

```
>> figure(fig_lambda)
>> subplot(3,1,3)
>> compass(lambda)
```

This produces Figure 6.27, which shows all three cases for comparison.

These basic predator-prey models show that the carrying capacity of the prey must be large enough to support a predator population, and that the predator and prey populations have an inherent tendency to oscillate. This oscillation is expressed only if the carrying capacity in the prey is high enough, because density dependence tends to stabilize the oscillations, and the density dependence is weakened if the carrying capacity is high relative to the prey's equilibrium abundance in the presence of the predator. The models also show that the abundance of the prey is largely controlled by properties of the predator, and the abundance of the predator is largely controlled by properties of the prey. Moreover, simultaneously increasing the death rate of the predators and lowering the intrinsic rate of increase in the prey leads to a net increase in the prey.

6.5 Host-Parasitoid Interaction

As we pan our theoretical camera from lions, tigers, and eagles to insects, and zoom in on wasps and other insect predators, we see a biology sufficiently different from that envisioned in the generic predator-prey models that a separate literature has been developed especially for them. Some insects are predators in the usual sense that they find prey, attack and kill them, and either eat them directly or bring them to nests as food for their young. But many other insect predators are called "parasitoids" because they have a lifestyle intermediate between a parasite and a usual predator. A parasitoid searches for prey and then lays its eggs on, or in, the prey. The prey is called a "host" and the eggs laid on, or in it, hatch into larvae. The larvae then slowly consume selected parts of its host, while leaving the host alive, although probably somewhat weakened. As the larvae mature, they finally consume their host's vital organs, the host dies, and the larvae metamorphose and emerge as adults. So, predation is brought about by an adult parasitoid who locates a host in which to lay eggs, and a host who escapes detection escapes predation. Therefore, the population dynamics of host and parasitoid has long been concerned with the searching behavior of the parasitoids relative to the spatial distribution of the hosts. If the hosts are themselves herbivores, then their distribution more or less coincides with the distribution of their food plant. Hence, the spatial distribution of a plant indirectly influences the predation pressure on its own herbivores.

The literature on the host-parasitoid interaction resembles that on generic predator-prey theory because it also starts out with a famous model, to which many authors have made additions. These additions concern what might be termed the "escapement function," which plays a role similar to functional response in generic predator-prey theory. The host-escapement function is the probability that a host escapes detection as a function of the number of parasitoids that are looking for it. Host-parasitoid models are written in discrete time, because life cycles of both host and parasitoid are usually synchronized with the seasons. So the time period over which the host must escape detection is one time step in the model; it is typically a season length.

Two functional forms for the host-escapement function that have been widely used are called the Poisson and negative binomial, named after statistical distributions. These are defined as

$$
\begin{aligned}
F(P) &= e^{-aP} && \text{Poisson} \\
F(P) &= \frac{1}{[1+(a/k)P]^k} && \text{Negative Binominal}
\end{aligned}
$$

where P is the number of parasitoids at the start of the season, $F(P)$ is the probability of a host's failing to be detected during the season, and a and k are parameters. k is known as the "clumping parameter." It's possible to devise statistical justifications for these functions, but for present purposes it's sufficient to show how they differ from each other, because these are reasonable functions irrespective of their connections with statistical distributions.

To define the escapement functions for MATLAB, type

```
poisson = 'exp(-a*p)'
negbinomial = '(1+a*p/k)^(-k)'
```

To graph them, go on to type

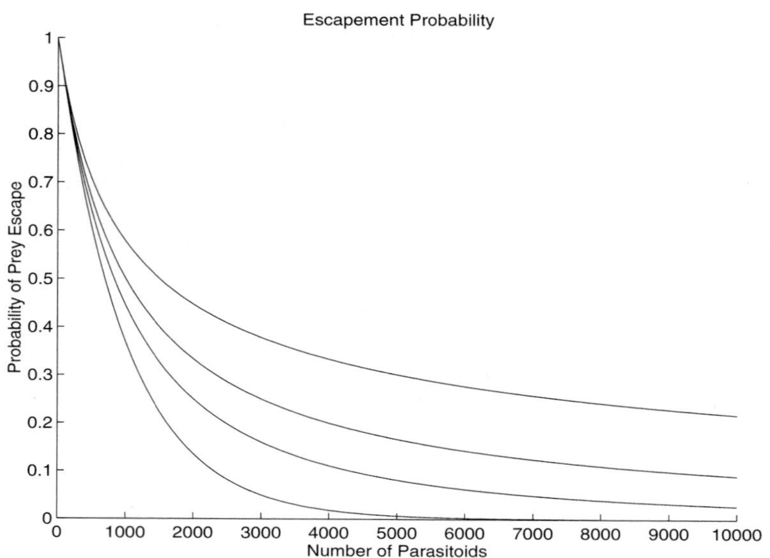

Figure 6.28: Host-escapement functions for host-parasitoid population-dynamic models. The bottom curve is the Poisson version, and the next three curves are the negative binomial forms, with the clumping parameter, k, equal to 2, 1, and 1/2, from lower to upper.

```
>> a = .001;
>> p=0:100:10000;
>> figure
>> hold on
>> plot(p,eval(sym2ara(poisson)),'y');
>> k=2;
>> plot(p,eval(sym2ara(negbinomial)),'r');
>> k=1;
>> plot(p,eval(sym2ara(negbinomial)),'g');
>> k=1/2;
>> plot(p,eval(sym2ara(negbinomial)),'c');
```

and Figure 6.28 appears. The lowest curve is for the Poisson version of an escapement function. The probability of a host's avoiding detection declines exponentially with the number of parasitoids. This means that as the parasitoid population increases, all hosts will quickly be found. The negative binomial forms are more gentle. The next lowest curve is for $k = 2$, and shows that host detection is less effective than in the Poisson form. With $k = 1$, which is the next curve up, the hosts are even less likely to be detected, and with $k = 1/2$, which is the top curve, the hosts are least likely to be detected for a given number of parasitoids. The interpretation of the negative binomial with $k < 1$, is that the host distribution is spread out over various patches of plants, so that some hosts have lots of company in their patch and others have nearly a patch to themselves. If the parasitoid's searching effort is concentrated on places where the hosts are most abundant, the parasitoids will ignore plant patches where the hosts are relatively rare. Therefore, the particular hosts who are lucky enough to have almost a whole patch of food plants to themselves are not going to be found by parasitoids, and have in effect a refuge from predation. Therefore, the parasitoids can't hunt them to extinction. Let's see then how these escapement functions influence the dynamics of host and parasitoid populations.

6.5.1 The Nicholson-Bailey Model

The general form of host-parasitoid models is

$$\begin{aligned} N_{t+1} &= RN_tF(P_t) \\ P_{t+1} &= cN_t[1 - F(P_t)] \end{aligned}$$

Here R is the factor of geometric increase in the hosts, N. In the absence of the parasitoid, the hosts simply grow geometrically, and increase by a factor R during each time step. But in the presence of the parasitoid, only a fraction, $F(P)$, escape detection, and these are the only ones that get to live and reproduce. Meanwhile, for the parasitoids, $[1 - F(P_t)]$ is the probability a host is discovered, and each discovered host gives rise to c new parasitoids for the next time step. From a biological standpoint, this model is conceptually equivalent to the Volterra model in that there is no density dependence in either host or parasitoid population, and the focus is completely on the interaction terms, $NF(P)$ and $N[1 - F(P)]$.

The first model of this form to be analyzed is known as the Nicholson-Bailey model. It was analyzed in the early 1930's and uses the Poisson form for the host-escapement function. To acquaint MATLAB with it, type

```
>> nprime = 'r*n*exp(-a*p)'
>> pprime = 'c*n*(1-exp(-a*p))'
```

The equilibria are then found by our typing

```
>> [nhat,phat] = solve(symop(nprime,'-','n'),symop(pprime,'-','p'),'n,p');
>> nhat = simple(nhat)
>> phat = simple(phat)
```

which yields collectively

```
nhat = [                    0]
       [log(r)*r/c/a/(r-1)]

phat = [       0]
       [log(r)/a]
```

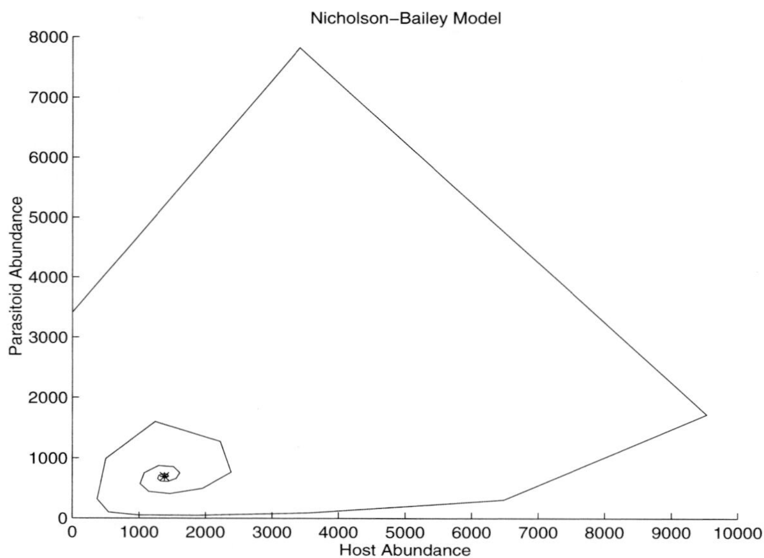

Figure 6.29: Trajectories of the Nicholson-Bailey model for host-parasitoid dynamics.

There are two equilibria, one for total extinction and the other for coexistence of host and parasitoid. If the parasitoids are extinct, the host increase to infinity, which is not counted as an equilibrium. The coexistence equilibrium is positive if the geometric factor of increase in the hosts, R, is greater than 1. If R is less than 1, the hosts will go extinct regardless of the parasitoids, so this is a trivial requirement. The number of hosts is reduced by the parasitoid if the parasitoid is a good searcher (high a), and if the parasitoid can produce lots of new parasitoids from each host that is found (high c). So, for a parasitoid to act as a good biological control of a host that is a pest to humans, it must be a good searcher and good reproducer. Whoa, you may be saying now. We haven't checked whether this equilibrium is stable. Well, it's not, and you're right to demand that we look into this a little more carefully.

Let's create a function to iterate these equations and then graph the results. Using a text editor, type and save the following as `nichbail.m`

```
function [n,p]=nichbail(runlen,no,po)
 global r a c
 n = [no]; p = [po];
 for t = 1:runlen
   nprime = r*n(t)*exp(-a*p(t));
   pprime = c*n(t)*(1-exp(-a*p(t)));
   if nprime < 0 nprime = 0; end;
   if pprime < 0 pprime = 0; end;
```

```
  n = [n nprime];
  p = [p pprime];
end
```

This function, like that to iterate the discrete-time logistic equation in Chapter 4, `logist_d`, takes as input the number of steps to be iterated, `runlen`, and the initial conditions for host and parasitoid, `no` and `po`. It returns vectors for the host and parasitoid through time. It also checks if a population size becomes negative along the way, and restores any negative abundance to zero.

To use this function, let's start off a trajectory near the equilibrium. To calculate the equilibrium, type

```
>> global r a c
>> r = 2; a = .001; c = 1;
>> nhat_n = eval(sym(nhat,2,1))
>> phat_n = eval(sym(phat,2,1))
```

yielding

```
nhat_n = 1.3863e+03
phat_n = 693.1472
```

Then Figure 6.29 is made by our typing

```
>> figure
>> hold on
>> plot(nhat_n,phat_n,'*')
>> [n p] = nichbail(50,1.01*nhat_n,1.01*phat_n);
>> plot(n,p)
```

The figure shows that the equilibrium is indeed unstable. The trajectories quickly spiral counterclockwise, with the parasitoid going extinct, and the host population then expanding to infinity.

The predictions of the Nicholson-Bailey model have been most perplexing, because parasitoids do in fact persist with their hosts in nature for long times, perhaps enduring continuing fluctuations, but persisting nonetheless. In the lab though, host-parasitoid oscillations leading to parasitoid extinction have been observed. The need to explain the long-term coexistence of parasitoids and their hosts in nature has therefore motivated changes or extensions to the Nicholson-Bailey model; these are intended to confer stability to the equilibrium. One could add density dependence to the host or parasitoid, and expect this to stabilize the dynamics. However, insect ecologists have for the most part preferred to explore various assumptions about how the parasitoids detect their hosts, and to look within the host-parasitoid interaction itself, for an explanation of stability.

6.5.2 Stability Analysis in Discrete Time

Because so much of the discussion of the host-parasitoid interaction involves the criterion for stability of an equilibrium, it's worth mentioning briefly how the criterion for stability of an equilibrium is formulated in multispecies discrete-time models. Recall that when we analyzed the stability of a single species model of the form $N_{t+1} = G(N_t)$, we computed dG/dN and evaluated this derivative at the equilibrium being analyzed, $\hat{N}$. If this derivative, called λ, is between -1 and 1, then the equilibrium is stable. Now, considering two species, we'll need to take all the possible derivatives, and let's arrange the derivatives in a matrix so that the first row is the derivative of $N_{1,t+1}$ with respect to N_1 and N_2, and the second row the derivative of $N_{2,t+1}$ with respect to N_1 and N_2. Yes, this is the Jacobian again, as I'll bet you guessed. The eigenvalues of the Jacobian, its two λ's, hold the key to the equilibrium's stability. Now, we can't require that each eigenvalue lie between -1 and 1, because they may be complex numbers, so we require instead that, *if the eigenvalues of the Jacobian of a multispecies discrete-time dynamical model lie in the unit circle centered on the origin, then the equilibrium is stable.* The way to tell if a complex number lies within the unit circle centered around the origin is to see if its absolute value is less than one[7].

To see if the equilibrium is predicted to be unstable, let's apply this criterion to the Nicholson-Bailey model. Let MATLAB compute the jacobian and its eigenvalues symbolically; type

```
>> jacob = jacobian(sym([nprime '; ' pprime]), sym('[n,p]'));
>> jacob = simple(subs(jacob,sym(nhat,2,1),'n'));
>> jacob = simple(subs(jacob,sym(phat,2,1),'p'));
>> lambdas = simple(eigensys(jacob));
```

The formulas for the λ's are pretty messy, but they turn out to depend only on R, the geometric growth factor in the hosts. So, let's plot them as a function of R and see what they look like. Typing

```
>> r = 1.1:.1:10;
>> lambda_fig = figure;
>> hold on
>> plot(r,abs(eval(sym2ara(sym(lambdas,1,1)))),'y')
>> plot(r,abs(eval(sym2ara(sym(lambdas,2,1)))),'y')
```

brings up Figure 6.30[8]. The top curve is the one for the Nicholson-Bailey model. Notice that it is above 1, and this signifies that the equilibrium is unstable.

---

[7]This requirement for stability is derived for two species in the same way as it was for one species in Chapter 4. Specifically, we consider small deviations from the equilibrium, and expand both $N_{1,t+1}$ and $N_{2,t+1}$ as Taylor series around the equilibrium point $\hat{N_1}$ and $\hat{N_2}$. Then, keeping only first order terms in the Taylor series, we obtain a pair of linear equations with constant coefficients for the deviations from equilibrium. The deviations from equilibrium are observed to go to zero as time tends to infinity, provided that the eigenvalues of the Jacobian are less than one in absolute value.

[8]Two `plot` statements are used because if the eigenvalues are real, two lines will be evident; whereas if they are complex conjugates, the two lines will coincide. Thus, seeing whether there are one or two lines will indicate whether the eigenvalues are complex or real.

Let's turn now to the models using the negative binomial host escapement function, in which the hosts have more of a chance of avoiding detection than in the Nicholson-Bailey model.

6.5.3 The Negative-Binomial Model

The negative-binomial-model host-escapement function was introduced by May in 1978 as a phenomenological description of host-parasitoid dynamics. To tell MATLAB of this new model, type

```
nprime = 'r*n*(1+a*p/k)^(-k)'
pprime = 'c*n*(1-(1+a*p/k)^(-k))'
```

The equilibria are found by our typing

```
>> [nhat,phat] = solve(symop(nprime,'-','n'),symop(pprime,'-','p'),'n,p');
>> nhat = simple(nhat)
>> phat = simple(phat)
```

which yields collectively

```
nhat = [                              0]
       [(-1+r^(1/k))*k/c/a/(r-1)*r]


phat = [                   0]
       [(-1+r^(1/k))*k/a]
```

As in the Nicholson-Bailey model, there are again two equilibria, and again the equilibrium is positive if the factor of geometric increase in the host population is greater than one, which it has to be anyway for it to live in the habitat in the absence of the parasitoid. The parasitoid can serve as a good biological control over the host if it is a good searcher (high a) and/or can produce many parasitoids per discovered host (high c). This equilibrium may also be stable. Let's compute the eigenvalues of the Jacobian for this system. We type

```
>> jacob = jacobian(sym([nprime '; ' pprime]), sym('[n,p]'));
>> jacob = simple(subs(jacob,sym(nhat,2,1),'n'));
>> jacob = simple(subs(jacob,sym(phat,2,1),'p'));
>> lambdas = simple(eigensys(jacob));
```

The eigenvalues turn out to depend only on R and k. Add plots of the absolute values of these eigenvalues as a function of R, for three values of k. These appear as in Figure 6.30 when you type

```
>> r = 1.1:.1:10;
>> figure(lambda_fig)
>> hold on
>> k = 2;
```

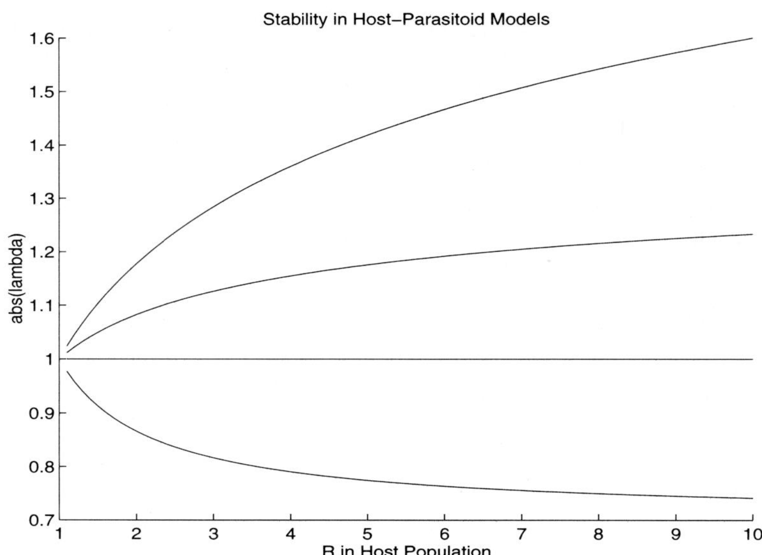

Figure 6.30: Absolute value of eigenvalues of the Jacobian matrix for host-parasitoid population dynamic models, as a function of R, the geometric increase factor in the host population. The top curve is for the Nicholson-Bailey model. The next curve down is for the negative binomial model with $k = 2$, followed by the flat curve for $k = 1$, and ending with the bottom curve for $k = 1/2$.

```
>> plot(r,abs(eval(sym2ara(sym(lambdas,1,1)))),'r')
>> plot(r,abs(eval(sym2ara(sym(lambdas,2,1)))),'r')
>> k = 1;
>> plot(r,abs(eval(sym2ara(sym(lambdas,1,1)))),'g')
>> plot(r,abs(eval(sym2ara(sym(lambdas,2,1)))),'g')
>> k = 1/2;
>> plot(r,abs(eval(sym2ara(sym(lambdas,1,1)))),'c')
>> plot(r,abs(eval(sym2ara(sym(lambdas,2,1)))),'c')
```

The figure shows that the absolute value of the eigenvalues is greater than 1 for $k = 2$, equal to 1 for $k = 1$, and is less than 1 for $k = 1/2$. Therefore, the equilibrium should be stable for $k = 1/2$.

To display trajectories of the host-parasitoid model with a negative-binomial host-escapement function, type and save the following function as `negbinom.m`

```
function [n,p]=negbinom(runlen,no,po)
 global r a c k
 n = [no]; p = [po];
```

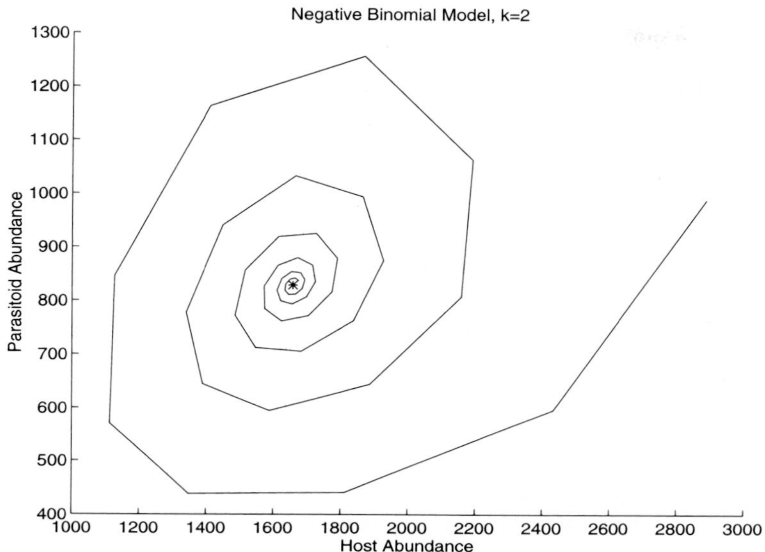

Figure 6.31: Trajectories of a model of host-parasitoid population dynamics with a negative binomial host escapement function, and clumping parameter, $k = 2$.

```
for t = 1:runlen
  nprime = r*n(t)*(1+a*p(t)/k)^(-k);
  pprime = c*n(t)*(1-(1+a*p(t)/k)^(-k));
  if nprime < 0 nprime = 0; end;
  if pprime < 0 pprime = 0; end;
  n = [n nprime];
  p = [p pprime];
end
```

Then in rapid succession, let's generate three figures for $k = 2$, 1, and 1/2. To set things up, type

```
>> global r a c k
>> r = 2; a = .001; c = 1;
```

The first case, when $k = 2$, is produced with

```
>> k = 2;
>> nhat_n = eval(sym(nhat,2,1))
>> phat_n = eval(sym(phat,2,1))
>> figure
>> hold on
```

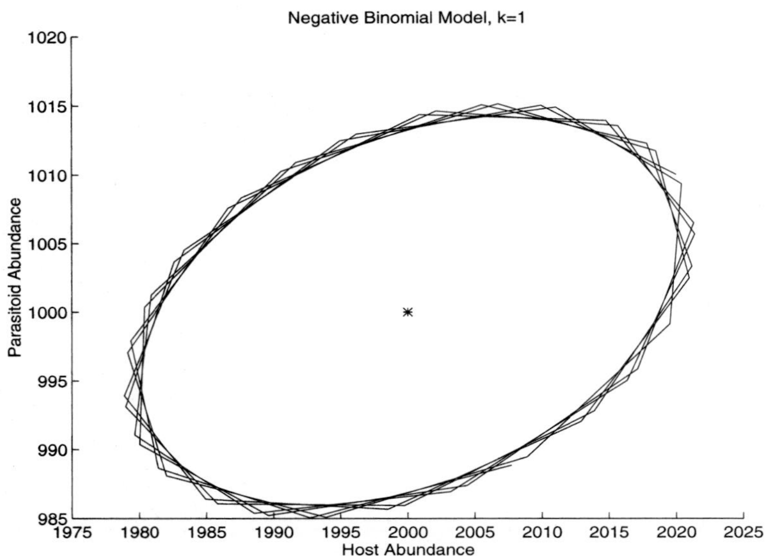

Figure 6.32: The trajectories of a model of host-parasitoid population dynamics with a negative-binomial host-escapement function, and clumping parameter, $k = 1$.

```
>> plot(nhat_n,phat_n,'*')
>> [n p] = negbinom(50,1.01*nhat_n,1.01*phat_n);
>> plot(n,p)
```

Figure 6.31 shows that the equilibrium at

```
nhat_n = 1.6569e+03
phat_n = 828.4271
```

is unstable, as anticipated.

Next, type

```
>> k = 1;
>> nhat_n = eval(sym(nhat,2,1))
>> phat_n = eval(sym(phat,2,1))
>> figure
>> hold on
>> plot(nhat_n,phat_n,'*')
>> [n p] = negbinom(50,1.01*nhat_n,1.01*phat_n);
>> plot(n,p)
```

and Figure 6.32 appears, showing the equilibrium at

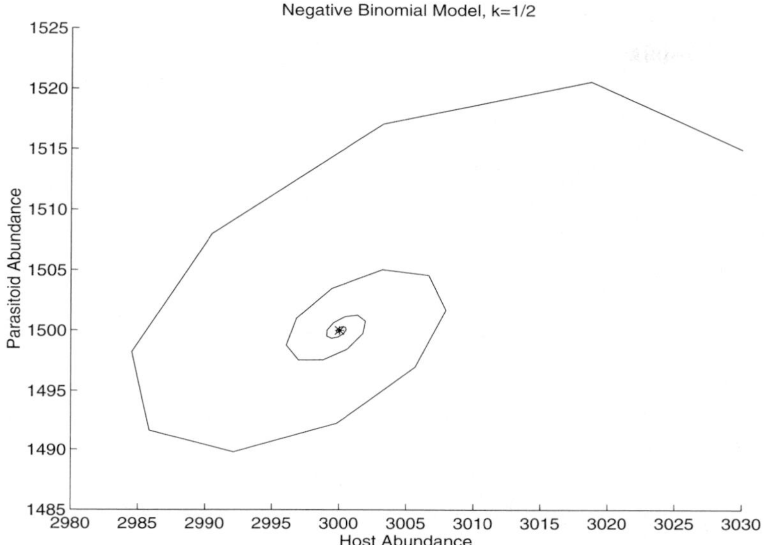

Figure 6.33: The trajectories of a model of host-parasitoid population dynamics with a negative-binomial host-escapement function, and clumping parameter, $k = 1/2$.

```
nhat_n = 2000
phat_n = 1000
```

is neutrally stable. The trajectories simply cycle around the equilibrium to rejoin a neighborhood of the initial condition. $k = 1$ is the demarcation between the stable and unstable equilibria of $k > 1$ and $k < 1$ respectively.

To see a stable host-parasitoid equilibrium, type

```
>> k = 1/2;
>> nhat_n = eval(sym(nhat,2,1))
>> phat_n = eval(sym(phat,2,1))
>> figure
>> hold on
>> plot(nhat_n,phat_n,'*')
>> [n p] = negbinom(50,1.01*nhat_n,1.01*phat_n);
>> plot(n,p)
```

Figure 6.33 now pops up. It shows a trajectory converging to the equilibrium at

```
nhat_n = 3000
phat_n = 1500
```

Figures 6.31–6.33 show that decreasing the clumping parameter, k, in the negative-binomial host-escapement function confers stability on the host-parasitoid interaction. What does this all mean for the future of biological control? Can we count on natural enemies of our pests to keep them under control for us for free, so that we don't have to buy pesticides?

To be good at biological control, a parasitoid must have both a good searching ability and reproductive capacity (high a and c), so that the equilibrium abundance of the hosts is as low as possible. But also requiring that k be low means that a host's chance of escaping detection can't go to zero too quickly as the number of parasitoids increase. Therefore, the parasitoid can't be too good either. It has to leave some hosts undiscovered. These undiscovered hosts are then food for the next generation of parasitoids. Only by failing to discover all the hosts can a parasitoid continue to exert control in the future. Otherwise the parasitoid will eat itself to extinction, and leave the host population to explode in its absence. So, biological control doesn't come for free. The hosts that the parasitoids must fail to detect are left to do the damage that has led them to be classified as pests to begin with. Look at the numerical values of `nhat` for $k = 2$, 1, and 1/2. The equilibrium number of hosts increases as k decreases. Therefore, stability is being purchased at a cost of maintaining a larger number of hosts at equilibrium. Ecologists tend to prefer biological control over pesticide control. There are two reasons for this preference. First, as the Volterra principle states, pesticides can lead to a net increase in the number of pests, because pesticides cause the evolution of pesticide resistance in the pest populations and therefore increasing doses of the pesticide are required. And second, pesticides are toxic to innocent creatures and to humans. Nonetheless, biological control is not free.

6.5.4 Host and Parasitoid as Weeds

Hey host, where you gonna hide? If the answer is nowhere, then is coexistence between host and parasitoid ruled out? On a local scale, yes, according to the previous model. But on a larger scale, a metapopulation scale, the hosts and parasitoids may coexist as weeds. The overall region is assumed to consist of H patches of habitat for the hosts. Both hosts and parasitoids can migrate among the patches. A patch can be classified as having neither host nor parasitoid (vacant), having hosts only, or having both hosts and parasitoids. The vacant patches are colonized by hosts from the host-only patches. The host-only patches are colonized by parasitoids from the patches with both host and parasitoid. The patches with both host and parasitoid become vacant because the parasitoids detect all the hosts there and both become extinct. It's also possible for a host-only patch to lose its population of hosts from causes not related to the arrival of a parasitoid. This setup is modeled with

$$\frac{dP_1}{dt} = m_1 P_1 (H - P_1 - P_2) - \mu_1 P_1 - m_2 P_2 P_1$$

$$\frac{dP_2}{dt} = m_2 P_1 P_2 - \mu_2 P_2$$

where P_1 is the number of host-only patches, P_2 the number of patches with both host and parasitoid. In these equations $m_1 P_1 (H - P_1 - P_2)$ is the rate vacant patches are being colonized from host-only patches, $mu_1 P_1$ is the rate at which hosts naturally become extinct in host-only patches for reasons not related to parasitoids, $m_2 P_2 P_1$ the rate at which host-only patches are colonized from patches that have both hosts and parasitoids, and $\mu_2 P_2$ is the rate at which patches with both host and parasitoids become vacant, because of the parasitoid driving the host to local extinction there and then becoming extinct itself. It's understood in this model that $\mu_2 > \mu_1$ and μ_1 may typically be zero or nearly so. The model for the weedy host and parasitoid can be rewritten as

$$\begin{aligned} \frac{dP_1}{dt} &= (m_1 H - \mu_1)P_1 - m_1 P_2^2 - (m_1 + m_2)P_2 P_1 \\ \frac{dP_2}{dt} &= m_2 P_1 P_2 - \mu_2 \end{aligned}$$

After gazing at this for a while, you can see that this model is the same as one we've already solved—namely, that for a prey with logistic density dependence together with a predator (without predator satiation). We can translate these parameters into the parameters we used before, according to

$$\begin{aligned} r &= m_1 H - \mu_1 \\ K &= H - \frac{\mu_1}{m_1} \\ a &= m_1 + m_2 \\ b &= \frac{m_2}{m_1 + m_2} \\ c &= \mu_2 \end{aligned}$$

As we already know, if K is larger than $d/(ba)$, there is a positive equilibrium, and it is stable. The approach to this equilibrium may or may not involve oscillations, according to how big K is, but the equilibrium is stable in any case, provided it is positive. If we reverse-translate the requirement that $K > d/(ba)$ back into the parameters for host-parasitoid weeds, we obtain

$$H > \frac{\mu_1}{m_1} + \frac{\mu_2}{m_2}$$

So, if there are enough habitat patches, H, then host and parasitoid can coexist in the region, even though the parasitoids always drive the hosts to extinction within each local patch they get into.

The lessons from the model of host-parasitoid weeds for biological control of pests and for conservation are stark. Habitat loss frees pests from biological control. Without enough habitat, the parasitoid goes extinct and the host population will explode in its absence. However, the maintenance of biological controls on a metapopulation scale is not any more free than it was on a local scale. The costs are first, to maintain a large enough region with enough patches to support the metapopulations; and second, to abide by the damage caused by the hosts within those patches that the parasitoids do not colonize. But unless you own stock in a pesticide company, biological control may well work out as the cheapest way to go.

6.6 Host-Pathogen Interaction

The predator-prey interaction conjures up images of big things catching little things—big fish eating little fish, and the host-pathogen interaction is, if anything, a triumph for the little guy. The host is a big fish, and the predator a tiny single-celled bacterium or protozoan called the pathogen. The predator-prey interaction between multicellular prey and single-celled predators enjoys a medical vocabulary and research agenda. When a prey is captured by a group of predators, the prey is said to be "diseased." When the predator population increases, an "epidemic" is said to occur. When research funding for this predator-prey interaction is sought, the medical discipline of epidemiology is the source. This medical vocabulary and research style often obscures the essentially ecological nature of the host-pathogen interaction. In medicine, the word disease covers any cause of poor health, of which "infectious diseases" are really predator-prey interactions. In medicine, diseases are said to "develop" resistance to drugs, whereas in reality, pathogen populations are evolving drug resistance. Schemes of multiple drug treatments to avoid producing drug-resistance amount to schemes of alternating selection pressures upon the pathogenic bacteria. The main difference, other than body size, between the host-pathogen interaction and a generic predator-prey interaction is that the pathogen lives in the host for some time. Therefore it's often not in the best interest of the pathogen to kill the host, and this may temper its virulent appetite. Many pathogens don't hurt their hosts very much, and instead just view them as mobile habitat patches to be colonized as often as possible.

6.6.1 Kermack-McKendrick SIR Model

Like both generic predator-prey theory and host-parasitoid theory, the theory of communicable diseases has a long history. The classic model assumes that the hosts (people) are divided into three groups, those without the disease, those with the disease, and those who have already had the disease and have now become resistant. A model with these components is called an SIR model, which stands for "susceptible," "infected," and "resistant." The classical formulation for this is known as the Kermack-McKendrick model, as proposed in 1927. Let x_1 be the number of hosts who are susceptible, x_2 the number who are presently infected, and x_3 the number who are resistant. The total number of hosts, N, is $x_1 + x_2 + x_3$. The classic formulation is for a communicable disease that has negligible impact on the host's survival; think of a common cold. The total population size, N is assumed to remain constant, and the death rate per individual, d, also equals the birth rate. The equations for the three categories of hosts are

$$\begin{aligned}
\frac{dx_1}{dt} &= -bx_1x_2 + px_3 + dN - dx_1 \\
\frac{dx_2}{dt} &= bx_1x_2 - ux_2 - dx_2 \\
\frac{dx_3}{dt} &= ux_2 - px_3 - dx_3
\end{aligned}$$

Taking these equations term by term, we find that $-bx_1x_2$ is the rate at which infected hosts contact susceptible hosts times what is called the transmission parameter, b, which measures the contagiousness of the disease; px_3 is the rate at which resistant hosts lose their resistance; dN is the rate at which new individuals are born, with all newborn assumed to be susceptible; $-dx_1$ is the death rate of susceptible individuals; and $-ux_2$ is the speed at which infected individuals heal themselves and become resistant. Because the total population is constant, the three equations can be boiled down to two; it is figured that $x_3 = N - x_1 - x_2$. Therefore, we will analyze

$$\begin{aligned}\frac{dx_1}{dt} &= -bx_1x_2 + p(N - x_1 - x_2) + dN - dx_1 \\ \frac{dx_2}{dt} &= bx_1x_2 - ux_2 - dx_2\end{aligned}$$

Condition for an Epidemic

Suppose a new pathogen pops up, either by jumping to humans from another species of host, say a domesticated animal, or by being carried by a traveler from a place where the disease is widespread to a place where it hasn't been experienced before. Will the new pathogen cause an epidemic? With the SIR model the answer is easily found. Just look to see if the infected pool, x_2, can increase when rare. That is, suppose that initially, $x_1 = N$, and ask if $dx_2/dt > 0$. By substituting $x_1 = N$ into the equation for dx_2/dt, we get

$$\begin{aligned}\frac{dx_2}{dt} &= bNx_2 - ux_2 - dx_2 \\ &= [bN - (u + d)]x_2\end{aligned}$$

So, the epidemic will take off if the expression in brackets is greater than zero. The important result here is that, in order for this expression to be positive, N must be big enough. Thus, there is a threshold host population size, and if the host population size is smaller than this threshold, the epidemic will not take off. This is basically why there is a higher risk of disease when there is a high population.

The expression in brackets above may be rearranged as

$$\frac{b}{u + d}N > 1$$

It's customary to define

$$R_o = \frac{b}{u + d}$$

as the reproductive rate of the disease. It is the net number of new infections arising from one infected individual who is surrounded by susceptibles. The host population size condition is therefore customarily written as

$$R_oN > 1$$

The threshold host population size, is $1/R_o$, and the host population must be larger than this for the epidemic to take off.

Now let's alert MATLAB to the model. For the record, let's define R_o for later evaluation as

```
>> ro = 'b/(u+d)'
```

We'll also want to use one of MATLAB's differential-equation-solvers to illustrate trajectories of x_1 and x_2. So, with a text editor, type in the model and save it as `sir.m`:

```
function xdot = sir(t,x)
 global n b p d u;
 xdot(1,1) = -b*x(1)*x(2)+p*(n-x(1)-x(2))+d*n-d*x(1);
 xdot(2,1) = b*x(1)*x(2)-u*x(2)-d*x(2);
```

We're now ready to go.

Progression to Pandemic

Suppose the epidemic can get itself started, what happens next? If everyone winds up infected, the epidemic turns into what is called a "pandemic." A pandemic results from a special case in which p, d, and u are zero. The model then becomes, in MATLAB notation,

```
>> dx1 = '-b*x1*x2'
>> dx2 = 'b*x1*x2'
```

To ask MATLAB for the equilibria, we type

```
>> [x1hat,x2hat] = solve(dx1,dx2,'x1,x2')
```

yielding

```
x1hat = [ 0]
        [x1]

x2hat = [x2]
        [ 0]
```

Thus, there are two equilibria: one with everyone infected and no susceptibles, and one with everyone susceptible with no one infected. The first of these is stable and the second unstable, so that trajectories go from the second to the first of these equilibria. To illustrate the progression of an epidemic culminating in a pandemic, type

```
>> global n b p d u;
>> n = 1000; b = .01; p = 0; d = 0; u = 0;
>> figure
>> [time x] = ode45('sir',0,3,[n-1;1]);
>> plot(time,x)
```

Here we assume that the host population is 1000 people, and we start out with 1 infected person. Figure 6.34 shows how everyone winds up infected.

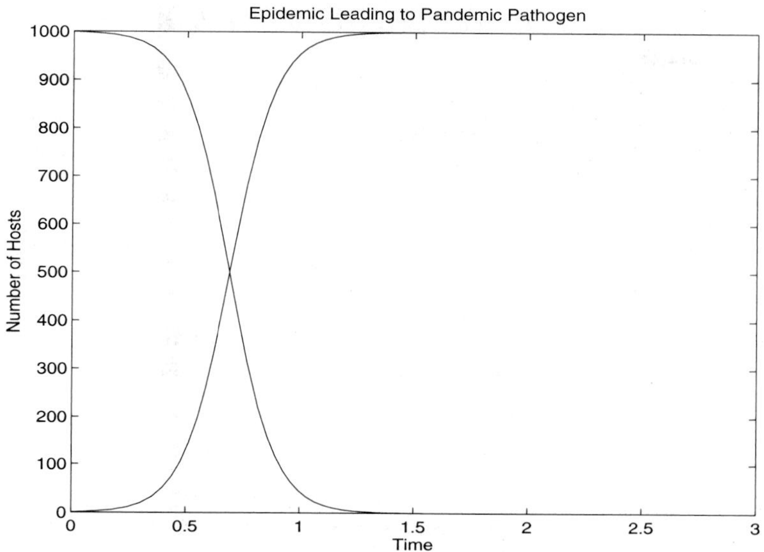

Figure 6.34: An epidemic's progression of an epidemic culminating in a pandemic disease. The curve of susceptibles starts at 1000 and drops to 0. The curve of infected hosts starts at 0 and rises to 1000.

Rise and Fall of an Epidemic

A happier outcome to the epidemic occurs when the hosts can develop resistance. Consider the special case where u is no longer zero, but p and d still are. The model now becomes, in MATLAB's notation,

```
>> dx1 = '-b*x1*x2'
>> dx2 = 'b*x1*x2-u*x2'
```

The equilibria are found with

```
>> [x1hat,x2hat] = solve(dx1,dx2,'x1,x2')
```

yielding

```
x1hat = [ 0]
        [x1]


x2hat = [0]
        [0]
```

These equilibria correspond, first, to no one being susceptible or infected (i.e., everyone is resistant), and the second, to everyone being susceptible. The first of these equilibria

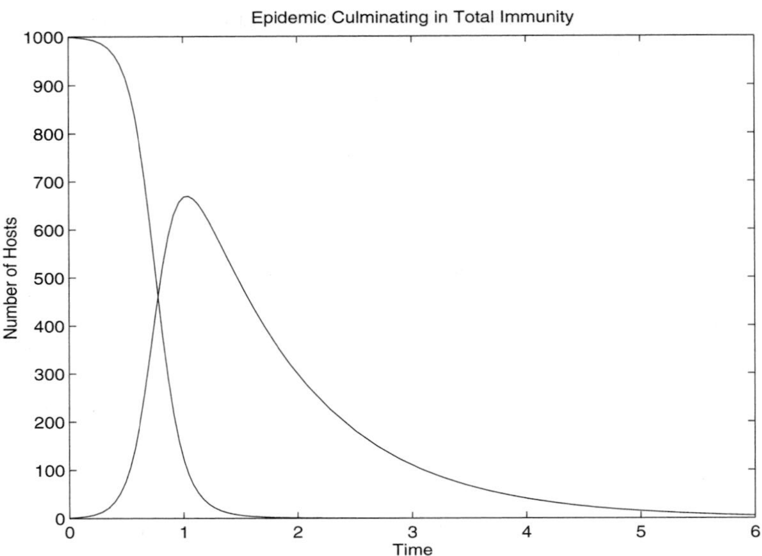

Figure 6.35: The progression of an epidemic that eventually disappears because all hosts become resistant. The curve of susceptibles starts at 1000 and drops to 0. The curve of infected hosts rises and then falls as the hosts all gradually become resistant.

is stable and the second unstable, so trajectories go from the second to the first of these equilibria. To illustrate the progression of this epidemic, type

```
>> n = 1000; b = .01; p = 0; d = 0; u = 1;
>> ro_n = eval(ro)
>> figure
>> [time x] = ode45('sir',0,6,[n-1;1]);
>> plot(time,x)
```

yielding

```
ro_n = 0.0100
```

together with Figure 6.35. Because R_o is 0.01, N must be greater than 100 for the epidemic to start off. The figure shows that, starting with 1 infected individual, the epidemic spreads but eventually dies out altogether, because everyone becomes resistant and the disease becomes extinct.

Progression to Endemic Disease

Alas, resistance to a disease is not forever, and the pool of susceptibles is eventually renewed, so that the disease does not completely die out. When the disease is established at a low steady-state level, it is said to be "endemic." This is the general case, and, to tell MATLAB about the full model, we'll type

```
>> dx1 = '-b*x1*x2+p*(n-x1-x2)+d*n-d*x1'
>> dx2 = 'b*x1*x2-u*x2-d*x2'
```

The equilibria are found by typing

```
>> [x1hat,x2hat] = solve(dx1,dx2,'x1,x2');
>> x1hat = simple(x1hat)
>> x2hat = simple(x2hat)
```

yielding collectively

```
x1hat = [       n]
        [(u+d)/b]

x2hat = [                              0]
        [(p+d)*(n*b-u-d)/b/(d+p+u)]
```

The first of these equilibria is for the condition in which everyone is susceptible, and no one is infected. This is the state that exists before the disease is introduced. The second equilibrium represents coexistence of the pathogen and host. To see if the coexistence-equilibrium is stable, we'll inspect the Jacobian. Type

```
>> jacob = jacobian(sym([dx1 '; ' dx2]), sym('[x1,x2]'))
>> jacob = subs(jacob,sym(x1hat,2,1),'x1');
>> jacob = subs(jacob,sym(x2hat,2,1),'x2');
>> jacob = simple(jacob)
>> trace = symop(sym(jacob,1,1),'+',sym(jacob,2,2))
>> determinant = determ(jacob)
```

yielding collectively

```
jacob = [-b*x2-p-d,  -b*x1-p]
        [      b*x2, b*x1-u-d]

jacob = [ -(p+d)*(p+n*b)/(d+p+u), -d-p-u]
        [(p+d)*(n*b-u-d)/(d+p+u),      0]

trace = -(p+d)*(p+n*b)/(d+p+u)
determinant = -(-d-p-u)*(p+d)*(n*b-u-d)/(d+p+u)
```

Look closely at both the trace and determinant of the Jacobian. The trace is negative, and the determinant is positive provided $R_o N > 1$. Therefore, the equilibrium with an endemic disease among the hosts is stable. Trajectories go from the first equilibrium to the second. For later use, let's calculate the discriminant too, to tell us more about the approach to the stable equilibrium:

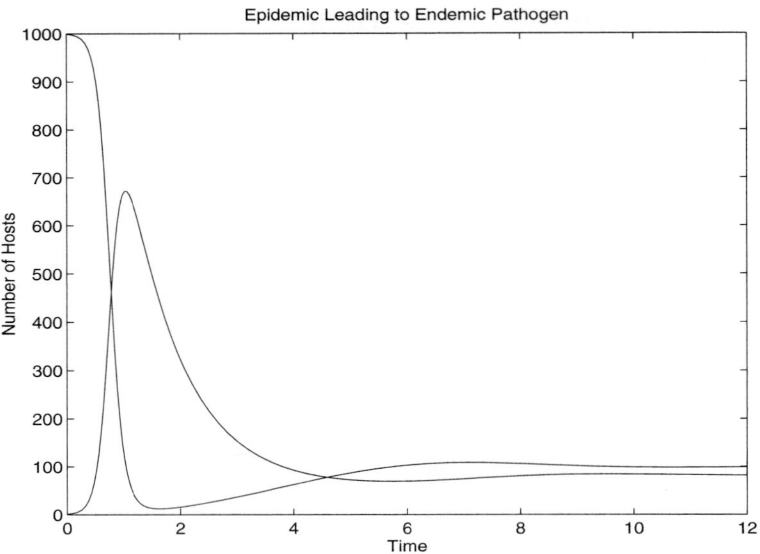

Figure 6.36: The progression of an epidemic culminating in an endemic disease. The curve of susceptible hosts starts at 1000 and drops and then rises to its equilibrium value. The curve of infected hosts starts, rises, and then drops to its equilibrium level.

```
discriminant = symop(trace,'^','2','-','4','*',determinant)
```

Now let's illustrate the progression of an epidemic that culminates in an endemic disease. Typing

```
>> n = 1000; b = .01; p = .1; d = 0; u = 1;
>> ro_n = eval(ro)
>> x1hat_n = eval(sym(x1hat,2,1))
>> x2hat_n = eval(sym(x2hat,2,1))
>> x3hat_n = n - x1hat_n - x2hat_n
>> discriminant_n = eval(discriminant)
```

collectively yields

```
ro_n = 0.0100
x1hat_n = 100
x2hat_n = 81.8182
x3hat_n = 818.1818
discriminant_n = -2.7569
```

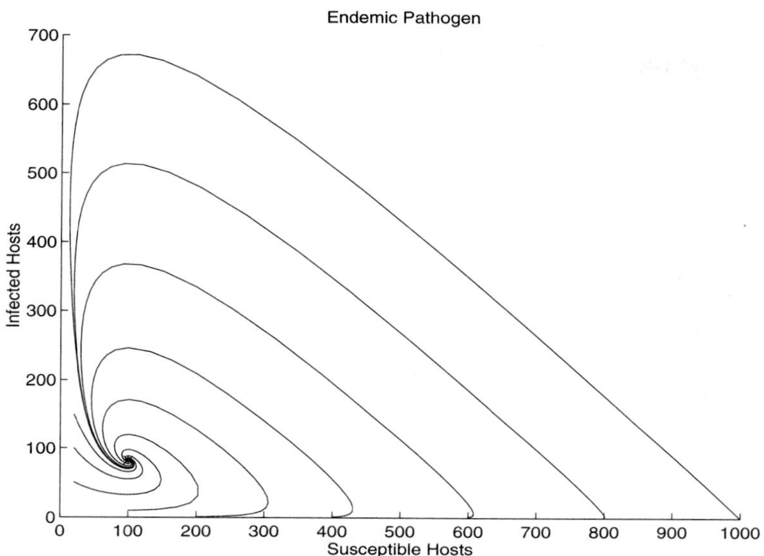

Figure 6.37: The trajectories of an epidemic culminating in an endemic disease started from an assortment of initial conditions.

Because R_o is 0.01, the number of hosts must be greater than 100 for the epidemic to start. With 1000 hosts, the endemic equilibrium at steady state will have 100 susceptibles, 82 infected individuals, and 818 resistant individuals. The progression of the epidemic starting with one infected individual is illustrated by our typing

```
>> figure
>> [time x] = ode45('sir',0,12,[n-1;1]);
>> plot(time,x)
```

as shown in Figure 6.36. The trajectory shows an overshoot of the equilibrium, which represents an oscillatory approach. Because the discriminant is negative, the equilibrium is a focus. And we already know it is stable because the trace is negative and determinant positive. To show the progression of the epidemic from an assortment of initial conditions, type

```
>> figure
>> hold on
>> plot(x1hat_n,x2hat_n,'*')
>> [time x] = ode45('sir',0,20,[20;50]);
>> plot(x(:,1),x(:,2))
>> [time x] = ode45('sir',0,20,[20;100]);
>> plot(x(:,1),x(:,2))
```

```
>> [time x] = ode45('sir',0,20,[20;150]);
>> plot(x(:,1),x(:,2))
>> [time x] = ode45('sir',0,20,[999;1]);
>> plot(x(:,1),x(:,2))
>> [time x] = ode45('sir',0,20,[800;1]);
>> plot(x(:,1),x(:,2))
>> [time x] = ode45('sir',0,20,[600;1]);
>> plot(x(:,1),x(:,2))
>> [time x] = ode45('sir',0,20,[400;1]);
>> plot(x(:,1),x(:,2))
>> [time x] = ode45('sir',0,20,[200;1]);
>> plot(x(:,1),x(:,2))
>> [time x] = ode45('sir',0,20,[100;10]);
>> plot(x(:,1),x(:,2))
```

This will result in Figure 6.37. This figure clearly illustrates the tendency of the epidemic to spiral in to the endemic equilibrium.

The tendency of a host-pathogen interaction to spiral into equilibrium, as depicted in Figure 6.37, is the basis for extending the model to explain cyclic epidemics as well as the seemingly chaotic spontaneous appearance of epidemics. If the transmission parameter, b, itself varies seasonally (think about measles among school children when the school year starts), then an external cycle is mixed in with the inherent tendency to cycle that comes from the host-pathogen interaction itself. Because the host-pathogen model is nonlinear, combining two cycles in the host-pathogen model does not simply add together, like combining two notes to produce a chord, but instead can produce still more cycles, and chaotic trajectories, as though striking only two notes on a piano caused still more notes to play, and thus produced a chord of many notes. All in all, one can think of greatly extending host-pathogen models to many types of diseases, each with its own special characteristics. This is the subject of mathematical epidemiology, a subject of growing importance in society.

6.7 Mutualism and Symbiosis

What about species helping one another? We've delved into competition, predation, and disease—all interactions in which at least one species is hurting another. Is nature just tooth and claw? Well, obviously not. "Mutualism" is an interaction wherein each of two or more species helps the others. Perhaps the most conspicuous mutualism is insect-plant pollination. There are many more, though. Many involve a partnership between sessile and mobile species, in which the sessile species pays the mobile species to move it around. Animals who carry pollen and fruits for plants are examples, as are animals that are paid by plants to attack their herbivores. The mode of payment is often candy, such as nectar or syrup, or the tasty fruit surrounding a hard pit. Other mobile-sessile partnerships seemingly are designed by the mobile party, who uses the sessile party as its home, and who feeds and tends the sessile party as though a garden. Damselfish who live in the protection of an anemone's tentacles feed their anemone to keep it healthy. Other mutualisms are based on biochemical exchange. Bacteria living in little bumps called root nodules on the roots of legume plants, such as peas, fix nitrogen which they give to their plant, and receive carbon products in return. Perhaps less known, but of huge importance, are mutualisms between chloroplasts and mitochondria and the cells in which they live as organelles.

"Symbiosis" is defined as a relationship between two species in which individuals of the two species live together in a long-term physical association. The bigger party is called the "host" and the smaller the "guest." If the individuals living together help one another, then their symbiosis is a mutualistic symbiosis. However, symbiosis also includes "parasitism" in which the guest exploits the host to its net detriment, and "commensalism" in which the guest has almost no net effect on the host. Sometimes symbiosis is used as a synonym of mutualism, but this usage is discouraged. The plant-pollinator interaction is a mutualism, but not a symbiosis, whereas the damselfish-sea anemone association is a mutualistic symbiosis, as are root nodules in peas, and mitochondria in eukaryotes.

A symbiotic mutualism should be distinguished from a nonsymbiotic mutualism, called a "mass-action mutualism," in that a symbiotic mutualist shares "property rights" with its partner. In a symbiotic association, an individual who helps its partner will itself benefit from its partner's improved condition. In contrast, a mass-action mutualism such as pollination is like an open-market transaction between the selling and the buying species. An economist has suggested that a symbiotic mutualism such as the zooxanthellae algae that live in the tissues of corals represents a "merger" of two species, and that the disaggregation of a symbiosis might be thought of as a "spin off," as when the coral release their zooxanthellae under low-light conditions, presumably because they no longer return a net energetic profit to the coral.

Not only do corals abandon their zooxanthellae, but many plants do not bother with insect pollination at all, and prefer the wind instead; these plants include the grasses, of which the world's major grain crops such as corn, wheat and rice are examples. And a plant that depends on pollinators does fine without them when it lands on oceanic islands where the pollinators are absent; the plant evolves other ways to obtain pollination. Nor is it all sweetness and light between mutualists. A plant would like nothing better than to dribble out nectar-lite instead of the real thing. They would both save energy and keep the pollinators hungry, so they are forced to scramble among as many flowers as possible for a decent meal, and pollinate as they go. Meanwhile, a pollinator would like nothing better than to visit one flower, fill up, and take a nap; and is certain to avoid restaurants that skimp on dessert. Thus, the theoretical issues raised by mutualism and symbiosis are, when do the associations form, when do they dissolve, and how good should each party be for the other?

Well, there's no hiding the fact that these issues have not been investigated very well in theoretical ecology. The competitive and predator-prey interactions enjoy a wealth of literature, but mutualism and symbiosis languish in comparative neglect. Theoreticians know this, of course, and for decades books reviewing ecological theory begin their discussion of mutualism and symbiosis with an apology for the relative absence of models for this interaction. And matters are not improving quickly. Why then, is an ecological topic known to be important progressing so slowly? The main reason is that simple models for mutualism don't stand alone and quickly turn into models that must include other types of interactions too, all of which complicates the matter enormously.

Here's why simple models of mutualism don't stand alone. Let's consider mutualism within a species—a hypothetical species where individuals help each other rather than compete. Suppose the intrinsic rate of increase, r increases with population, N, rather than decreases as usual. To be specific, suppose the intrinsic rate of increase increases linearly with population size, as $r(N) = aN$. Then the population's growth is

$$\frac{dN}{dt} = aN^2$$

This is called an "autocatalytic" process, and it's bad news. The solution, $N(t)$, goes to infinity in finite time. That is, with good ol' exponential growth, N only reaches infinity as t reaches infinity. But the solution of an autocatalytic process explodes even faster, and reaches infinity at a specific time in the future. If you were scared of the population bomb, you should be terrified now. One theoretician has indelicately called the autocatalytic population bomb an orgy of mutual benefaction.

Because of the tendency of mutualism to lead to autocatalysis, models of mutualism wind up building in a lot of density dependence to keep the autocatalysis under control. This may be considered the main result of the population dynamics of mutualism. Because mutualistic associations are not covering the earth even faster than *E. coli* could at its best, there must be some stabilizing factor, such as competition, present somewhere in the system as well. Thus, pure mutualism cannot exist by itself, but must be embedded in other interactions that provide boundedness and perhaps even stability.

Mass-Action Mutualism

A population-dynamic model of mass-action mutualism has often been written as a set of Lotka-Volterra competition equations with negative competition coefficients. But because the Lotka-Volterra parameters, especially the competition coefficients, α_{ij}, and carrying capacity, K_i, have special interpretations in relation to competition that don't naturally transport to mutualism, I'd rather pose a Lotka-Volterra-like model as

$$\begin{aligned}\frac{dN_1}{dt} &= (r_1 - b_1 N_1 + \beta_{12} N_2) N_1 \\ \frac{dN_2}{dt} &= (r_2 - b_2 N_2 + \beta_{21} N_1) N_2\end{aligned}$$

This assumes that an individual's density-independent component of growth, r_i, is diminished by intraspecific competition according to $-b_i N_i$ and is augmented by interspecific mutualism according to $+\beta_{ij} N_j$. For convenience, let's analyze a symmetric example. We type in to MATLAB

```
>> dn1 = '(r-b*n1+beta*n2)*n1'
>> dn2 = '(r-b*n2+beta*n1)*n2'
```

The equilibria are found by our typing

```
>> [n1hat n2hat] = solve(dn1,dn2,'n1,n2')
```

yielding

```
n1hat = [           0]
        [           0]
        [         r/b]
        [r/(b-beta)]

n2hat = [           0]
        [         r/b]
        [           0]
        [r/(b-beta)]
```

The first equilibrium is total extinction, the second and third are for each species alone, and the fourth equilibrium is for the coexistence of the two mutualists. The coexistence equilibrium is positive if $b > \beta$, that is, if the interindividual effect of intraspecific competition exceeds the interindividual effect of interspecific mutualism. This is the condition that allows competition to check the autocatalysis of mutualism. If this condition is satisfied, then each species has a higher abundance when both are present than either does when it is alone. To see if the coexistence equilibrium is stable, type

```
>> jacob = jacobian(sym([dn1 '; ' dn2]), sym('[n1,n2]'));
>> jacob = subs(jacob,sym(n1hat,4,1),'n1');
>> jacob = subs(jacob,sym(n2hat,4,1),'n2');
>> jacob = simple(jacob);
>> trace = simple(symop(sym(jacob,1,1),'+',sym(jacob,2,2)))
>> determinant = simple(determ(jacob))
>> discriminant = simple(symop(trace,'^','2','-','4','*',determinant))
```

yielding collectively

```
trace = -2*r*b/(b-beta)
determinant = r^2*(b+beta)/(b-beta)
discriminant = 4*beta^2*r^2/(b-beta)^2
```

The trace is negative and the determinant positive if $b > \beta$, which was also the condition for the equilibrium to be positive to begin with. So a positive equilibrium is stable. Moreover, the discriminant is positive, so the equilibrium is a stable node. This type of model might conceivably be tailored for a plant-pollinator system of nonsymbiotic mutualism.

To use a text editor to develop a numerical illustration of mass-action mutualism, type in a function and save as `mutual.m`,

```
function ndot = mutual(t,n)
 global r b beta;
 ndot(1,1) = (r-b*n(1)+beta*n(2))*n(1);
 ndot(2,1) = (r-b*n(2)+beta*n(1))*n(2);
```

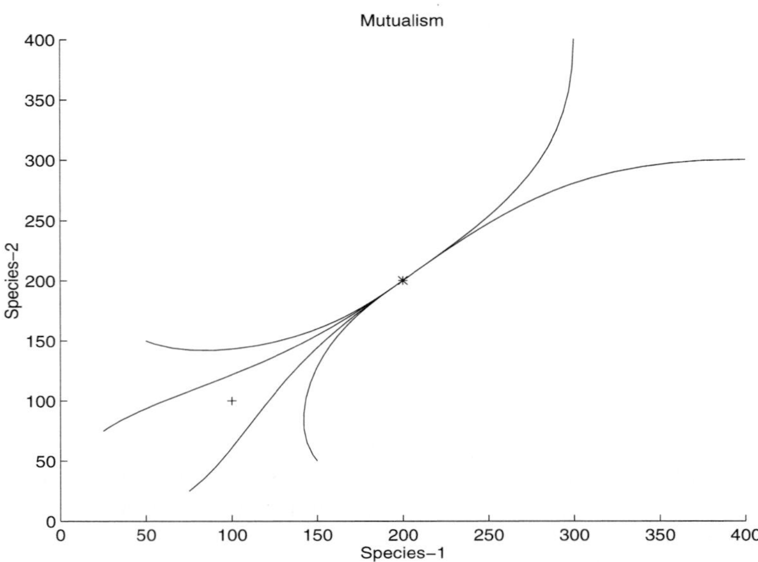

Figure 6.38: A coexistence of mass-action (nonsymbiotic) mutualists. In the absence of mutualism, the species would coexist at the point marked with a +. When there is mutualism, both species coexist at a higher abundance than either would alone.

Then let's evaluate the quantities we calculated above. We'll type

```
>> global r b beta;
>> r = 1; b = 0.01; beta = 0.005;
>> n1hat_n = eval(sym(n1hat,4,1))
>> n2hat_n = eval(sym(n2hat,4,1))
>> trace_n = eval(trace)
>> determinant_n = eval(determinant)
>> discriminant_n = eval(discriminant)
```

yielding collectively

```
n1hat_n = 200
n2hat_n = 200
trace_n = -4
determinant_n = 3
discriminant_n = 4
```

Because the trace is negative and the determinant positive, the equilibrium is stable, and because the discriminant is positive, the equilibrium is a node. For comparison, the equilibrium abundance of each species if it were not a mutualist condition (i.e., $\beta = 0$), is

```
>> r/b
```

yielding

```
ans = 100
```

To illustrate a pair of species coexisting with mass-action mutualism, type

```
>> figure
>> hold on
>> plot(n1hat_n,n2hat_n,'*')
>> plot(r/b,r/b,'+')
>> [time n] = ode45('mutual',0,10,[0.75*(r/b);0.25*(r/b)]);
>> plot(n(:,1),n(:,2))
>> [time n] = ode45('mutual',0,10,[0.25*(r/b);0.75*(r/b)]);
>> plot(n(:,1),n(:,2))
>> [time n] = ode45('mutual',0,10,[0.75*n1hat_n;0.25*n2hat_n]);
>> plot(n(:,1),n(:,2))
>> [time n] = ode45('mutual',0,10,[0.25*n1hat_n;0.75*n2hat_n]);
>> plot(n(:,1),n(:,2))
>> [time n] = ode45('mutual',0,10,[1.50*n1hat_n;2.00*n2hat_n]);
>> plot(n(:,1),n(:,2))
>> [time n] = ode45('mutual',0,10,[2.00*n1hat_n;1.50*n2hat_n]);
>> plot(n(:,1),n(:,2))
```

and Figure 6.38 appears. It depicts a stable node located beyond the point at which non-mutualistic species would coexist.

Symbiotic Mutualism

Symbiotic mutualism requires changing the model to represent the physical union between both symbionts. The benefit received by one partner depends on the other partner being alive. In mass-action mutualism, the total benefit received by one party depends only on the number of the other parties with which it can interact. When there is symbiotic mutualism, the total benefit received by one party depends on the number of potential partners, and also on whether an intact symbiotic state persists, which in turn depends on the average lifespan of the partner. One possible model to incorporate this feature may be formed from the previous model when we decompose r_i into the birth and death components, and assume that competition and mutualism affect birth and not death. Thus,

$$\begin{aligned}\frac{dN_1}{dt} &= (B_1 - D_1 - b_1 N_1 + \frac{\beta_{12}}{D_2} N_2) N_1 \\ \frac{dN_2}{dt} &= (B_2 - D_2 - b_2 N_2 + \frac{\beta_{21}}{D_1} N_1) N_2\end{aligned}$$

In this model, the interindividual mutualistic effect, β_{ij}, is weighted by the average lifespan of the partner, which is $1/D_j$, and is multiplied by the number of partners, N_j. To see what this model predicts, let's again look at a symmetric illustration. We can type

```
dn1 = '((bo-d)-b*n1+beta*n2/d)*n1'
dn2 = '((bo-d)-b*n2+beta*n1/d)*n2'
```

Solving for the equilibria and checking the stability, exactly as before, yields

```
n1hat = [                    0]
        [                    0]
        [           -(d-bo)/b]
        [-d*(d-bo)/(b*d-beta)]

n2hat = [                    0]
        [           -(d-bo)/b]
        [                    0]
        [-d*(d-bo)/(b*d-beta)]

trace = 2*b*d*(d-bo)/(b*d-beta)
determinant = (d-bo)^2*(b*d+beta)/(b*d-beta)
discriminant = 4*beta^2*(d-bo)^2/(b*d-beta)^2
```

Positivity now requires that $B > D$, as is obvious, and also that $bD > \beta$. As for the Jacobian, the trace is negative, the determinant positive, and the discriminant positive also if both $B > D$ and $bD > \beta$. Thus, when there is a positive equilibrium, the dynamics of the symbiotic mutualism model is the same as that of the mass-action model. But the starting conditions for a positive equilibrium are different. Specifically, $bD > \beta$ means that if the partners have a high death rate, D, then the impact of mutualism is low, because the symbiotic state doesn't last very long, and so stability can be achieved with less intraspecific competition, b. The symbiotic-mutualism model allows for the coevolution of mutually helpful traits in both species. The fitness in species-i is raised when the death rate in species-j, D_j, is lowered, and vice versa. Interestingly though, the net effect is to increase the abundance of both species but also to destabilize the relationship, because as coevolution leads to lower D_i's, the condition that $bD > \beta$ comes closer to being violated.

Endosymbiosis

The model for symbiotic mutualism as written above assumes that individuals of the two species retain their identity, whether they are in a symbiotic state or not. A damselfish is still a damselfish whether or not it is living in a sea anemone, and the identity of a sea anemone isn't affected by whether a damselfish is living in it. But endosymbionts, such as the algae that live in corals, raise the issue of a change in identity. Do the symbionts function together so that they become in effect a new type of organism, different from either living apart? Moreover, is this third type a self-replicating entity of its own? Mitochondria are maternally inherited, passed from mother to offspring in the cytoplasm of the egg, along with the mother's genes located in the nucleus of the egg. So a eukaryote is a self-replicating symbiotic entity, distinct and independent of the population dynamics of its partners who, conceptually, might also exist in free-living states.

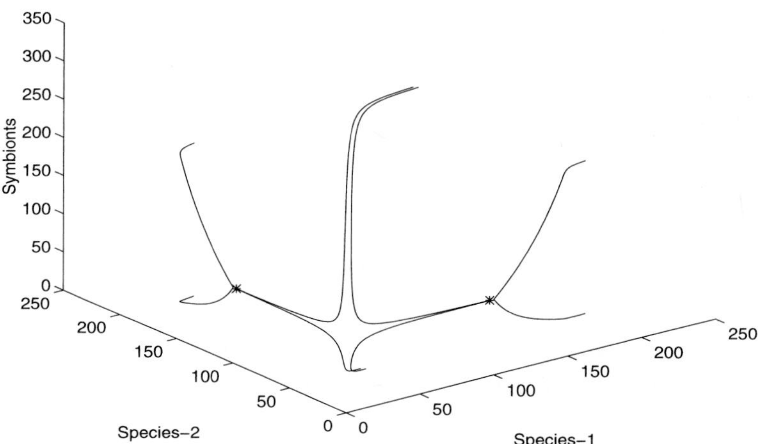

Figure 6.39: The formation of endosymbiosis. Two stable equilibria are indicated as asterisks.

The idea that a symbiotic association is a third self-replicating type, that is distinct from, and can live independently from, either its free-living partners, can be modeled as

$$\begin{aligned}\frac{dN_1}{dt} &= -aN_1N_2 + cN_3 + rN_1(1 - N_1/K) \\ \frac{dN_2}{dt} &= -aN_1N_2 + cN_3 + rN_2(1 - N_2/K) \\ \frac{dN_3}{dt} &= aN_1N_2 - cN_3 + sN_3(1 - N_3/L)\end{aligned}$$

The term aN_1N_2 means that an association is formed at rate a when species-1 and species-2 contact. This quantity is debited from both species-1 and species-2 and added to species-3, which is the symbiont. Also, symbionts may dissociate from each other, with the rate of collapse being c. Meanwhile, each species, including the symbiont, may grow logistically on their own. For simplicity, I've let both species-1 and species-2 have the same intrinsic rate of increase and carrying capacity, whereas the symbiont is different in this regard from either of its constituents. Even this simple model is difficult to solve symbolically, but some of how it works can be discerned from a numerical example. With a text editor, type in the following and save it as `symbios.m`:

```
function ndot = symbios(t,n)
 global r k a c s l;
```

```
ndot(1,1) = r*n(1)*(1-n(1)/k)-a*n(1)*n(2)+c*n(3);
ndot(2,1) = r*n(2)*(1-n(2)/k)-a*n(1)*n(2)+c*n(3);
ndot(3,1) = s*n(3)*(1-n(3)/l)+a*n(1)*n(2)-c*n(3);
```

Now, Figure 6.39 is produced by our typing

```
>> global r k a c s l;
>> r = 1; k = 100; a = 1; c = .01; s = r; l = k;
>> m = 2*k; n = k/2; tstop = 7.5;
>> [timea na] = ode45('symbios',0,tstop,[m+1;m-1;m]);
>> [timeb nb] = ode45('symbios',0,tstop,[m-1;m+1;m]);
>> [timec nc] = ode45('symbios',0,tstop,[m;n;m]);
>> [timed nd] = ode45('symbios',0,tstop,[m;n;0]);
>> [timee ne] = ode45('symbios',0,tstop,[n;m;m]);
>> [timef nf] = ode45('symbios',0,tstop,[n;m;0]);
>> [timeg ng] = ode45('symbios',0,tstop,[n+1;n-1;0]);
>> [timeh nh] = ode45('symbios',0,tstop,[n-1;n+1;0]);
>> figure
>> plot3(na(:,1),na(:,2),na(:,3),'y',nb(:,1),nb(:,2),nb(:,3),'r',...
>>       nc(:,1),nc(:,2),nc(:,3),'y',nd(:,1),nd(:,2),nd(:,3),'y',...
>>       ne(:,1),ne(:,2),ne(:,3),'r',nf(:,1),nf(:,2),nf(:,3),'r',...
>>       ng(:,1),ng(:,2),ng(:,3),'y',nh(:,1),nh(:,2),nh(:,3),'r',...
>>       na(length(na),1),na(length(na),2),na(length(na),3),'*c',...
>>       nb(length(nb),1),nb(length(nb),2),nb(length(nb),3),'*c')
```

The figure shows two stable equilibria, marked with asterisks. Their locations are found by our typing

```
nhat1_n = [na(length(na),1),na(length(na),2),na(length(na),3)]
nhat2_n = [nb(length(nb),1),nb(length(nb),2),nb(length(nb),3)]
```

yielding

```
nhat1_n = 97.0551    0.0104  100.0579
nhat2_n = 0.0104   97.0551  100.0579
```

Both equilibria represent "take-overs." In the first of the stable equilibria, a free-living species-1 and the symbiotic state exist near carrying capacity, whereas a free-living species-2 is nearly absent. The second of the stable equilibria features a free-living species-2 and a symbiont near carrying capacity, and a rare free-living species-1. If we were to encounter this symbiosis near one of the stable equilibria, we would say that the rare free-living species was an obligate symbiont, and the other free-living species a facultative symbiont. Which of the species emerges as obligate, and which facultative, depends on the initial condition.

Interest in symbiosis is slowly growing; it is stimulated in large part by Margulis' studies concerning endosymbiosis. Symbiosis also offers an interesting angle on cancer. If organelles are symbionts, it is conceivable that they attempt to break the partnership, and a malfunctioning organelle that could be diagnosed as a cancer might in fact be a selfish organelle. If medical interest also shines on symbiosis, research in this area will greatly expand.

Researchers on mutualism and symbiosis have tended to focus on coevolution and coadaptation, such as how much good each party should do for the other. Another topic is the degree of species-specificity between partners. Does species-a help *only* species-b, and vice-versa, or do species help one another in broad groupings? Evidence so far points to mutualism being shared across species, rather than being narrowly confined within a particular pair of species. For more in this fascinating area, consult the readings listed at the end of the chapter.

We've come to the end of this chapter that has focussed on two, and sometimes three, interacting populations. We now want to reach out for the whole thing, not just two or three populations, but all of them at once. It's gonna be fun, so stick around.

6.8 Further Readings

Anderson, R. (1989). Populations and infectious diseases: Ecology or epidemiology? *Journal of Animal Ecology*, 60:1–50.

Anderson, R. and May, R. (1982). Coevolution of hosts and parasites. *Parasitology*, 85:411–426.

Anderson, R. and May, R. (1992). *Infectious Diseases of Humans: Dynamics and Control.* Oxford University Press.

Bailey, N. (1975). *The Mathematical Theory of Infectious Diseases and their Application.* Griffin, London.

Dugatkin, L. and Mesterton-Gibbons, M. (1996). Cooperation among unrelated individuals: reciprocal altruism, by-product mutualism and group selection in fishes. *BioSystems*, 37:19–30.

Futuyma, D. and Slatkin, M. (1983). *Coevolution.* Sinauer Associates.

Gause, G. F. (1934). *The Struggle for Existence.* Hafner Press (Reprint 1964).

Godfray, H. (1994). *Parasitoids: Behavioral and Evolutionary Ecology.* Princeton University Press.

Hassell, M. (1978). *The Dynamics of Arthropod Predator-Prey Systems.* Princeton University Press.

Hastings, A. and Harrison, S. (1994). Metapopulation dynamics and genetics. *Annual Review of Ecology and Systematics*, 25:163–184.

Horn, H. and MacArthur, R. H. (1971). Competition among fugitive species in a harlequin environment. *Ecology*, 53:749–752.

Huffaker, C. (1958). Experimental studies on predation: dispersion factors and predator-prey oscillations. *Hilgardia*, 27:343–383.

Iwasa, Y. and Roughgarden, J. (1986). Interspecific competition among metapopulations with space-limited subpopulations. *Theoretical Population Biology*, 30:260–214.

Kermack, W. and McKendrick, A. (1927). A contribution to the mathematical theory of epidemics. *Proceedings of the Royal Society of London A*, 115:700–721.

Kermack, W. and McKendrick, A. (1932). A contribution to the mathematical theory of epidemics. II. the problem of endemicity. *Proceedings of the Royal Society of London A*, 138:55–83.

Kostitzen, V. A. (1939). *Mathematical Biology.* Harrap, London.

Levins, R. (1968). *Evolution in Changing Environments.* Princeton University Press.

Levins, R. (1969). Some demographic and genetic consequences of environmental heterogeneity for biological control. *Bulletin of the Entomological Society of America*, 15:237–240.

Levins, R. and Culver, D. (1971). Regional coexistence of species and competition between rare species. *Proceedings of the National Academy of Sciences (USA)*, 68:1246–1248.

Lotka, A. J. (1925). *Elements of Physical Biology.* Williams and Wilkins (Reprinted 1956, Elements of Mathematical Biology, Dover).

MacArthur, R. H. and Levins, R. (1964). Competition, habitat selection, and character displacement in a patchy environment. *Proceedings of the National Academy of Sciences (USA)*, 51:1207–1210.

MacArthur, R. H. and Levins, R. (1967). The limiting similarity, convergence and divergence of coexisting species. *American Naturalist*, 101:377–385.

Margulis, L. (1976). Genetic and evolutionary consequences of symbiosis. *Experimental Parasitology*, 39:277–349.

Margulis, L. and Fester, R. (1991). *Symbiosis as a Source of Evolutionary Innovation: Speciation and Morphogenesis.* MIT Press.

May, R. (1973). *Stability and Complexity in Model Ecosystems.* Princeton University Press.

May, R. (1978). Host-parasitoid systems in patchy environments: a phenomenological model. *Journal of Animal Ecology*, 47:833–843.

May, R. M. (1976). *Theoretical Ecology, Principles and Applications.* Saunders.

May, R. M. and Leonard, W. J. (1975). Nonlinear aspects of competition between three species. *SIAM J. Appl. Math.*, 29:243–253.

Murdoch, W. W., Reeve, J. D., Huffaker, C. B., and Kennett, C. E. (1984). Biological control of olive scale and its relevance to ecological theory. *American Naturalist*, 123:371–392.

Nardon, P. and Grenier, A. (1993). Symbiose et evolution. *Ann. Soc. Entomol. Fr. (N.S.)*, 29:113–140.

Nicholson, A. and Bailey, V. (1935). The balance of animal populations. Part 1. *Proceedings of the Zoological Society of London*, 3:551–598.

Pacala, S. W. and Tilman, D. (1994). Limiting similarity in mechanistic and spatial models of plant competition in heterogeneous environments. *American Naturalist*, 143:222–257.

Poulin, R. and Vickery, W. (1995). Cleaning symbiosis as an evolutionary game: to cheat or not to cheat? *Journal of Theoretical Biology*, 175:63–70.

Renaud, F. and deMeeus, T. (1991). A simple model of host-parasite evolutionary relationships. parasitism: compromise of conflict? *Journal of Theoretical Biology*, 152:319–327.

Roughgarden, J. (1975). Evolution of marine symbiosis—a simple cost-benefit model. *Ecology*, 56:1201–1208.

Roughgarden, J. (1979). *Theory of Population Genetics and Evolutionary Ecology: An Introduction.* Macmillan (Reprinted 1996, Prentice Hall).

Roughgarden, J. (1995). Anolis *Lizards of the Caribbean: Ecology, Evolution, and Plate Tectonics.* Oxford University Press.

Schaffer, W. and Kot, M. (1985). Nearly one dimensional dynamics in an epidemic. *Journal of Theoretical Biology*, 112:403–427.

Strobeck, C. (1973). N-species competition. *Ecology*, 54:650–654.

Thompson, J. (1982). *Interaction and Coevolution.* Wiley.

Thompson, J. (1994). *The Coevolutionary Process.* University of Chicago Press.

Tilman, D. (1977). Resource competition between planktonic algae: an experimental and theoretical approach. *Ecology*, 58:338–348.

Tilman, D. (1982). *Resource Competition and Community Structure.* Princeton University Press.

Tilman, D. (1988a). On the meaning of competition and the mechanisms of competitive superiority. *Functional Ecology*, 1:304–315.

Tilman, D. (1988b). *Plant Strategies and the Dynamics and Structure of Plant Communities.* Princeton University Press.

Tilman, D. (1990). Constraints and tradeoffs: toward a predictive theory of competition and succession. *Oikos*, 58:3–15.

Tilman, D. (1994). Competition and biodiversity in spatially structured habitats. *Ecology*, 75:2–16.

Tilman, D., May, R. M., Lehman, C. L., and Nowak, M. A. (1994). Habitat destruction and the extinction debt. *Science*, 371:65–66.

Volterra, V. (1926). Variazioni e fluttuazioni del numero d'individui in specie animali conviventi. *Mem. Acad. Lincei*, 2:31–113.

Wilson, D. S. (1980). *The Natural Selection of Populations and Communities.* Benjamin/Cummings.

6.9 Application: Abundance is No Guarantee of Safety

Traditionally biologists think of extinction as a problem for species with small populations[9]. David Tilman has pointed out, however, that if habitat destruction is the process threatening species (and it often is; in fact, habitat loss is the primary source of endangerment for species throughout the world), then the most common species may be those most at risk. This paradoxical prediction follows from a simple model of species competing in a patchy environment, where only one species is able to occupy any particular site. One key to this model is the assumption that species form a rigid competitive hierarchy, such that the species at the top of the hierarchy always displaces species #2, #3, and so on. The second species in the hierarchy (#2) loses to #1, but always displaces species #3, #4, and so forth. A second key assumption is that there is a negative correlation between competitive ability and colonization ability. Good competitors are poor colonizers and vice-versa. We can use these assumptions as the foundation for a simple model that keeps track of the fraction of sites or habitat patches occupied by each species. We start off with a function that evaluates the fraction of sites occupied by species `i`, denoted `p(i)`. The key differential equations represent changes in `p(i)` as a function of colonization of vacant sites `(1-p(i))` and extinction from occupied sites.

Enter the following function, and save it as `six1.m`:

```
function pdot = six1(t,p)
 global m c n;
 for i = 1:n
   sum1 = 0.0;
   sum2 = 0.0;
   for j =1:i ;
     sum1 =sum1 +p(j);
   end
```

[9] This section is contributed by Peter Kareiva.

```
    for j = 1:(i-1)
      sum2 = sum2 +c(j)*p(i)*p(j):
    end
    pdot(i,1) = c(i)*p(i)*(1-sum1)-m(i)*p(i)-sum2;
  end
```

To apply this model, we need to specify three species-specific parameters and the total number of species competing for sites, where `m(i)` is the mortality/year for species-i, `c(i)` is the colonization rate for species-i, `p(i)` is the proportion of patches occupied by species-i, and `n` is the number of species to be evaluated (we will use 5).

Now we are ready to run the model, and we'll see what type of dynamics it predicts. At the command line, type

```
>> global m c n
```

We will assume the simplification that all species have the same mortality rates.

```
>> m = [.02;.02;.02;.02;.02];
```

We use colonization rates that will produce nonzero equilibria for all species:

```
>> c = [.025;.039;.061;.095;.149];
```

We declare the number of species to be used for a model run,

```
>> n = 5;
>> [time p] = ode45('six1',0,1000,[ .1;.2;.3;.4;.5]);
```

and the output is then graphed by our typing

```
>> figure
>> plot(time,p(:,1),'r',time,p(:,2),'y',time,p(:,3),'g', ...
        time,p(:,4),'c',time,p(:,5))
```

Figure 6.40 is produced. By viewing this plot, we can see that after some initial fluctuations, species settle down to a stable equilibrium. Now we are ready to simulate habitat destruction. To do that, we subtract an extra term, `q`, from each `p(i)`. Modify the previous function, `six1.m`, as follows, and save it as `six2.m`:

```
function pdot = six2(t,p)
 global m c q n;
 for i = 1:n
   sum1 = 0.0;
   sum2 = 0.0;
   for j =1:i ;
     sum1 =sum1 +p(j);
   end
   for j = 1:(i-1)
```

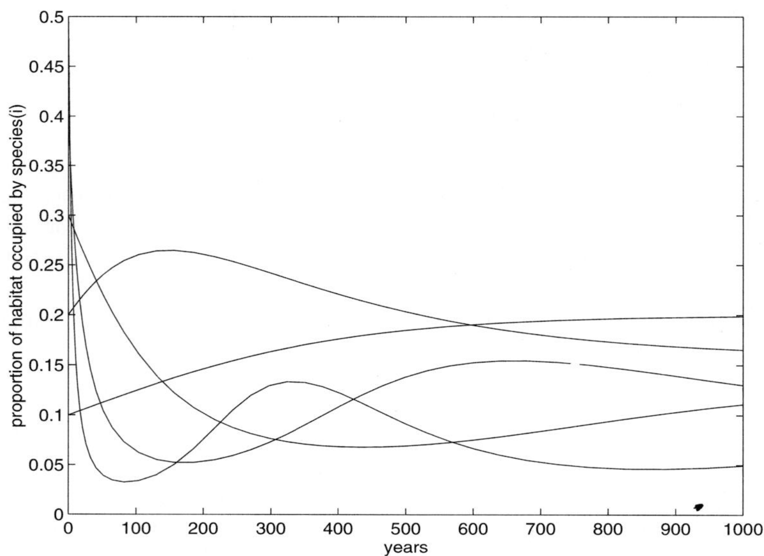

Figure 6.40: An undisturbed system of five competing species shows each species reaching a steady equilibrium as the result of a balance between colonization rates and competitive abilities.

```
      sum2 = sum2 +c(j)*p(i)*p(j)
    end
    pdot(i,1) = c(i)*p(i)*(1-q(1)-sum1)-m(i)*p(i)-sum2;
  end
```

Now redeclare the variables. This time use the initial abundance rates and colonization rates as set out by Tilman (1996), with $p(1) = 0.2$, which is supposedly realistic for temperate forests. The abundance of other species follows a geometric series such that the ith species has abundance $p(i) = p(1)(1 - p(1))^{i-1}$. Colonization rates are those that produce equilibrium in a non-disturbed system with equal mortality rates for all species. To declare variables and add disturbance rate `q`, type

```
>> global c m n q
>> m = [.02;.02;.02;.02;.02];
>> c = [.025;.039;.061;.095;.149];
>> n = 5;
```

We destroy 30% of the habitat and observe the consequences.

```
>> q = [.3;.3;.3;.3;.3];
>> [time p] = ode45('six2',0,2000,[.2;.16;.128;.1024;.082]);
```

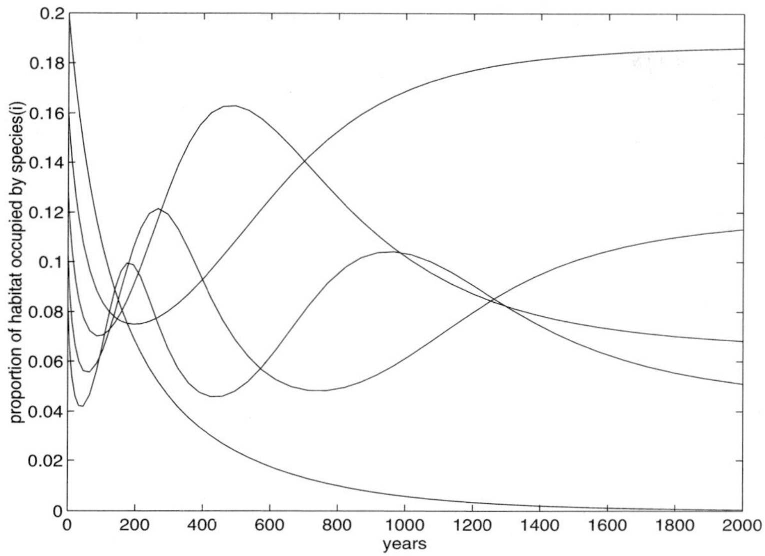

Figure 6.41: The distribution of five competing species when 30% of the habitat is destroyed. The initial abundance is calculated from a geometric series, with the proportion of habitat initially occupied by the most dominant competitor set at .2, as has been estimated for a temperate forest. Colonization rates for the first to the fifth most dominant species are .025, .039, .061, .095, .149. Only the most dominant competitor becomes extinct as a result of habitat destruction.

```
>> plot(time,p(:,1),'r',time,p(:,2),'y',time,p(:,3),'g', ...
        time,p(:,4),'c',time,p(:,5))
```

Figure 6.41 shows how the superior competitor and most abundant species goes extinct, while scarcer species persist.

The time course and outcome vary if we alter the initial frequency of the superior competitor, say to $p(1) = 0.03$, which is supposed to be more typical of a tropical rainforest (where no single species achieves a high initial frequency). But regardless of details, there is a substantial timelag before the extinction occurs; Tilman has labeled the phenomenon an extinction debt.

This model clearly suggests that abundance itself is not a guarantee of immunity to extinction risk. The driving factor here is a tradeoff between colonization and competitive ability, such that common organisms tend to be poor dispersers.

Tilman, D. et al. 1996. Habitat destruction and the extinction debt. *Nature* 371: 65–66.

6.10 Application: How to Control a Pest

All species that attack pests will not be equally effective as biological control agents[10]. One way of identifying those species most likely to control pest populations is to examine models for host-parasitoid or predator-prey interactions. The idea behind this exercise is to associate predator and parasitoid attributes with parameters that yield the best pest control. To do this, we simulate population dynamics for a pest, and then introduce a control agent and follow the ensuing pest depression. A suitable model for pest-parasitoid or pest-predator interactions is the Nicholson-Bailey model discussed in section 6.5.1; we'll add the realistic notion that pests could experience resource-limited population growth when they are at very high densities. (We add a $(1 - N/K)$ term to express that possibility.) The idea is that pests could so thoroughly destroy their crops that their populations crash to zero, a phenomenon sometimes seen in nature.

Type in the following function, to be saved as `six3.m`:

```
function [n,p] = six3(runlen,intro,no,po)
 global r a c k
 n = [no]; p = [0];
 for t= 1:runlen
   if t < intro
     nprime = r*n(t)*(1-n(t)/k);
     if nprime < 0
       nprime =0;
     end;
   n = [n nprime];
   p = [p 0];
   if t == (intro-1)
     avg1 = mean(n)
     maxload1 = max(n)
   end;
   else
   if t==intro
     p(t)=po;
   end;
   nprime = r*n(t)*(1-n(t)/k)*exp(-a*p(t));
   pprime = c*n(t)*(1-exp(-a*p(t)));
   if nprime < 0
     nprime = 0;
   end;
   if pprime < 0
     pprime =0;
   end;
```

[10]This section is contributed by Peter Kareiva.

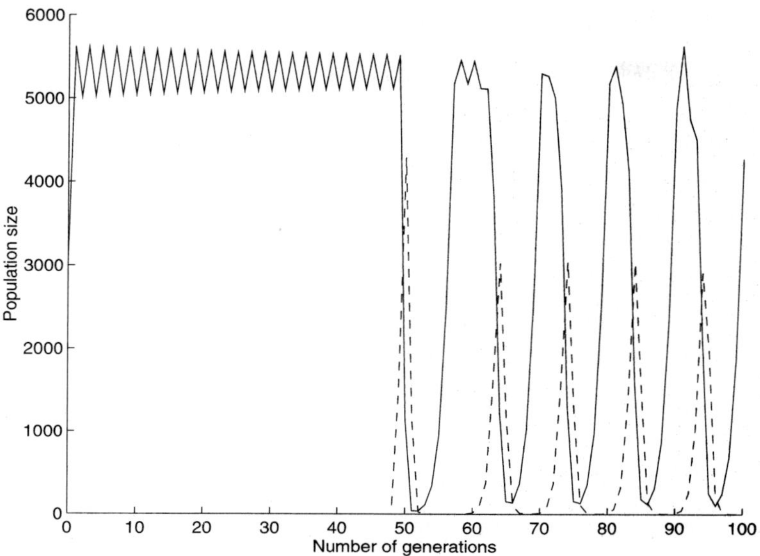

Figure 6.42: A time series showing how a pest population is reduced by the introduction of a biocontrol agent. The growth rate for the pests is 3, the attack rate for the control agent is 0.001, and the conversion rate for pests into new control agents is 1.

```
    n = [n nprime];
    p = [p pprime];
    if t == runlen
      avg2 = mean(n(intro:runlen))
      maxload2 = max(n(intro:runlen))
      avgdiffer = (avg1-avg2)
      maxdiffer = (maxload1-maxload2)
      end;
    end;
  end;
```

In the above function nprime is the pest population, pprime is the biocontrol agent population, r is the maximum rate of population growth for the pest, k is a population density for the pest, which if exceeded causes the pest population to crash to zero, c quantifies the rate at which consumed pests are converted into new biocontrol agents, and a is a measure of the biocontrol agent's searching and hunting efficiency (see discussion under section 6.5.). In addition, runlen is the run time (in generations), intro is the generation of hosts at which the biocontrol agent is introduced, no is the initial host population size, and po is the initial parasite-population size (when it was introduced). When it is done, MATLAB will print out these terms: avg1 is the average pest population before the introduction of

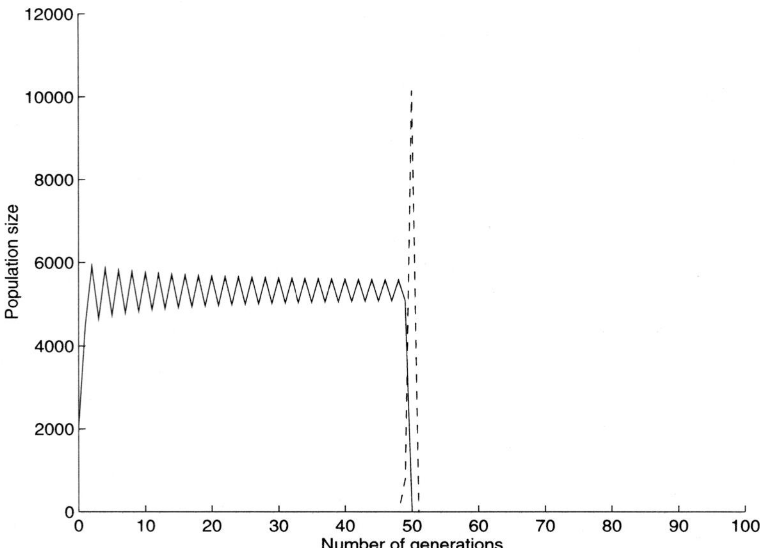

Figure 6.43: A time series showing that a pest population is eliminated with the introduction of a biocontrol agent. The growth rate for the pests is 3, the attack rate for the control agent is 0.2, and the conversion rate for pests into new control agents is 2.

biocontrol agent, **maxload1** is the maximum pest population before the introduction of the biocontrol agent, **avg2** is the average pest population after the introduction of the biocontrol agent, **maxload2** is the maximum pest population after the introduction of the biocontrol agent, **avgdiffer** is the difference between averages, and **maxdiffer** is the difference between maximums. The following variables are needed to call the **six3** function: **r** is the host growth, **a** is the attack rate, **c** is the parasitoid attack rate, and **k** is the carrying capacity.

So, to run our first biocontrol experiment, we type

```
>> global r a c k
>> r=3; a=.2; c=2; k = 8000;
```

Now, to call the function **six3** with an iteration of 100 generations, a biocontrol agent introduced at the pest's 50th generation, an initial **n** of 2000, and an initial **p** of 800, type

```
>> [n,p] = six3(100,50,2000,800)
>> plot(0:100,n,'g')
>> plot(0:100,p,'r')
```

and inspect the resulting figure, Figure 6.42.

We can try a different life history for the biocontrol agent. Let's use a higher searching and hunting capacity and a higher conversion rate for making new biocontrol agents out of consumed pests:

```
>> r=3; c=1; a = .001; k = 8000;
>> [n,p] = six3(100,15,3000,1500)
>> plot(0:100,p,'r-')
>> plot(0:100,n,'g-')
```

This gives us Figure 6.43, which shows the population crashing shortly after the biocontrol agent is introduced. It would seem that this second agent is the better control agent because it drives the pest quickly to zero.

The answers are not always so clearcut. For example, if one biocontrol agent that was efficient at hunting and killing, was compared to an agent that was not such a good hunter, but that converted consumed pests very efficiently into new biocontrol agents, the answer would be more complex. This type of simple model can be used to guide researchers in the selection of biocontrol agents.

Beddington, J., Free, C. and J. Lawton. 1978. Modeling biological control: on the characteristics of successful natural enemies. *Nature* 273: 513–519.

Murdoch, W. and Briggs, C. 1996. Theory for biological control: recent developments. *Ecology* 77: 2001–2013.

Chapter 7

Ecosystems at Work

Vrooom! The bulldozer awakens, gazes at a stand of old-growth forest, and quivers with anticipation. Should the driver shift gears from neutral to forward? Should the driver clear the land of vegetation that's waiting to burn at the next strike of lightning? And should the driver pave the way for playgrounds and schools that will gladden the hearts of children and keep their parents employed? Well, what do you get, and what do you lose, if that old-growth forest is clear-cut? The answer depends on the goods and services that the forest provides compared with the goods and services provided by whatever the forest is replaced with. This chapter is about the theory for the biological side of this issue—the goods and services that ecological systems provide. We leave others to present the case for development, and hope the decision is democratically resolved according to an impartial economic analysis that fully considers both sides.

The two major categories of goods and services provided by ecological systems are those that depend on particular species and those that are carried out by the aggregate of living things. While there is some overlap between these categories, it's convenient to distinguish them conceptually.[1] Today, these distinctions are not rigidly followed, and the many goods that depend on particular species include harvests of fish and timber where people distinguish the kind of fish by its taste and texture, and the kind of tree by the color, hardness and grain of its wood. People also find certain species attractive or interesting, such as parrots, giraffes, and gorillas, or symbolically significant such as the American eagle. And people also gain biologically reactive chemicals from species, such as spices and pharmaceuticals. There is an endless list of species that people value and want to keep around. Services that depend on the aggregate of living things include the gas exchange of oxygen and carbon dioxide by plants with the atmosphere, the role of plants in holding soil for agriculture and

[1]Traditionally, an ecological "community" is a set of interacting populations, such as a forest, and an "ecosystem" is a community plus the physical environment it lives in, such as the forest plus the soil and any streams and ponds in the woods. I use the phrase "ecological system" to refer generically to either a community or ecosystem. Also, an ecosystem is usually defined with a higher degree of aggregation than is a community. A "landscape" is a spatially, or geographically, structured ecosystem. These distinctions are not rigid, however.

of controlling how hot and dry the soil becomes, and the role of worms in mud flats that control how fast sediment is moved by tides and storms. People benefit from endless functions carried out by the aggregate of living things, and will pay dearly if they are somehow lost.

The models here divide naturally into those examining of how many species are in ecological systems and what they do, and those examining aggregate chemical and geological processing by ecological systems. Of course, species are made up of individual organisms, and chemical processing is also carried out by individual organisms, so in principle a model could deal simultaneously with both the species composition and geochemical processing. But in practice, the degree of aggregation is so different in species-oriented and geochemically-oriented models that they are usually formulated with different state variables from the beginning, and treated separately. A state variable in a species-oriented model is typically the number of organisms in each populations or group of populations, and the state variable in a geochemically-oriented model might be the total weight of the organisms, expressed in units of tons of carbon. To determine the overall value of an ecological system to be compared with, say, the value of a development project, one adds together the values of the species-based services with the values of the geochemical function-based services and this comes up with a bottom line.

The present-day urgency to determine the value of ecological goods and services joins a research program underway in ecology since the 1920's. Ecologists have long wondered about the nature of ecological systems. Early on, two views were enunciated—that an ecological system is like a superorganism whose species are its vital organs, contrasted with the view that an ecological system is an unstructured collection of successful immigrants, no more an integrated community than is Grand Central Station at rush hour. The Grand Central Station view is sometimes know as the "individualistic concept" of an ecological system. These polarities persist to this day because of evidence on both sides. Some species are vital to the geochemical function of ecological systems, and others are not, though they may be valuable for other reasons. The venerable dispute between whether an ecological system is an integrated superorganism or an unstructured assembledge of immigrants is important to one's intuition about whether ecological systems are fragile, and to one's fear that humanity is about to destroy the habitability of the planet for all time. If an ecological system is tightly integrated, then it is fragile and the amputation of its pieces is certain to kill it; whereas if an ecological system is a loose grab bag of species, then it will persist through many indignities. Thus, the models of this chapter all tend to refer somehow to the essential question, What is the nature of ecological systems?

7.1 Biodiversity in Ecological Communities

People have been known to say, "If you've seen one, you've seen 'em all." Is this true for the species of trees in a forest, of fish on a coral reef, or of sparrows in a grassland? Do the species of trees in an acre of Panamanian rain forest, the species of sharks at Shoal Bay, the species of pigeons at Katmandu, all do about the same thing or are they functionally different from one another? If they all do about the same thing, then they are substitutable for one another, and in a sense, redundant. If they are all functionally different from one another, then some collective property of all the species taken together is diminished by the loss of one of them. The truth is somewhere in between. Paul Ehrlich has described the consequences of species extinction by analogy to popping the rivets on an airplane's wing. A plane will fly with some of its rivets missing, but after enough pop loose, the wing falls off and the plane crashes. How many species can pop loose from an ecological community before the collective properties of the community are devastated?

The connection between biodiversity and the collective properties of a community have been studied theoretically by simulating numerically what happens to a community when a random species is added to it, or alternatively, what happens when a random bunch of species are all thrown together. The construction of a community by the sequential addition of species is called "community assembly." The analysis of communities resulting from throwing a bunch of species together is called "statistical community dynamics." Let's start with a model of community assembly.

7.1.1 Niche Spacing in Guilds

A "guild" is a collection of species of the same general type, such as the various species of peas, of pigeons, of parrots, or of parrot fish living together. A guild is a good candidate for analysis by the model of nearest-neighbor competition, based on similarity along a niche axis. A common niche axis in animals is body size, or more specifically, the logarithm of body size. The strength of competition depends on the ratio of the body sizes of the competitors or equivalently, on the difference between the logarithms of the body sizes. In plants, the axis might be the extent of the tradeoff between requirements for different nutrients, or between growth rates and dispersal rates. This situation of nearest-neighbor competition along a niche axis was discussed in the preceding chapter, when the concept of limiting similarity was introduced. That discussion used three species. Here we extend this setup to look at how a guild with more than three species is assembled.

Imagine a "bookcase" of competitors, as illustrated in Figure 7.1. The horizontal axis is the niche axis. The carrying capacity, K, equals a constant, K_o, for any species whose position on the niche axis, x, is between $-e_w$ and e_w. We'll call e_w the "environment width." In symbols,

$$\begin{aligned} K(x) &= K_o|x| \quad (|x| < e_w) \\ & \quad\; 0 \qquad\; (|x| > e_w) \end{aligned}$$

Similarly, the intrinsic rate of increase, r, for a species is

$$r(x) = r_o K(x)/K_o$$

so that r and K vary together. Finally, the nearest neighbor competition is modeled as a triangle. The competition coefficient between a species at position x_1 and a species at x_2 on the niche axis is

$$\begin{aligned} \alpha(x_1, x_2) &= 1 - |x_1 - x_2| \quad (|x_1 - x_2| < n_w) \\ & \quad\; 0 \qquad\qquad\qquad (|x_1 - x_2| > n_w) \end{aligned}$$

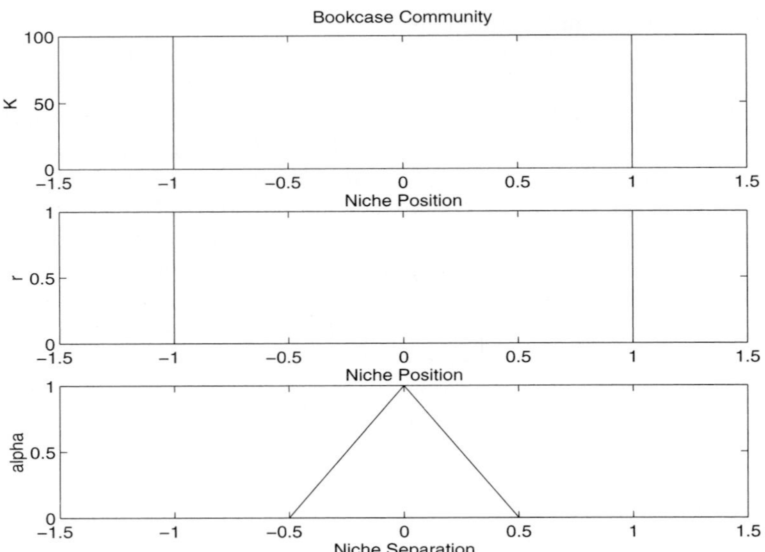

Figure 7.1: Setup for a "bookcase" of competitors. Each species is a "book" that tries to invade at a random spot on the shelf. The top and middle panels show that the carrying capacity and intrinsic rate of increase, as functions of niche position, are uniform from $-e_w$ to e_w. The bottom panel shows a triangle-shaped curve to describe "nearest neighbor" competition, in which the strength of competition between two species depends on how close they are to each other on the niche axis.

The distance that the competition extends along the niche axis is thus $2n_w$, where n_w is called the "niche width." Figure 7.1 illustrates these three functions, and was made by my typing

```
>> ew = 1; nw = 1/2;
>> ko = 100; ro = 1;
>> figure
>> subplot(3,1,1)
>> x = [-1.5*ew -ew -ew ew ew 1.5*ew];
>> k = [      0   0  ko ko  0      0];
>> plot(x,k)
>> subplot(3,1,2)
>> r = [      0   0  ro ro  0      0];
>> plot(x,r)
>> subplot(3,1,3)
>> x = [-1.5*ew -nw 0 nw 1.5*ew];
>> a = [      0   0 1  0      0];
```

```
>> plot(x,a)
```

Let's imagine that no member of this guild is presently in the habitat. The habitat might be a newly formed oceanic island, a mountain top just exposed by glacial retreat, or an area of formerly clear-cut forest being allowed to recover. Let's further imagine that there is a pool somewhere filled with species ready and willing to colonize this newly opened habitat, but the habitat is assumed to be far enough away from the source pool that the colonizing species arrive one at a time. We'll assume that each species is chosen at random from a niche position between $-e_w$ and e_w. What sort of community forms from this process?

Let's develop a computer program to simulate how the community on this habitat is assembled. Let's suppose that we'll look at a total of 100 invaders, so type

```
>> invade_max = 100;
```

Next, initialize the random number generator with

```
>> rand('seed',0);
```

Then, let's define an initially empty vector to contain the names of all the successful colonists:

```
>> x = [];
```

Let's show what goes on as the community is assembled in a figure, so let's initialize the figure with

```
>> figure
>> hold on
```

The main loop for the invaders is then commenced with

```
>> for invasion=1:invade_max
```

We choose the invader's location at random, plot it in the figure as a red asterisk on the horizontal axis, and pause 1/2 second, with

```
>>  x_invade = ew*(2*rand - 1);
>>  plot([x_invade],[0],'*r');
>>  pause(.5)
```

So, at this point we have an invader, and the question is whether it can enter the community. To find out, we introduce the invader to the community. To append it to the list of the community's residents, we use

```
>>  x = [x x_invade];
```

Next, we determine the vector of K's and r's for the community with

```
>>  k = ko*ones(length(x),1);
>>  r = ro*k/ko;
```

The matrix of all the competition coefficients is then determined with

```
>>  a = zeros(length(x),length(x));
>>  for i=1:length(x)
>>    for j=1:length(x)
>>     if abs(x(i) - x(j)) < nw
>>      a(i,j) = 1 - (1/nw)*abs(x(i) - x(j));
>>     end
>>    end
>>  end
```

The first invasion is assumed automatically to succeed, so the population size of the invader simply equals K_o, and the vector of population sizes for all the species is then simply `[ko]`.

```
>>  if (invasion == 1)
>>    n = [ko];
```

We'll also want to keep a log of what happens after each invasion. Specifically, we'll want to remember what population size the successful invader attained, and store this in the vector `invas_his`. Also, we want to remember how many residents, if any, were kicked out by the invader, and store this in the vector `resid_his`. These logs are maintained, for the very first invasion, by our typing

```
>>    invas_his(invasion) = ko;
>>    resid_his(invasion) = 0;
>>  else
```

For all the invaders after the first, we have to test whether they can increase when rare. If there are presently s species in the community, the last one of which is the invader, the condition for increase-when-rare for the invader is

$$K_s - \Sigma_{i=1}^{s-1} \alpha(x_s, x_i) N_i > 0$$

In computerese, this condition is written as

```
>>    if (k(length(x)) - a(length(x),1:length(x)-1)*n) > eps
```

where `eps` is the name of the smallest positive floating-point number that the computer recognizes. Assuming that this condition is met, the invasion succeeds, and the population sizes of all the species now becomes

```
>>        n = a\k;
```

where the \ sign (backwards division symbol) is a special symbol in MATLAB to solve a system of linear equations with its preferred method. Having obtained the vector of population sizes following the invasion, we check for the population sizes that are greater than one, and assume that any populations whose sizes are less than one are extinct. Therefore, the niche positions of the species remaining after the invasion are

```
>>         s_after = find(n>1);
```

To update the log of what happened, we record the abundance of the invader as

```
>>         invas_his(invasion) = n(length(x));
```

and the number of residents who were kicked out is

```
>>         resid_his(invasion) = ...
>>                          (length(x)-1) - length(find(n(1:length(x)-1)>1));
```

Meanwhile, if the invasion did not succeed because the invader could not increase when rare, we delete the invader from the end of the list of niche positions, record that the invader's abundance was zero, and that the number of residents kicked out was also zero. To do all this, we type

```
>>      else
>>         s_after = 1:(length(x)-1);
>>         invas_his(invasion) = 0;
>>         resid_his(invasion) = 0;
>>      end
```

Now that the invasion has played out, we can reassign the vectors of niche positions, abundances, carrying capacities, intrinsic rates of increase, and competition coefficients to reflect the new reality, all by our typing

```
>>      x = x(s_after);
>>      n = n(s_after);
>>      k = k(s_after);
>>      r = r(s_after);
>>      a = a(s_after,s_after);
>>   end
```

Now that the invasion is over we can plot the outcome. First, to erase the screen, type

```
>>   hold off
>>   plot([])
```

Then, we can plot each species now in the community as a vertical line at its niche position, and with a height to indicate its abundance. We type

```
>>   hold on
>>   for i=1:length(x)
>>     plot([x(i) x(i)],[0 n(i)])
>>   end
>>   pause(.25)
```

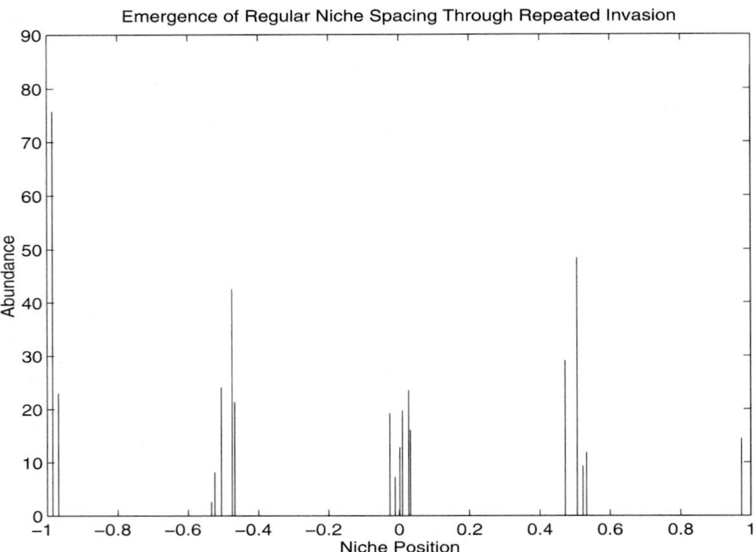

Figure 7.2: The community's composition resulting after 100 invasion attempts. The horizontal axis is niche position, and the vertical axis is abundance. Each vertical line represents one species in the final community.

The result of all 100 invasions appears in Figure 7.2. If you run this simulation, you'll see species after species invading. At first they are spread out all over the niche axis, but gradually they come to be concentrated in five clusters. But before finishing the loop and proceeding to the next invasion, let's record some further summary information about the collective properties of the community we have assembled so far. The number of species in the community is recorded as

```
>>  s_tot(invasion) = length(n);
```

The total abundance of all species, the biomass, is recorded as

```
>>  n_tot(invasion) = sum(n);
```

The most complicated collective property of the community is its stability. As you now know, the stability of the community is determined by the real part of the eigenvalues of its Jacobian evaluated at the equilibrium. Although I didn't mention it in the last chapter because it wasn't needed, the Jacobian for a Lotka-Volterra competition system can be written in a particularly simple form; namely, as the product of a certain diagonal matrix with the α matrix. Specifically, the Jacobian works out to be

$$\text{Jacobian} = \begin{pmatrix} -r_1\hat{N}_1/K_1 & 0 \\ 0 & -r_2\hat{N}_2/K_2 \end{pmatrix} \begin{pmatrix} 1 & \alpha_{12} \\ \alpha_{21} & 1 \end{pmatrix}$$

The formula above is for two species, but the same pattern is true for any number of species. You may want to return to the last chapter to check out this identity with the Jacobian's we calculated previously. Anyway, this identity makes life easy. We can calculate the Jacobian numerically in one line:

```
>>  jac = diag(-r.*n./k)*a;
```

In this command, the `-r.*n./k` computes a vector whose elements are $r_i \hat{N}_i / K_i$. Then `diag(-r.*n./k)` puts this vector on the diagonal of a matrix. Finally, multiplying by `a`, which is the α-matrix, yields the diagonal. Next, we want the magnitude of the real part of all the eigenvalues of this matrix. This is also done in one line:

```
>>  eigs = flipud(sort(-real(eig(jac))));
```

The interesting features of these eigenvalues are first, the one with the smallest magnitude, and second, the number of eigenvalues needed to account for 95% of the sum of the eigenvalues. The eigenvalue with the smallest magnitude tells us how slowly the system returns to equilibrium along its slowest direction. The number of eigenvalues needed to account for 95% of the total tells us the number of dimensions along which 95% of the action is occurring—it is a measure of the system's dynamic dimensionality. The eigenvalue with the smallest magnitude is recorded with

```
>>  eig_min(invasion) = eigs(length(n));
```

To record the number of eigenvalues needed to explain 95% of the total, type

```
>>  eigs_cum = cumsum(eigs)/sum(eigs);
>>  eig_dim(invasion) = min(find(eigs_cum>0.95));
```

The loop for the invasion is then finished up by typing

```
>> end
```

That completes the program to simulate the assembly of a guild, by successive invasions at random positions along the niche axis.

Figure 7.2 illustrates the niche positions created in the final community after 100 invasions have been attempted. As already noted, the species cluster into five groups even though the total diversity is much higher. There is much redundancy. In each group species can be removed without affecting the community much, but if all the species from a group are removed, the community will clearly be diminished.

We can plot more figures to illustrate how the collective properties of the community has developed during the sequence of invasions. The total abundance is illustrated with

```
>> figure
>> plot(n_tot);
```

resulting in Figure 7.3. The figure shows a monotonic increase in the total abundance in the community, but 85% of the final total is attained after the first five species have invaded. To display the number of species occurring in the community through the invasion history, type

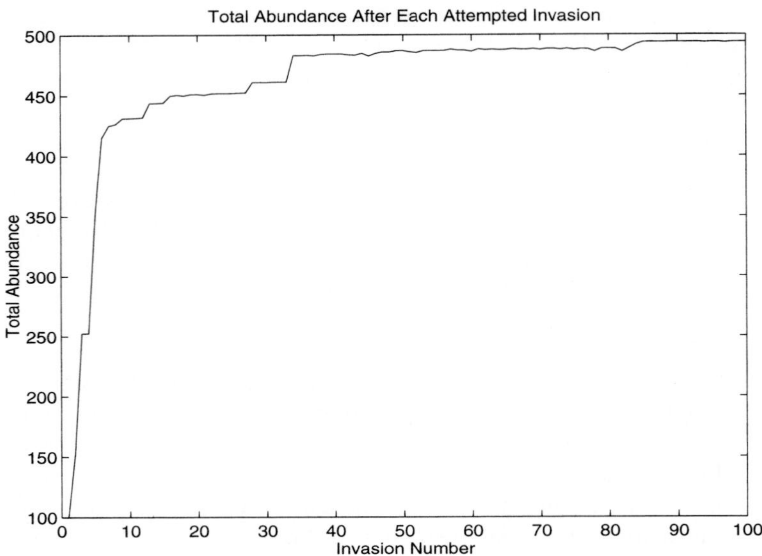

Figure 7.3: The total abundance of all populations in the community, as a function of the invasion number.

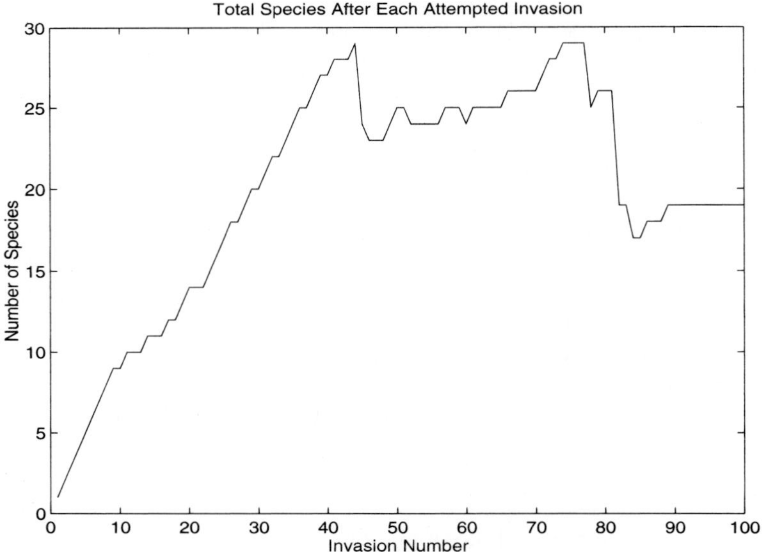

Figure 7.4: The number of species in the community, as a function of the invasion number.

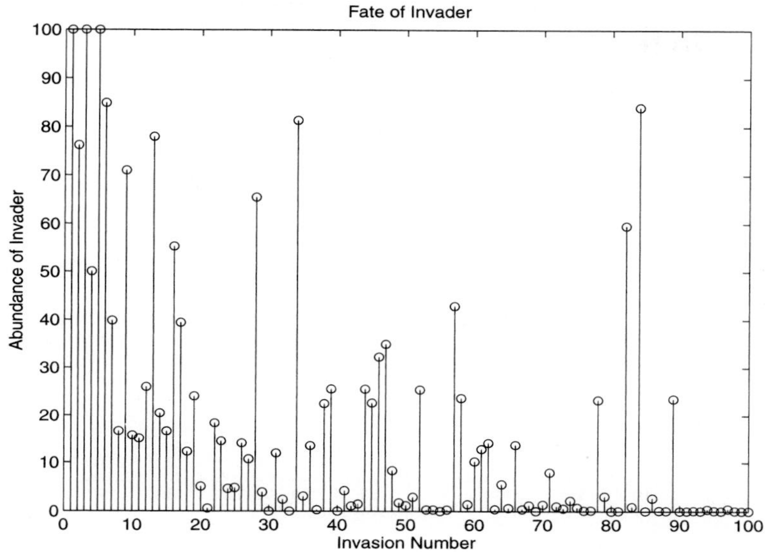

Figure 7.5: The invader's abundance, as a function of the invasion number.

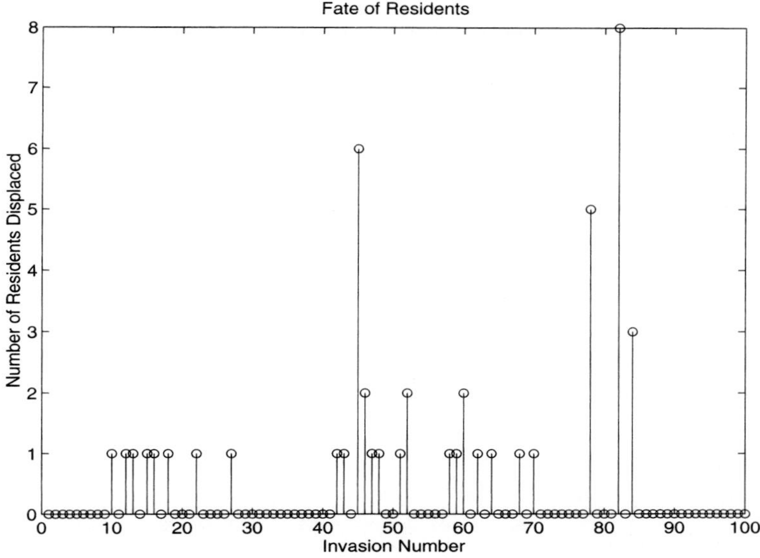

Figure 7.6: The number of residents displaced by an invader, as a function of the invasion number.

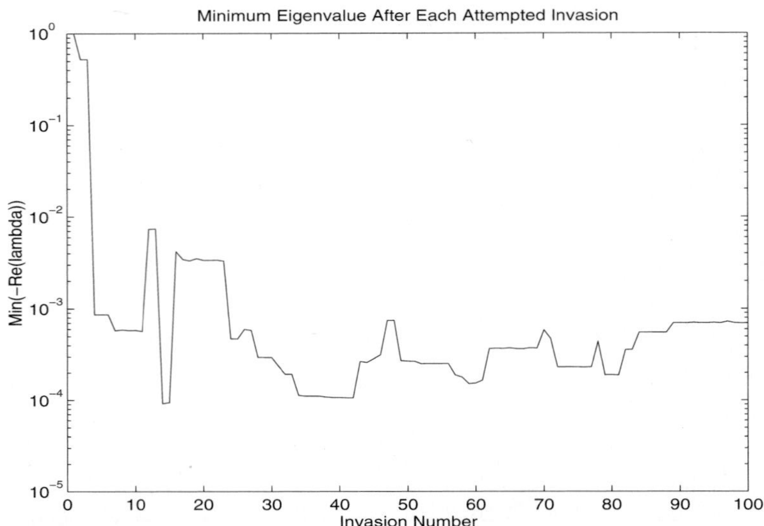

Figure 7.7: The minimum magnitude of the real part of the community's eigenvalues, as a function of invasion number. This indicates the speed of the community's response to a perturbation along its slowest direction.

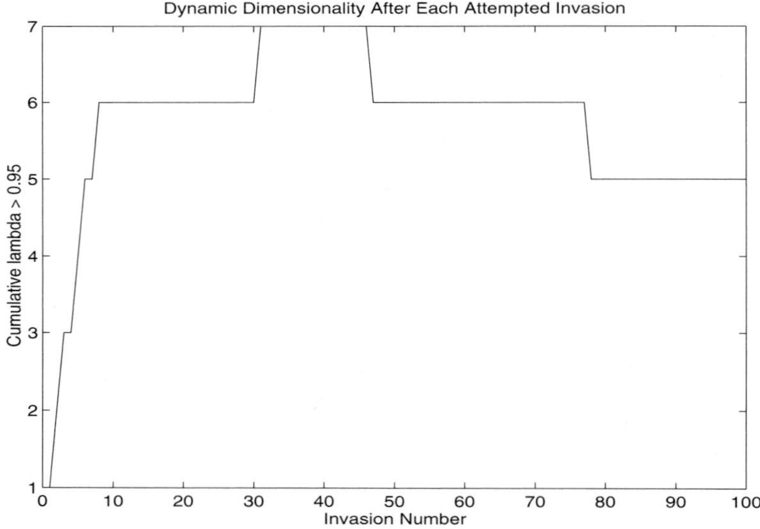

Figure 7.8: The number of eigenvalues to account for 95% of the community's dynamic response to a perturbation. This number is showed as a function of invasion number.

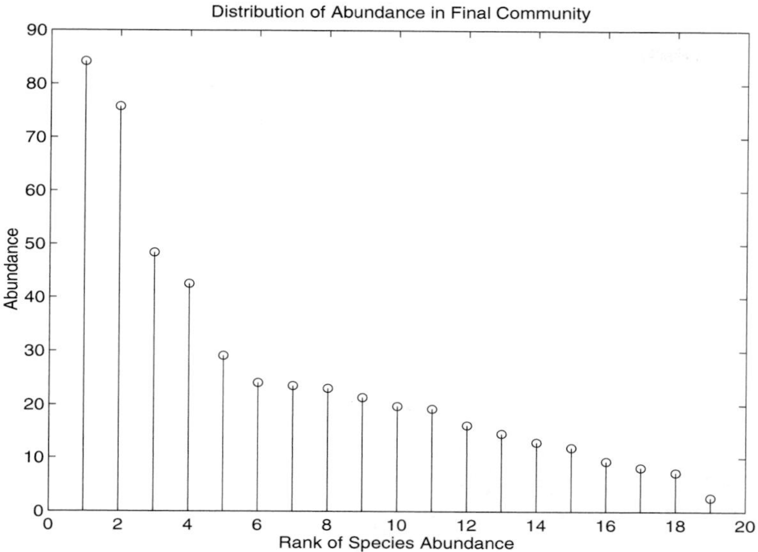

Figure 7.9: The distribution of abundance in the final community after 100 invasion attempts. This is displayed in rank order.

```
>> figure
>> plot(s_tot);
```

This results in Figure 7.4. After 40 invasion attempts, the total number in the community is about 25 species. This drops to about 20 species near invasion 85, when a couple of invaders kick out several residents. The community is clearly filling up after it has around 20 species. The size of an invading population is plotted by our typing

```
>> figure
>> stem(invas_his);
```

resulting in Figure 7.5. The abundance of successful invaders generally drops as the the invasion sequence progresses, although the couple of species near invasion 85 did very well for themselves. The fate of the residents generally becomes more endangered by invaders as the invasion sequence progresses. Typing

```
>> figure
>> stem(resid_his);
```

yields Figure 7.6. Residents are more likely to be knocked out by invaders as the community develops.

Turning to the stability of the community, we can display the minimum magnitude of the real part of the eigenvalues of the community's Jacobian. Typing

```
>> figure
>> semilogy(eig_min);
```

yields Figure 7.7. Early on, the minimum eigenvalue is still quite high, indicating a fast rate of return following a perturbation even in the slowest direction. By about 5 species, the minimum eigenvalue stabilizes as a value between about 0.001 and 0.0001. Thus, after about 5 species, the response to a perturbation in the slowest direction is about 10^{-3} to 10^{-4} times slower than the intrinsic rate of increase, r, which equals 1. The dynamic dimensionality during the sequence of invasions is displayed with

```
>> figure
>> plot(eig_dim);
```

which yields Figure 7.8. This shows that about five to six eigenvalues account for most of the dynamic response of the community to a perturbation. This observation shows that both the structural and dynamic capabilities of the community are attained after 5 to 10 species have invaded. The additional 10 to 20 species don't contribute much to either total biomass or dynamic responsiveness.

Finally, the distribution of abundances of the species in the final community are displayed in rank order by our typing

```
>> figure
>> stem(flipud(sort(n)))
```

as shown in Figure 7.9.

So, how many rivets can we pop and still keep this community flying? To realize its structural and dynamic potential, this community clearly needs 5 to 10 species covering the 5 clusters on the niche axis. Additional species are redundant, but may still be valuable as insurance. If the community is thinned to 5 species, 1 species for each cluster, and 1 of them is attacked by an epidemic, then the aggregate function is impacted. By carrying say, 15 species, structure and function are both conserved and insured. Of course, all possible species, including the 15 needed for ecological function and insurance, may be valuable to people for other reasons. What emerges is that, when species are added to a community, the purely ecological value of each species drops as the community saturates. Moreover, as the community saturates, it becomes increasingly hard to protect all the species in it, because they are increasingly susceptible to being kicked out by an invader, even though the probability that a random invader can enter is becoming lower as the biodiversity increases. Taken together, these results suggest that an economically optimal level of biodiversity for a community exists, a level that takes into account the declining ecological value of incremental biodiversity, and the increasing difficulty of conserving incremental biodiversity, as the community saturates. So, some intermediate degree of biodiversity between species impoverishment and full saturation willwork out as optimal.

The policy implications of determining an economically optimal degree of biodiversity depend on whether the biodiversity of natural communities is currently below or above the optimal level. If natural communities are undersaturated to begin with, then every species probably matters a lot, whereas if natural communities are saturated, some species can be lost with little ecological impact. It's therefore important to determine empirically on which side of the biodiversity optimum natural communities occur.

Consideration of an economically optimal level of biodiversity between species impoverishment and ecological saturation assumes that the non-ecological value of species do not change as the biodiversity changes. Non-ecological values include the value of pharmaceutical compounds produced by plants, and the existence value of beautiful, attractive, or symbolic organisms such as parrots, panda bears, and eagles, and so forth. The ecological value of a species shows what economists call a declining return to scale, which means that each species is worth less ecologically than the preceding one. Moreover the cost of keeping species in a community shows an increasing return to scale, which means that each additional species costs more to keep in the community. The effects of either of these premises alone is sufficient to ensure an optimal biodiversity level greater than zero and less than full saturation, and taken together, these effects even more strongly imply that an intermediate level of biodiversity is optimal. But there is a sneaky little catch here. The non-ecological value of a species is assumed to be independent of biodiversity—a species drawn from a low-diversity community is assumed to be just as valuable, on the average, as a species drawn from a high-diversity community. In fact though, the non-ecological value of a species from more diverse communities may be higher than that from depauperate ones. This is because the non-ecological features we tend to value in species are traits, like unique beauty and unique bioactive chemicals, that have evolved in the context of complex interactions. It's no coincidence that coffee and chocolate come from the tropics, where there are many species, and not the tundra, where there are few. So, the non-ecological value of biodiversity may show an increasing return to scale. If the non-ecological value is sufficiently more than the ecological value, then the optimum biodiversity level may be full saturation, and not some intermediate level. All in all then, the final calculation of the economically optimal level of biodiversity will require that both ecological and non-ecological values are taken into account.

7.1.2 Randomly Constructed Communities

Similar results about the value of biodiversity emerge from looking into communities that are assembled all at once from a bunch of randomly selected colonists. Robert May introduced the idea that by looking at randomly constructed communities, one could investigate the conventional wisdom that biodiversity begets ecological stability. The wisdom was, and is sometimes still claimed to be, that the overall impact of biodiversity is to provide species who can cover for each other when times are tough, and who can, though a system of checks and balances, prevent a renegade species from overtaking the others. But this is a best-case scenario. An alternative intuition is provided by the large electric-power-distribution networks in North America and Europe. When a tree falls and knocks down a power line in Idaho, the perturbation propagates through the power-distribution network, and such a perturbation has caused a power outage for the entire West Coast of the U.S. A diverse ecological community, with interconnections among many species, may be an unstable sort of thing, more like a large, complex power-distribution network subject to massive failures, than a harmonious basketball team firing on all cylinders.

A community constructed all at once from a random bunch of species is not necessarily an abstraction. Often when new habitats open up, they are colonized by many species at once. A fresh-water lake that suddenly develops a connection with the ocean after a storm is simultaneously colonized by many marine species. To explore the community properties of randomly constructed communities, we'll use two functions, one to produce the community and another to inspect its properties. The community will be a set of species competing according to the Lotka-Volterra competition equations. The matrix of competition coefficients is chosen at random. The function used to construct the community is

```
function [a,r,k]=mkcommun(s,ao,ro,ko)
 a = ao*rand(s);
 a = max(diag(ones(s,1)),a);
 r = ro*ones(s,1);
 k = ko*ones(s,1);
```

This function accepts as input the size of the community, `s`; the maximum value of the competition coefficient `ao`; the value of the intrinsic rate of increase for all the species, `ro`; and the value of the carrying capacity for all the species, `ko`. The function returns the competition matrix whose off-diagonal elements are chosen at random independently and uniformly between 0 and `ao` and whose diagonal elements equal one. The function also returns column vectors whose elements are `ro` and `ko`, respectively. With a text editor, type in the function above and save it as `mkcommun.m`.

High competition coefficients imply that the species are strongly "connected" to one another. The `ao` parameter controls the community connectance. If this is low, all the competition coefficients are low too, so that the species are largely independent of one another. If `ao` is high, then many of the competition coefficients are high, and many species are strongly connected to one another.

When `mkcommun` is called, it gives us a randomly constructed community. To see the properties of this community, we'll design what I'll call a "community stethoscope" to tell us all about it. The stethoscope is built from the pieces of the last section—it takes as input the `a` matrix and `r` and `k` vectors that `mkcommun` has provided, and returns information on whether all the species in the community have positive abundances, what is the total number of individuals of all species combined, whether the community is stable, what is the smallest magnitude of the real part of the eigenvalues, how many eigenvalues are needed to account for 95% of the total of the eigenvalues, and a measure of how many species in the community are "keystone" species. A keystone species is defined as one whose removal leads to the loss of additional species—its removal triggers a cascade of further extinctions.

With a text editor, type in the following function and save it as `stetho.m`. Its first two lines are the function's definition and initialization of variables.

```
function [pos,n_tot,stab,eig_min,eig_dim,keyness] = stetho(a,r,k)
 pos = 0; n_tot = 0; stab = 0; eig_min = 0; eig_dim = 0; keyness = 0;
```

The vector of equilibrium population sizes is computed with

```
n = a\k;
```

If all the abundances are positive, the variable `pos` is set equal to 1 (otherwise it's zero), and the total abundance is recorded in the variable `n_tot`:

```
if n > 0
  pos = 1;
  n_tot=sum(n);
```

Now that we know the community is positive, we want the eigenvalues of its Jacobian. As in the last section, we want the magnitude of the real parts of the eigenvalues, sorted from biggest to smallest, as obtained with

```
  jac = diag(-r.*n./k)*a;
  eigs = flipud(sort(-real(eig(jac))));
```

Because we've taken the negative of the real parts, stability is indicated if `eigs` is positive. So, if `eigs` is positive, we'll set `stab` equal to one, record the eigenvalue of smallest magnitude in `eig_min`, and the number of eigenvalues needed to account for 95% of the total in `eig_dim` with

```
  if eigs > 0
    stab = 1;
    eig_min = eigs(length(n));
    eigs_cum = cumsum(eigs)/sum(eigs);
    eig_dim = min(find(eigs_cum>0.95));
```

All this was done in the last section. The next piece of code is new. To calculate how many of the species in the community are keystones, we go through the community and see what happens if each species is deleted. If the remaining community is not positive, then we know we've found a keystone. To begin, we assume that all species are keystones, and start with species `i` in the community:

```
    if length(n)>1
      keys = length(n);
      for i=1:length(n)
```

We make a copy of the `a` matrix, and then delete its ith column and row with

```
        a_reduced = a;
        a_reduced(:,i) = [];
        a_reduced(i,:) = [];
```

Similarly, we make a copy of the `k` vector and delete its ith element with

```
        k_reduced=k;
        k_reduced(i) = [];
```

Now, to calculate the equilibrium abundances of the remaining species, we use

```
n_reduced = a_reduced\k_reduced;
```

If this community is positive, then species-i is not a keystone, and the total keystone count is decremented, with

```
if n_reduced > 0
  keys = keys - 1;
end
```

Actually, species-i theoretically still could be a keystone even if the community in its absence is positive, because we haven't checked if the community in its absence is also stable. So, this calculation yields a conservative estimate of the number of keystone species in the community. The fraction of species that are keystones is recorded in the variable `keyness`, and the function is concluded with

```
        end
        keyness = keys/length(n);
      end
    end
  end
```

Now that `mkcommun.m` and `stetho.m` reside on the disk, we can orchestrate our investigate of random communities. Let's look at 250 instances of communities with up to eight species. So, at the MATLAB command line, define

```
>> maxspecies = 8; maxtrials = 250;
```

Let's also define some vectors to contain the summary statistics, as follows:

```
>> ave_pos = zeros(maxspecies,2);
>> ave_n_tot = zeros(maxspecies,2);
>> ave_stab = zeros(maxspecies,2);
>> ave_eig_min = zeros(maxspecies,2);
>> ave_eig_dim = zeros(maxspecies,2);
>> ave_keyness = zeros(maxspecies,2);
```

We'll do all this for two sets of parameters, for comparison. The first set of parameters are

```
>> ao=1; ro=1; ko=100;
```

So, all we need now is some nested loops to go from `1` to `maxspecies` and from `1` to `maxtrials`; these will accumulate statistics as we go. To initialize variables and start the loops, type

```
>> for s=1:maxspecies
>>  tot_pos = 0; tot_n_tot = 0; tot_stab = 0;
>>  tot_eig_min = 0; tot_eig_dim = 0; tot_keyness = 0;
>>  for trial=1:maxtrials
```

The action is here. To make the community and apply the stethoscope, type

```
>>     [a r k] = mkcommun(s,ao,ro,ko);
>>     [pos,n_tot,stab,eig_min,eig_dim,keyness] = stetho(a,r,k);
```

Now, to accumulate the statistics for this community, type

```
>>     tot_pos = tot_pos + pos;
>>     tot_n_tot = tot_n_tot + n_tot;
>>     tot_stab = tot_stab + stab;
>>     tot_eig_min = tot_eig_min + eig_min;
>>     tot_eig_dim = tot_eig_dim + eig_dim;
>>     tot_keyness = tot_keyness + keyness;
>>   end
```

And now compute the averages for s species. Type

```
>>   ave_pos(s,1) = tot_pos/maxtrials;
>>   ave_n_tot(s,1) = tot_n_tot/tot_pos;
>>   ave_stab(s,1) = tot_stab/tot_pos;
>>   ave_eig_min(s,1) = tot_eig_min/tot_stab;
>>   ave_eig_dim(s,1) = tot_eig_dim/tot_stab;
>>   ave_keyness(s,1) = tot_keyness/tot_stab;
>> end
```

The results of these runs are now stored. Before drawing figures of the results, let's generate runs from an second set of parameters.

Consider now another set of parameters. Here ao $= 2/3$, to indicate a lower average degree of interaction and connectedness than was used in the first set of runs. Type

```
>> ao=0.66; ro=1; ko=100;
```

A set of runs for these parameters is generated as before, with

```
>> for s=1:maxspecies
>>   tot_pos = 0; tot_n_tot = 0; tot_stab = 0;
>>   tot_eig_min = 0; tot_eig_dim = 0; tot_keyness = 0;
>>   for trial=1:maxtrials
>>     [a r k] = mkcommun(s,ao,ro,ko);
>>     [pos,n_tot,stab,eig_min,eig_dim,keyness] = stetho(a,r,k);
>>     tot_pos = tot_pos + pos;
>>     tot_n_tot = tot_n_tot + n_tot;
>>     tot_stab = tot_stab + stab;
>>     tot_eig_min = tot_eig_min + eig_min;
>>     tot_eig_dim = tot_eig_dim + eig_dim;
>>     tot_keyness = tot_keyness + keyness;
>>   end
```

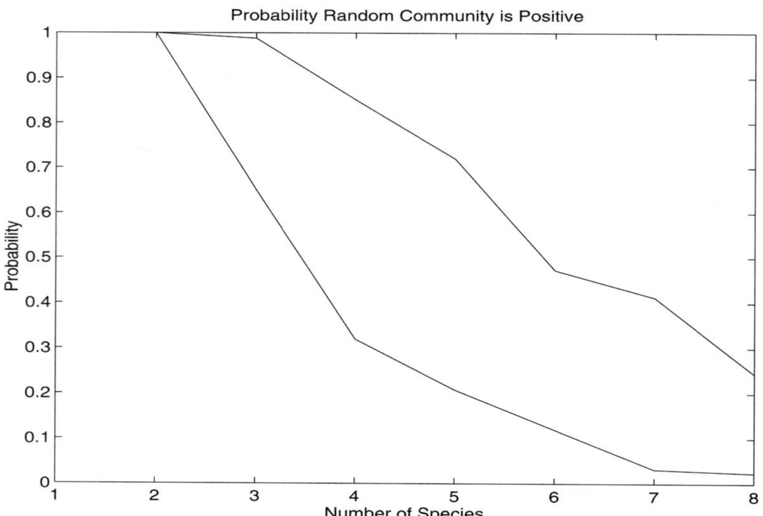

Figure 7.10: The probability that a community of randomly chosen competing species has a positive equilibrium. This probability is computed as a function of biodiversity. The top curve is for biodiversity with low connectance, the bottom curve for that with high connectance.

```
>>   ave_pos(s,2) = tot_pos/maxtrials;
>>   ave_n_tot(s,2) = tot_n_tot/tot_pos;
>>   ave_stab(s,2) = tot_stab/tot_pos;
>>   ave_eig_min(s,2) = tot_eig_min/tot_stab;
>>   ave_eig_dim(s,2) = tot_eig_dim/tot_stab;
>>   ave_keyness(s,2) = tot_keyness/tot_stab;
>> end
```

Now for the fun part. Let's see what the results look like. The probability that a random bunch of species can form a positive community is illustrated by our typing

```
>> figure
>> plot(ave_pos)
```

yielding Figure 7.10. The figure clearly shows that the chance of a bunch of species meshing together so that all the population sizes are positive declines as the biodiversity increases. Moreover, the stronger the connectance, `ao`, the harder it is to find species that will coexist. Throughout the following figures, we will find that biodiversity and strength of interaction go hand in hand. Here, a high probability of coexistence can be achieved with either low biodiversity and high connectance, or high biodiversity with low connectance.

The total abundance summed over all species is plotted by

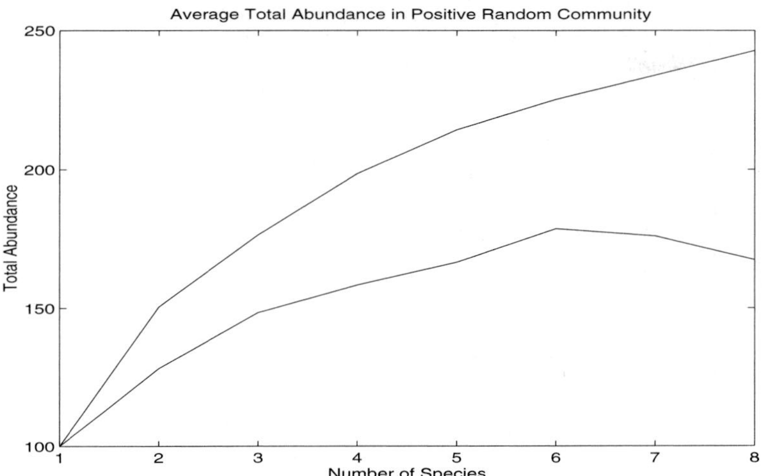

Figure 7.11: The total abundance summed over all species, in a positive community of randomly chosen competing species. Abundance is computed as a function of biodiversity. The top curve is for a community with low connectance, the bottom curve for one with high connectance.

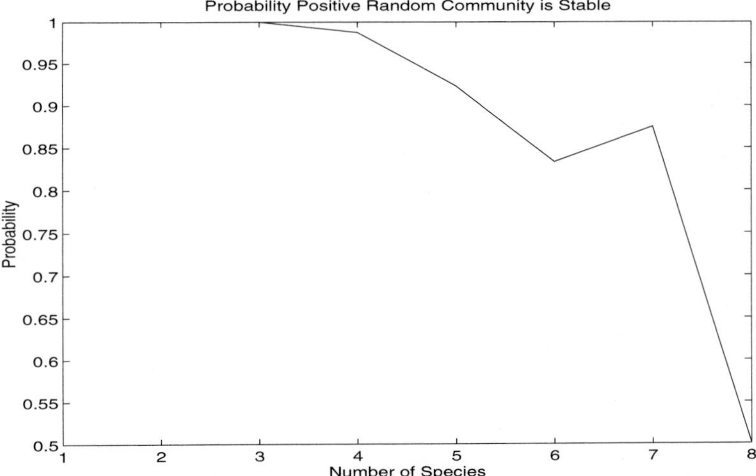

Figure 7.12: The probability that a positive community of randomly chosen competing species is stable. This probability is computed as a function of biodiversity. The top curve is for a community with low connectance, the bottom curve for one with high connectance. (The top curve is hard to see because it runs along the top border of the figure.)

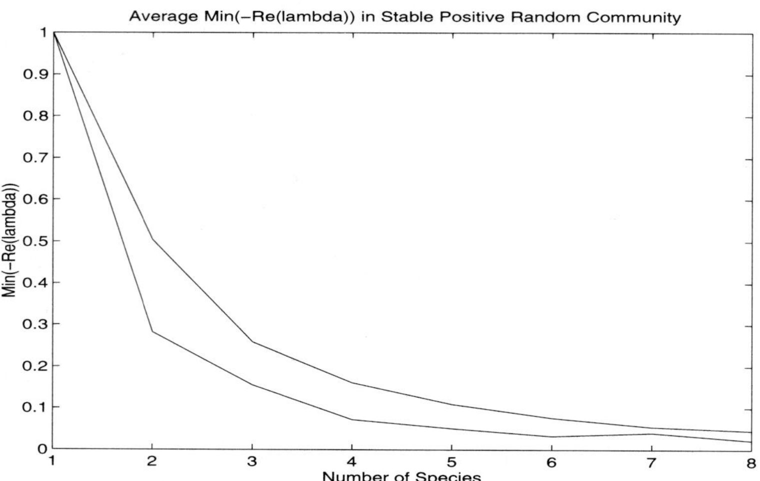

Figure 7.13: The minimum magnitude of the real part of the eigenvalues of a positive stable community of randomly chosen competing species. This is computed as a function of biodiversity. The top curve is for a community with low connectance, the bottom curve for one with high connectance.

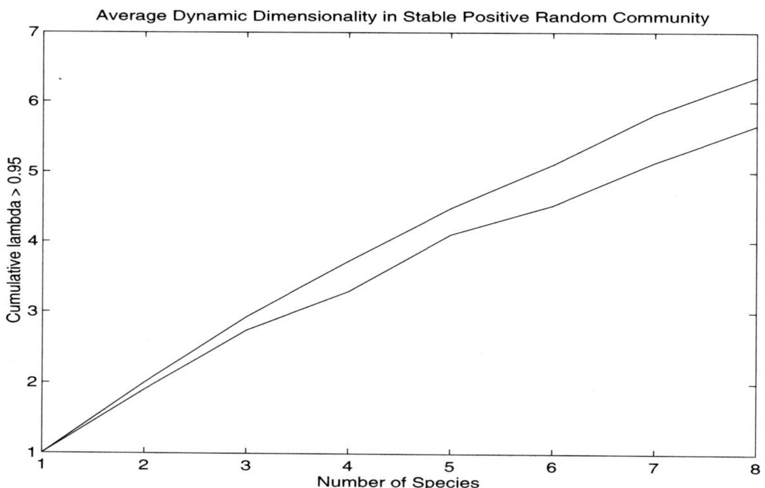

Figure 7.14: The number of eigenvalues accounting for 95% of the response to a perturbation by a positive stable community of random competing species. This is computed as a function of biodiversity. The top curve is for a community with low connectance, the bottom curve for one with high connectance.

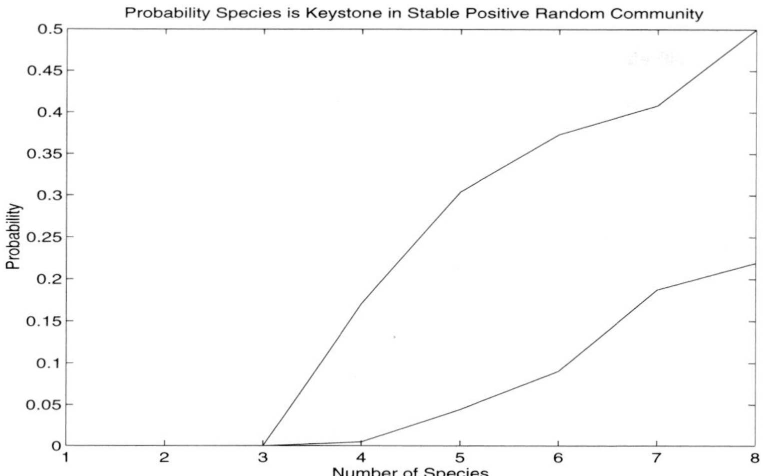

Figure 7.15: The probability that a species in a positive stable community of random competing species is a keystone. This is computed as a function of biodiversity. The bottom curve is for a community with low connectance, the top curve for one with high connectance.

```
>> figure
>> plot(ave_n_tot)
```

yielding Figure 7.11. As in the last section, we see the community saturating as the biodiversity increases. The total abundance shows a decreasing return to scale as biodiversity is increased.

Some of the communities in the preceding figures are not stable. The probability that a positive community is stable is plotted with

```
>> figure
>> plot(ave_stab)
```

This produces Figure 7.12. The probability that a positive community is stable decreases with increasing biodiversity and connectance. And if a positive community is stable, the rate of its return to equilibrium in the slowest direction also declines with biodiversity and connectance, as shown in Figure 7.13. This figure is made by our typing

```
>> figure
>> plot(ave_eig_min)
```

Here the competition is more diffuse, rather than nearest-neighbor as in the last section, and the species are less replaceable with one another. The relation between dynamic dimensionality and biodiversity is illustrated in Figure 7.14, and is made by our typing

```
>> figure
>> plot(ave_eig_dim)
```

The slope relating dynamic dimensionality to biodiversity is less than one, and the curve shows no clear tendency to deviate from linear. These findings suggest that species in positive, stable, diffuse-competition communities are less redundant dynamically than those in nearest-neighbor competition communities.

Finally, the probability that a species is a keystone increases with biodiversity, as shown in Figure 7.15. This figure is made by

```
>> figure
>> plot(ave_keyness)
```

This result illustrates the increasing precariousness of species in diverse, strongly connected communities. It also highlights the increasing difficulty of maintaining biodiversity at high levels.

All in all, the analysis of communities made from a random bunch of colonists shows a picture of community structure and function that more resembles a large electric power-distribution network than a harmonious basketball team. Since results like this began to appear in theoretical ecology in the 1970's, many workers have tried to construct some not-so-random communities instead. They were hoping that communities of harmonious groupings would emerge. With MATLAB it's easy to change the recipe by which the random community is generated. Just replace the `mkcommun` function with one of your choice, and then apply `stetho` to diagnose what you came up with. I'll bet you find that the negative relations between positivity and stability vs. biodiversity and connectance are hard to work around. Go ahead, check it out.

7.1.3 Species-Area Curves and Island Biogeography

The view of an ecological community as a Grand Central Station of immigrants has been most developed in studies of the biodiversity on islands. In fact, islands have long been influential in the intellectual development of evolution and ecology. Darwin visited the Galapagos Islands during his voyage on the H.M.S. Beagle and encountered the finches, now named after him, that provoked his thoughts on evolutionary change. In the 1940's, Lack observed that the body sizes of Darwin's finches varied among islands and he became the first to suggest the idea of nearest-neighbor competition, whereby species with different body (and beak) sizes consume different resources. In the 1960's, patterns that related island size and distance from the mainland to bird-species diversity were discovered by MacArthur and Wilson. Their work spawned the subfield of biogeography, which is the study of the geographical distribution of plants and animals; "island biogeography" focuses specifically on islands. This subfield continues to have great importance today, especially for conservation biology, wherein islands are viewed as prototypes of habitat fragments. Thus islands have historically been fundamental in evolution and ecology, and continue so today.

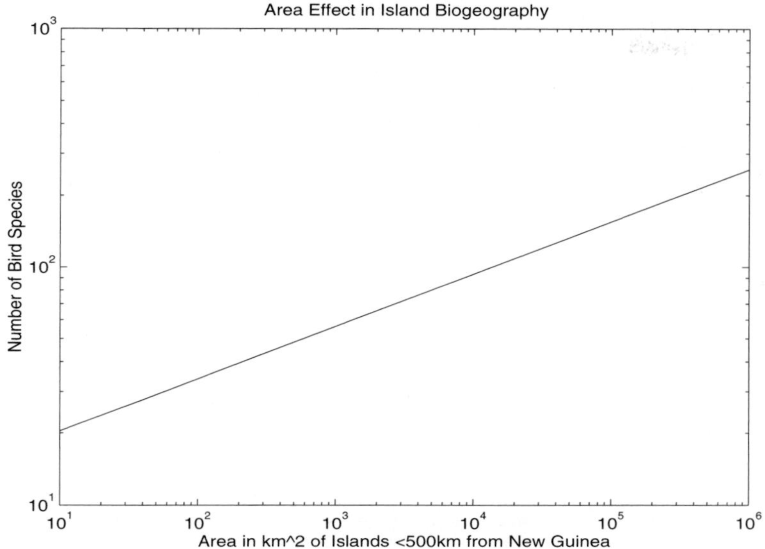

Figure 7.16: The "area effect" of island biogeography for bird species in the vicinity of New Guinea.

Biodiversity on Islands

The basic pattern of biodiversity on islands relates both the area of an island and its distance from a source pool to the number of species on it. If the island is sufficiently close to the source region, then the number of species on it increases as the island's area to some exponent. For islands that are not close to the source region, the number of species on it is compared to the number expected on an island of its area close to the source area. This ratio decreases as an exponential function of the island's distance from the source area. These patterns are called the "area effect" and "distance effect" of island biogeography.

Diamond has summarized the data on the numbers of bird species on islands in the region of New Guinea. He used the following formula, expressed in MATLAB's notation as

```
>> s_new_guinea = '12.3*(1+8.9e-5*l)*exp(-d/2600)*a^0.22'
```

where `a` is the island's area in km^2, `d` is its distance from New Guinea, and `l` is its maximum elevation. With these data, the exponent for the area effect is 0.22. This formula includes the island's elevation in addition to the standard area and distance variables.

The area effect may be illustrated by our typing

```
>> l = 0;
>> d = 0;
>> a = logspace(1,6);
```

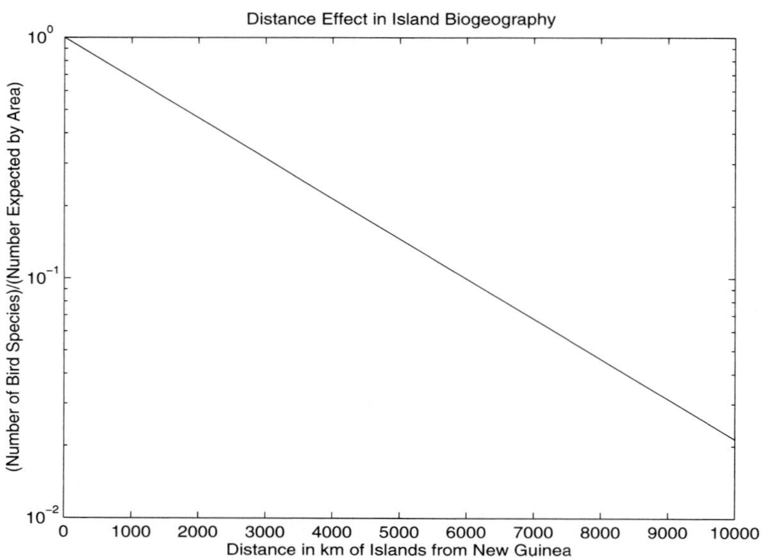

Figure 7.17: The "distance effect" of island biogeography for bird species in the vicinity of New Guinea.

```
>> figure
>> loglog(a,eval(sym2ara(s_new_guinea)))
```

which yields a straight line on a log-log plot, as shown in Figure 7.16.

The distance effect is illustrated by our typing

```
>> d = 0:2000:10000;
>> figure
>> semilogy(d,exp(-d/2600))
```

which is a straight line on a semi-log plot, as shown in Figure 7.17.

These patterns are purely empirical relationships. Regardless of why they are true, they have economic implications. If biodiversity is valuable, then for islands at a particular distance from the source, the value of habitat area shows decreasing return to scale because the exponent of `a` in the formula is less than one. That is, each additional sliver of area added to a piece of habitat yields less increase in biodiversity. Therefore, if each sliver of area has some fixed cost, an economically optimal level of biodiversity exists at which the value of the gain in biodiversity equals the cost of the sliver of land being used to acquire the additional biodiversity. Similarly, the distance effect implies that islands near the source are more valuable for producing biodiversity than are remote islands.

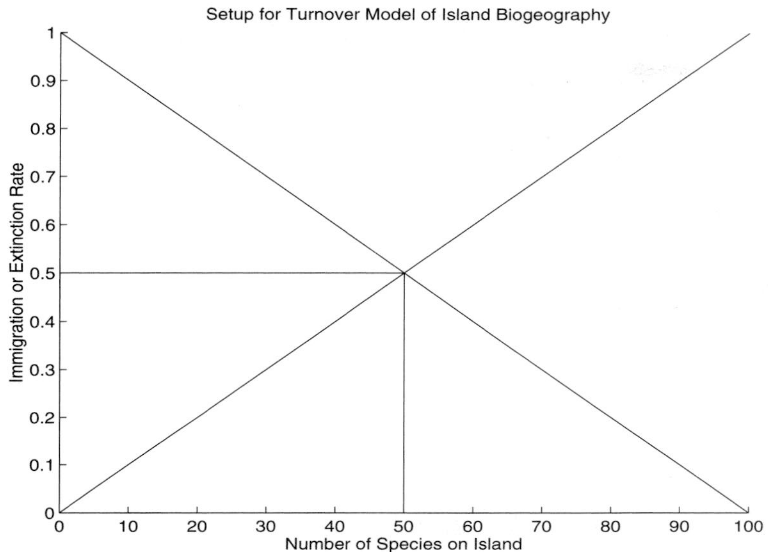

Figure 7.18: The setup for the turnover model of island biogeography uses linear curves for species immigration and extinction. The immigration rate declines with biodiversity, and the immigration rate increases with biodiversity. The curves intersect at the equilibrium level of biodiversity and equilibrium degree of turnover.

The Turnover Theory of Island Biogeography

To explain the area and distance effects of island biogeography, MacArthur and Wilson proposed in 1963 that the fauna on any island exhibit the equilibrium level of biodiversity set by a balance of immigration and extinction on the island. They visualized that an island is continually receiving propagules of new species, and that the arrival rate of new species to the island is a decreasing function of the number of species already on the island. They also supposed that the total rate at which species become extinct on the island increases with the number of species on it.

Graphically, the immigration rate of new propagules to the island can be drawn as a decreasing function of the number on it, and the extinction rate as an increasing function, so that the intersection of these curves represents the steady-state biodiversity on the island on the horizontal axis, and the steady-state turnover rate on the vertical axis. This hypothesis is known as the "turnover" theory of island biogeography.

Linear Immigration/Extinction Curves

To illustrate this theory in MATLAB, define linear immigration and extinction rates by typing

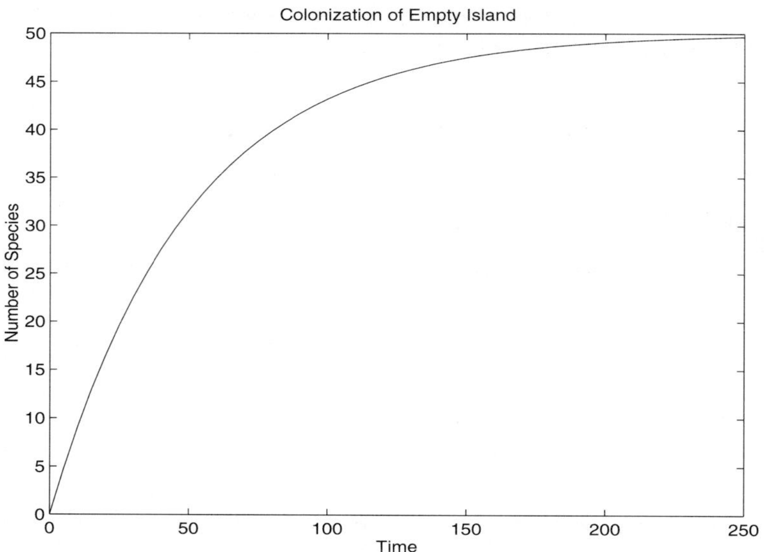

Figure 7.19: The accumulation of biodiversity through time on an island that was initially empty. Following the turnover theory of island biogeography, this representation uses linear species immigration and extinction curves.

```
>> i_rate = 'a*(1-s/p)'
>> e_rate = 'b*s/p'
```

where `s` is the number of species on an island, `p` is the number of species in the source pool, and `a` and `b` are constants of proportionality for immigration and extinction. (`a` here does not mean area, it's just a parameter.) The net rate of change of species on the island is then the immigration minus the extinction rates, which is

```
>> ds = symop(i_rate,'-',e_rate)
```

yielding

```
ds = a*(1-s/p)-b*s/p
```

The equilibrium level of biodiversity on the island is the value of `s` at which `ds` is zero, which is found by our typing

```
>> s_hat = simple(solve(ds,'s'))
```

yielding

```
s_hat = a*p/(a+b)
```

The turnover that takes place while this equilibrium is being maintained is found by our evaluating the extinction rate (or the immigration rate) at **s** equal to **s_hat**. This is found by our typing

```
>> t_hat = simple(subs(e_rate,s_hat,'s'))
```

yielding

```
t_hat = b*a/(a+b)
```

To produce a figure showing these relations, assign some parameters with

```
>> p=100; a=1; b=1;
```

Then type

```
>> s_hat_n = eval(s_hat)
>> t_hat_n = eval(t_hat)
```

which collectively yields

```
s_hat_n = 50
t_hat_n = 0.5000
```

This means there are 50 species on the island at equilibrium and that, on the average, one species enters and one becomes extinct for every two time units. To make the graph, type

```
>> figure
>> hold on
>> s=0:p/10:p;
>> plot(s,eval(sym2ara(i_rate)),'y')
>> plot(s,eval(sym2ara(e_rate)),'r')
>> plot([s_hat_n s_hat_n],[0 t_hat_n],'g')
>> plot([0 s_hat_n],[t_hat_n t_hat_n],'c')
```

This results in Figure 7.18.

The equation for the immigration minus extinction can predict how an empty island accumulates species through time. The equation, **ds**, is a linear differential equation, and MATLAB will find its integral when asked[2]

```
>> s_of_t = simple(dsolve(['Ds = ' ds],'s(0)=so','t'))
```

A graph of the number of species on an island through time is then made by our typing

```
>> so=0;
>> figure
>> t=0:5:250;
>> plot(t,eval(sym2ara(s_of_t)))
```

[2]The formula representing the differential equation to be solved has to be synthesized as a string, `['Ds = ' ds]`.

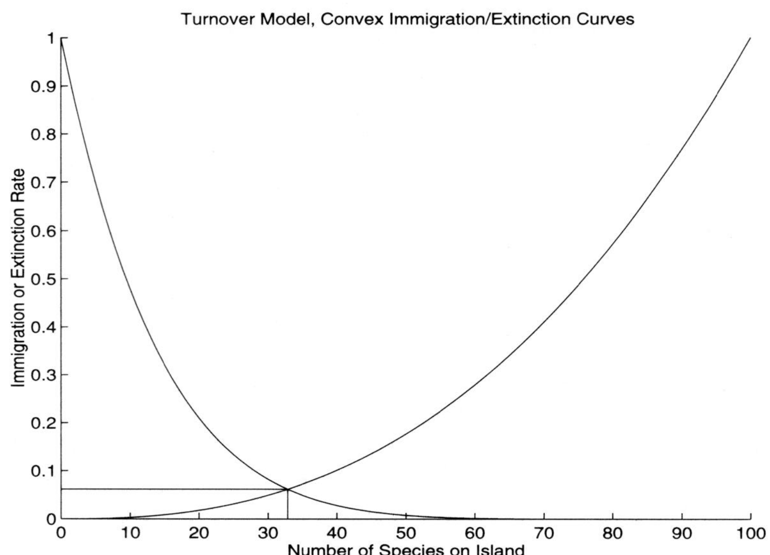

Figure 7.20: This setup for the turnover model of island biogeography has convex species immigration and extinction curves.

resulting in Figure 7.19.

By itself, the hypothesis that the biodiversity on an island is continually turning over, while maintaining a constant total biodiversity, says nothing about the area and distance effects of biogeography. To address these biogeographic patterns, MacArthur and Wilson added to the turnover model two additional assumptions: that the immigration rate to an island decreases with its distance from the source area, and that the extinction rate on an island increases as the island becomes smaller.

Thus, the classic theory of island biogeography is today identified with three propositions: the area and distance effects that solely describe empirical patterns; the hypothesis that the biodiversity on islands is in a continual state of turnover; and the further hypothesis that the area and distance effects are explained by the geographic variation of immigration and extinction rates in relation to the island's area and distance from the source fauna.

Convex Immigration/Extinction Curves

Is biodiversity on islands in a continual state of turnover? The alternative is that an island "fills up" to a level that depends on its area and variety of habitats, and that thereafter any propagules to the island fail because no ecological space is left. Williamson in 1983 termed turnover as "ecologically trivial" in application to birds on the island of Skokholm, because the core species were unaffected while the propagules going extinct never had an

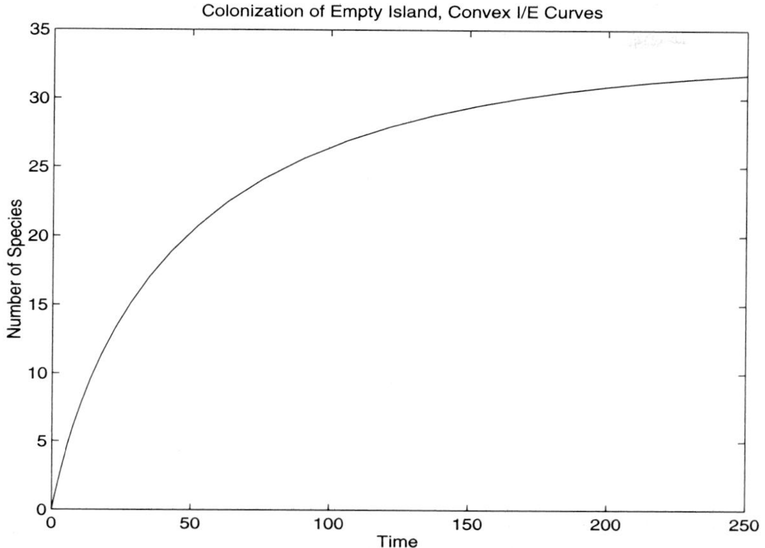

Figure 7.21: The accumulation of biodiversity through time on an island that was initially empty. Following the turnover theory of island biogeography, this prediction uses convex species immigration and extinction curves.

opportunity to establish; the island was already full.

Most textbooks draw the immigration and extinction curves as linear functions of the number of species on the island. In fact, in 1981 Gilpin and Diamond measured the extinction and immigration rates of birds in the Solomon Islands as convex curves: the decline in the immigration rate is slower than linear and the rise of the extinction rate is faster than linear. The produced shape is called convex, and is illustrated as sharply scalloped curves. The implication is that the islands tend to have a core fauna, and that any species beyond this core will have very high extinction rates and very low addition rates. Gilpin and Diamond view the situation as still consistent with the turnover model, when it is taken with the proviso that turnover is not uniformly distributed throughout the fauna. Williamson interprets essentially the same situation as consistent with the qualitatively different model wherein species inside the core have zero turnover and species outside the core all fail at the beach head.

To use the convex immigration and extinction curves found by Gilpin and Diamond to work out island-biogeographical theory, type

```
>> i_rate = 'a*(1-s/p)^(7)'
>> e_rate = 'b*(s/p)^(2.5)'
```

The equilibrium biodiversity is found numerically. Type the following function with a text editor, and save it as convex_e.m.

```
function sdot = convex_e(s)
 global p a b;
 sdot=a*(1-s/p)^(7)-b*(s/p)^(2.5);
```

The equilibrium biodiversity is then found by our typing

```
>> global p a b;
>> p=100; a=1; b=1;
>> s_hat_n = real(fzero('convex_e',0))
```

yielding

```
s_hat_n = 32.8243
```

The turnover corresponding to this equilibrium is then

```
>> t_hat_n = eval(subs(e_rate,'s_hat_n','s'))
```

yielding

```
t_hat_n = 0.0617
```

A graph of the convex immigration and extinction curves, together with the equilibrium biodiversity and turnover is made by our typing

```
>> figure
>> hold on
>> s=0:p;
>> plot(s,eval(sym2ara(i_rate)),'y')
>> plot(s,eval(sym2ara(e_rate)),'r')
>> plot([s_hat_n s_hat_n],[0 t_hat_n],'g')
>> plot([0 s_hat_n],[t_hat_n t_hat_n],'c')
```

This results in Figure 7.20. Because the curves are so scalloped, the equilibrium turnover is low.

MATLAB's numerical methods can predict the accumulation of biodiversity on an empty island if the immigration and extinction curves are convex. With a text editor, type the following function and save it as convex_t.m

```
function sdot = convex_t(t,s)
 global p a b;
 sdot=a*(1-s/p)^(7)-b*(s/p)^(2.5);
```

Then, typing

```
>> figure
>> [t s] = ode23('convex_t',0,250,0);
>> plot(t,s)
```

results in Figure 7.21.

So, is the biodiversity on islands truly in a state of continual turnover, albeit with convex curves for species immigration and extinction, or do islands just fill up, leaving no room in the inn for late arrivals?

For vertebrates, further evidence against turnover comes from the observation of "nested subset species-area relationships." This pattern is one wherein the smallest islands have a particular species, say A, then the next larger has that plus another, A and B, a still larger island has A, B, and C, and so forth. Each island's fauna is a proper subset of a larger island's fauna until the largest island is reached. This pattern occurs for the herpetofauna of the British Virgin Islands and for lizards in the Bahamas. This pattern is not consistent with the turnover model. According to the turnover model, each island small enough to have only one species from the source area should have a random species from the source pool of species, and each island with two species should have a random pair of species. The existence of a relation between biodiversity and island area does not itself imply that turnover is taking place as well. The Virgin Islands have an increasing species–area relation, but the nested subset pattern of species identities rules out the turnover theory of island biogeography there.

For invertebrates, the data for turnover are strong and convincing. In 1969–70, Simberloff and Wilson experimentally removed with insecticide the arthropods on small mangrove islands in the Florida cays. They directly observed two stages: the approach to a stable steady-state biodiversity as these mangrove islands were recolonized by arthropods; and the continual turnover of species within the islands even after the steady-state biodiversity was attained. The extinction of arthropod species within the mangrove islands was balanced by the arrival of new species that took their place. More recently, in the Bahamas, Schoener in 1983 directly quantified the relative propensity of arthropods and vertebrates to exhibit turnover and showed that, indeed, insects exhibit a higher turnover than lizards.

The implications for conservation seem quite clear. The conservation of vertebrates requires large islands, because small islands tend to be homogeneous, and even a great number of them will contain little biodiversity. For the conservation of invertebrates, one can trade a large island for lots of small islands with the same total area and come out ahead if the exponent in the species-area relation is less than one.

7.1.4 Succession and Disturbance

"Succession" is the sequential replacement of species in newly opened habitat. The first species in the sequence are called the "pioneers," the final species in the sequence comprise what is called the "climax community," and any species between the pioneers and the climax are called "transitional species." Ecological succession has been discussed at length in ecology since the 1920's, when the idea of a community as a superorganism was proposed. Succession for an ecological community was considered to be the counterpart of development in an individual. Two forms of succession are usually distinguished. "Primary succession" takes place on substrate that was previously unoccupied. Colonization of sand dunes, bare rock exposed by retreating glaciers, and newly cooled lava extruded from the sides of a volcano are examples. "Secondary succession" takes place on substrate that has recently contained organisms, and has still has its soil. Two examples of this process are the colonization of abandoned agricultural fields by propagules from the surrounding forest, which is called "old-field succession," and the recovery of habitat after a fire or hurricane. Fire and hurricanes are examples of what are called "disturbance" by ecologists because these events reset the community to an earlier stage in a successional sequence.

The rationale for thinking of succession as a counterpart to development is the hypothesis that the pioneer species are actually necessary for successional sequence to proceed. It is hypothesized that the transitional and climax species cannot enter the habitat unless their way has been paved by the pioneer species. Subsequent work has qualified this hypothesis substantially. In some cases a pioneer species is clearly needed. Beach grass that propagates with runners stabilizes the sand; the sand doesn't blow around and leaf litter, whose decomposition starts the development of soil, can accumulate. The transitional and climax species need the soil and a stable substrate before they can grow. In other cases, the climax species can do just fine if they happen to arrive at the habitat first, and the pioneering species, if anything, can retard the growth of the climax species if they compete for water and nutrients. Furthermore, climax communities may vary, according to the identity and sequence of the transitional species that happen to colonize the habitat. Here are two simple models that illustrate the main themes of succession theory.

Primary Succession

The main idea behind primary succession is that the pioneer species produces something in the environment, such as a soil, that the climax species needs. Here's how this idea may be modeled. Let x_1 be the amount of soil in the habitat, x_2 be the abundance of the pioneer species, and x_3 the abundance of the climax species. We assume that the rate of soil production is proportional to the abundance of the pioneer species, so

$$\frac{dx_1}{dt} = x_2$$

Next, the pioneer species can grow logistically in the habitat by itself, but also receives interspecific competition from the climax species,

$$\frac{dx_2}{dt} = x_2(1 - \frac{x_2 + x_3}{K_2})$$

Finally, we'll assume that the climax species also grows logistically, but with an intrinsic rate of increase and carrying capacity that are proportional to the amount of soil, x_1. So, if the climax species' intrinsic rate of increase equals x_1, its carrying capacity equals $x_1 K_3$, and it receives no interspecific competition from the pioneer species, its logistic equation works out to be

$$\frac{dx_3}{dt} = x_3(x_1 - \frac{x_3}{K_3})$$

Even this simple model is best analyzed numerically.

Using a text editor, type in the following function and save it as `psuccess.m`

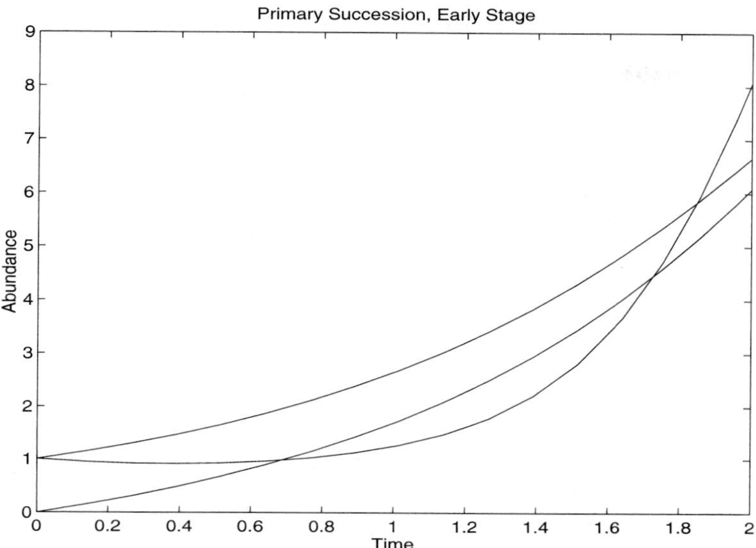

Figure 7.22: Primary succession in its early stage. Both the pioneer and climax species start with equal abundance, and there is no soil. During this early stage, the pioneer increases in abundance, the climax species declines slightly before increasing and overtaking the pioneer, and the soil continuously accumulates.

```
function dx=psuccess(t,x)
 global k2 k3;
 dx(1,1) = x(2);
 dx(2,1) = x(2)*(1-(x(2)+x(3))/k2);
 dx(3,1) = x(3)*(x(1)-x(3)/k3);
```

Then type

```
>> global k2 k3;
>> k2 = 100; k3 = 2;
>> [t x] = ode45('psuccess',0,2,[0;1;1]);
>> figure
>> plot(t,x)
```

This yields Figure 7.22, which models the early stage of succession, from time from 0 to 2. Initially there is no soil, and one unit each of the pioneer and climax species. Because there is no soil yet the climax species declines somewhat while the pioneer grows. And as the pioneer grows the soil accumulates. So, to run the model for a longer time, from 0 to 12, type

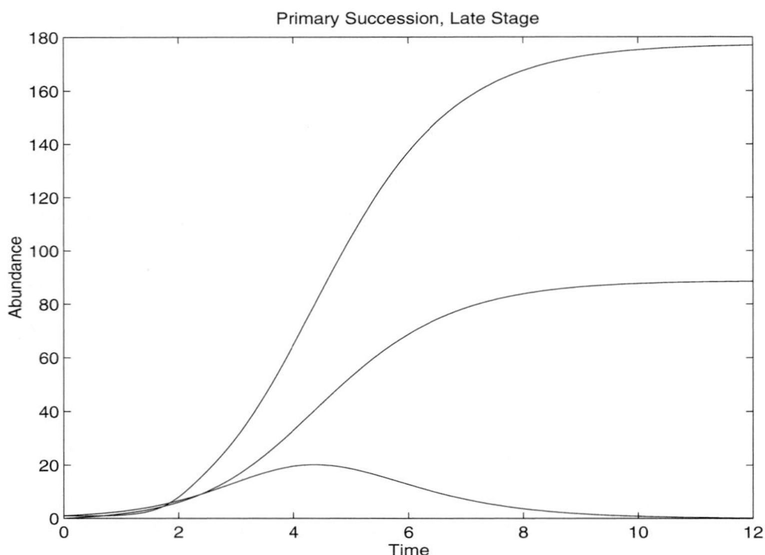

Figure 7.23: The full course of primary succession. The top curve is the abundance of the climax species, the middle curve is amount of accumulated soil, and the bottom curve is the abundance of the pioneer. The pioneer species is eliminated and both the climax species and soil approach asymptotic levels.

```
>> [t x] = ode45('psuccess',0,12,[0;1;1]);
>> figure
>> plot(t,x)
```

which yields Figure 7.23. This longer time-period shows the pioneer being eliminated, and the climax community consisting of the climax species together with the accumulated soil. The policy implications of primary succession have long been known. Obviously, if land use damages the soil so that primary succession has to start all over again, recovery is greatly prolonged in comparison to recovery from patterns of use that conserve the soil. Models like this may be tailored to predict the time needed to restore habitat when primary succession is involved, and then this time may be compared with that needed by less degraded habitat that is restored by the process of secondary succession.

Secondary Succession

Secondary succession is taking place all around us. The spatial mosaic of disturbances (such as when and where hurricanes hit and fires strike) leaves the regional landscape in a corresponding mosaic of recovery stages. If we focus on any one spot in this mosaic, we often find a cycle. After a fire, a place starts out with little vegetation on top of the soil, so that little fuel is left to burn should another flash of lighting immediately strike the site. After time, the vegetation grows back, and the susceptibility to another burn increases. An old, hot, and dry stand of chaparral is a fire waiting to happen. Thus, the climax community is more susceptible to disturbance than its preceding stages. When the disturbance comes, the community is reset to an earlier successional stage and the cycle starts all over again. A model can easily illustrate how such fire cycles recur.

We'll assume there is a "fire index" which describes the propensity of the vegetation to burn, and we'll call this index x_1. The vegetation is denoted as x_2, and we'll assume that the fire index of standing vegetation continuously increases as branches dry out and become increasingly brittle and prone to burn. Thus, we'll assume that dx_1/dt increases with x_2. But dry branches also fall, and decompose, so the fire index will decline by itself if it is not continuously resupplied. These considerations together suggest that

$$\frac{dx_1}{dt} = x_2 - x_1$$

Next, we'll imagine that the vegetation grows exponentially, except when there are fires, which occur in proportion to the fire index. These considerations suggest that

$$\frac{dx_2}{dt} = x_2 - x_1 x_2$$

MATLAB will tell us much about this model if we type

```
>> dx1 = 'x2-x1'
>> dx2 = 'x2-x1*x2'
```

and then type

```
>> [x1hat x2hat] = solve(dx1,dx2,'x1,x2')
```

which yields two equilibria

```
x1hat = [0]
        [1]

x2hat = [0]
        [1]
```

The stability of the nonzero equilibrium is then analyzed by typing

```
>> jacob = jacobian(sym([dx1 '; ' dx2]),sym('[x1,x2]'))
>> jacob = subs(jacob,sym(x1hat,2,1),'x1');
>> jacob = subs(jacob,sym(x2hat,2,1),'x2');
>> jacob = simple(jacob)
>> trace = simple(symop(sym(jacob,1,1),'+',sym(jacob,2,2)))
>> determinant = simple(determ(jacob))
>> discriminant = simple(symop(trace,'^','2','-','4','*',determinant))
```

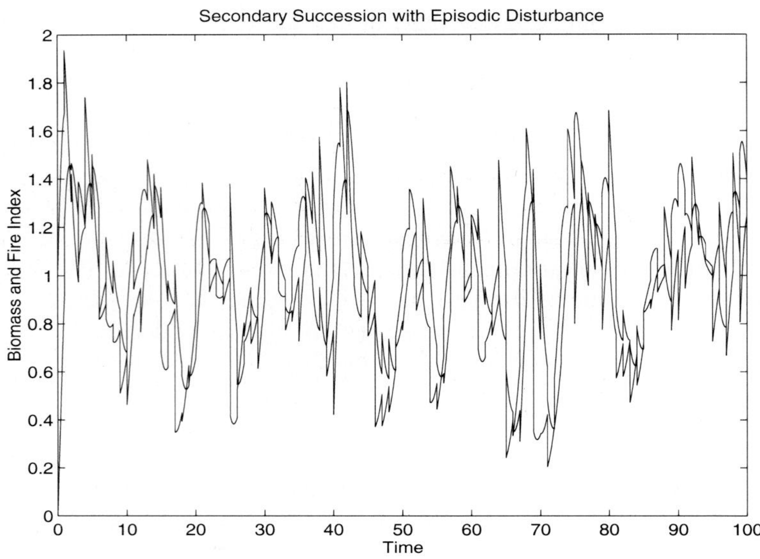

Figure 7.24: The fire cycle resulting from the combination of exponential growth and the density-dependent susceptibility to episodic disturbance. The curve for the fire index leads the curve for the vegetation level.

yielding

```
trace = -1
determinant = 1
discriminant = -3
```

Because the trace is negative, determinant positive, and discriminant negative, we know the equilibrium is a stable focus. This means that the exponential growth of the vegetation is checked by the fires that occur as the fire index builds up, and that there is an oscillatory approach to the equilibrium. The fire acts as a density-dependent factor that stabilizes the growth of the vegetation. As is, the model does not predict a recurring fire cycle because the system gradually settles into a steady state with a degree of continual burning that compensates for the continuous growth of fuel.

A real fire is episodic, not continuous, and starts when lightning strikes. A fire burns to a random extent, depending on the wind direction and speed, and on the air temperature. If we add random perturbations to the model above, then the system is prevented from coming to equilibrium and instead is always spiraling into the equilibrium without ever getting close to it for very long. To include the random aspects of fire, let's integrate the differential equations year by year. At the start of each year we'll add (or subtract) a random amount to the state left at the end of the preceding year, and then integrate for one year.

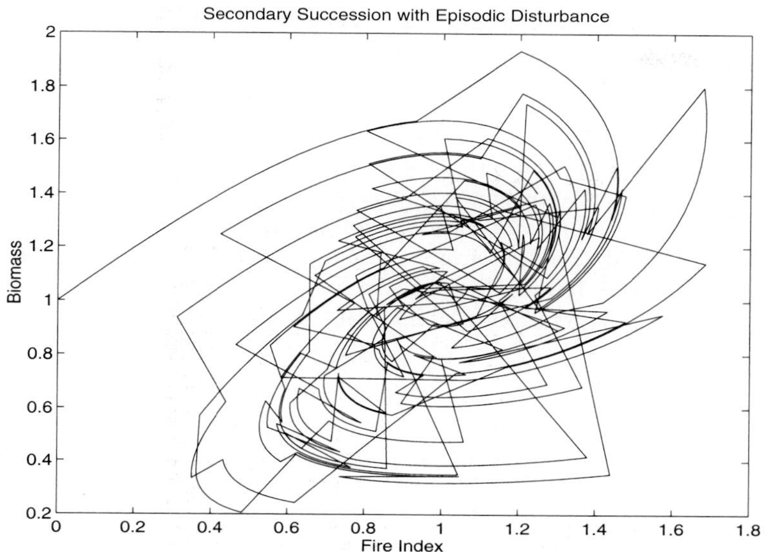

Figure 7.25: The fire cycle, when the fire index is plotted on the horizontal axis and the vegetation level is plotted on the vertical axis. The spiral starts on the vertical axis at the point (0,1), and spirals clockwise around an equilibrium point.

By splicing together the year-by-year trajectories, we obtain a long-term trajectory that includes random impulses inserted at each year.

Type in the following function, and save it as `ssuccess.m`.

```
function dx=ssuccess(t,x)
 dx(1) = x(2)-x(1);
 dx(2) = x(2)-x(1)*x(2);
```

This is now ready to be used with our trusty ol' numerical integrator, `ode45`. Let's store the long-term trajectory in `t_cum`, which contains all the time points, and `x_cum` which contains x_1 and x_2 at these time points. We'll initialize these as empty vectors, and also define the initial condition for x_1 as 0 and for x_2 as 1. This is all done by our typing at the MATLAB command line

```
>> t_cum = []; x_cum = []; x1init = 0; x2init = 1;
```

Then a trajectory spanning 100 years is assembled by our typing

```
>> for year=1:100
>>   [t x] = ode45('ssuccess',year-1,year,[x1init;x2init]);
>>   t_cum(length(t_cum)+1:length(t_cum)+length(t)) = t;
```

```
>>   x_cum(length(x_cum)+1:length(x_cum)+length(x),:) = x;
>>   x1init = (1+(0.25*randn))*x(length(x),1);
>>   x2init = (1+(0.25*randn))*x(length(x),2);
>> end
```

The first line in the loop does the integration of the model's differential equations during the year. The next two lines append the `t_cum` and `x_cum` vectors by the results from the year's integration. Then the initial conditions for the following year are obtained, when a bit of normally distributed random variable, `randn`, is added to the state variables at the end of this year. Doing this 100 times yields the long-term trajectory.

To plot a graph of the vegetation level and fire index through time, we type

```
>> figure
>> plot(t_cum,x_cum)
```

This results in Figure 7.24. The figure clearly reveals a cyclic character. Similarly, if we plot the fire index, x_1, on the horizontal axis and the vegetation level, x_2, on the vertical axis, we see the spiral nature of the trajectory. Figure 7.25 is made by our typing

```
>> figure
>> plot(x_cum(:,1),x_cum(:,2))
```

These figures illustrate the classic fire cycle that results when density-dependent susceptibility to episodic disturbance is coupled with exponential growth.

It's tempting to manage natural habitats such as national forests and parks to have continuous burning, rather than episodic burning. This will tend to make the community more spatially uniform, and greatly change the metapopulation structure of the environment. The fire cycle in time translates into a mosaic of successional stages in space, and this mosaic is home to many species, especially wild flowers and butterflies. The policy tradeoff, therefore, is between continuous burning, which produces a monotonous expanse of climax species with relative safety from fire damage, and episodic burning, which produces more biodiversity with more risk of damage from fire. Of course, the worst of all worlds is to prevent the fire altogether, because this both minimizes biodiversity and maximizes the risk of catastrophic fire.

7.1.5 Food Webs and Trophic Dynamics

A "food web" graphically represents all the who-eats-whom relations in a community. Competing species are viewed as being at the same horizontal level in a community, while a predator and prey are viewed as being at different vertical levels, with the predator being on top of the prey. A level in a food web is called a "trophic level." The base of the food web consists of plants and the top predator may be separated from the plant base by several trophic levels. Figure 7.26 is an example of a food web for the terrestrial community on the Caribbean island of St. Martin. *Anolis* lizards are important components of this web. The web is clearly quite complicated, even though it has some highly aggregated compartments,

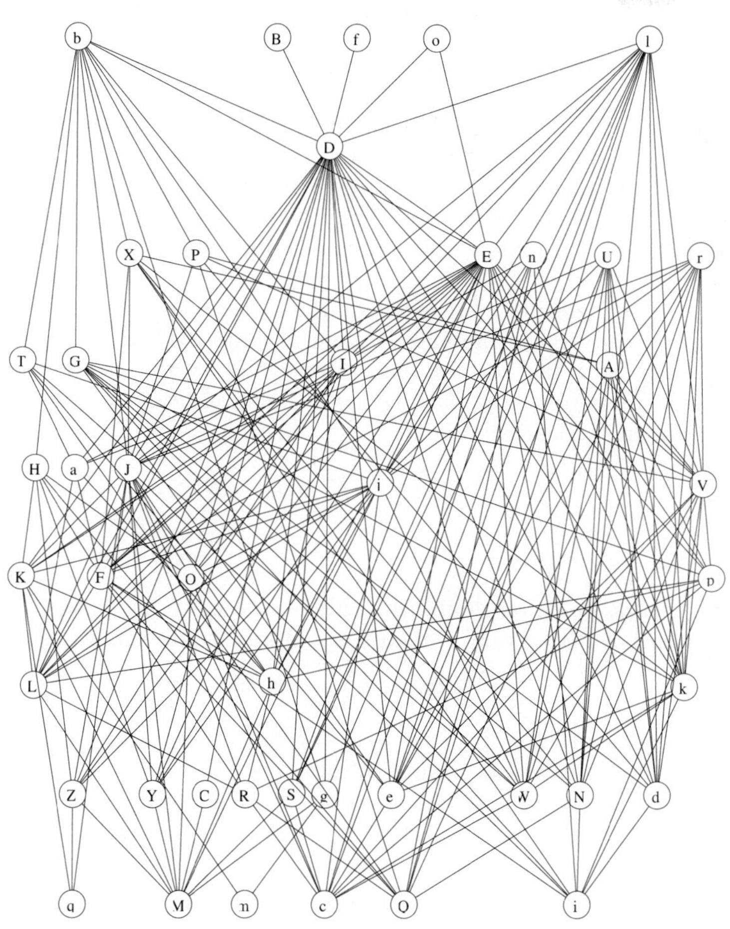

Figure 7.26: Terrestrial food web on St. Martin in the eastern Caribbean. Species codes: A. Adult spider, B. *Allogyptus crenshawi*, C. Annelid, D. *Anolis gingivinus*, E. *Anolis pogus*, F. Ants, G. Bananaquit, H. Bullfinch, I. Centipede, J. Coleoptera adult, K. Coleoptera larva, L. Collembola, M. Detritus, N. Diptera adult, O. Diptera larva, P. Elaenia, Q. Fruit&seeds, R. Fungi, S. Gastropoda, T. Grassquit, U. Gray Kingbird, V. Hemiptera, W. Homoptera, X. Hummingbirds, Y. Isopoda, Z. Isoptera, a. Juvenile spider, b. Kestrel, c. Leaves, d. Lepidoptera adult, e. Lepidoptera larva, f. *Mesocoelium* sp., g. Millipede, h. Mites, i. Nectar&floral, j. Orthoptera, k. Other hymenoptera, l. Pearly-Eyed Thrasher, m. Roots, n. Scaly-Breasted Thrasher, o. *Thelandros cubensis*, p. Thysanoptera, q. Wood, r. Yellow Warbler

such as "Leaves," "Mites," and so forth, that contain many species. At present, a couple hundred or so food webs have been drawn for ecological communities, and limited progress has been made at using graph-theoretic measures to find descriptive generalizations from the entire data set of webs. Perhaps the most successful approach to webs has focused on the number of trophic levels, and not paid much attention to how many species occur at each level.

Have you ever wondered why some lakes or ponds have clear fresh water that invites a good swim on a summer day, and why other ponds are pea-soup green leading one to stay on the beach even in sweltering heat? It turns out that the number of trophic levels in a community, whether there are an odd or even number of levels, greatly and visibly impacts the abundance of the lowest trophic level, which for a pond consists of algae and fresh water vegetation.

Let's investigate a sequence of models with an increasing number of trophic levels. The first two levels are already quite familiar. A one-trophic-level model is just the logistic equation. The productivity of the base level is specified by the r and K of this logistic equation. A two-level model is this base level with the addition of one predator. We've already analyzed this in the last chapter:

$$\begin{aligned}\frac{dN_1}{dt} &= rN_1(1 - N_1/K) - aN_1N_2\\ \frac{dN_2}{dt} &= baN_1N_2 - dN_2\end{aligned}$$

As you know, this system has three equilibria, only one of which has both species coexisting. The equilibrium with both coexisting, called the "interior equilibrium," is the solution of

$$\begin{aligned}(r/k)N_1 + aN_2 &= r\\ baN_1 &= d\end{aligned}$$

In matrix notation, the interior equilibrium solves

$$\begin{pmatrix} r/K & a \\ ba & 0 \end{pmatrix} \begin{pmatrix} N_1 \\ N_2 \end{pmatrix} = \begin{pmatrix} r \\ d \end{pmatrix}$$

MATLAB finds the root of this linear equation for us when we type

```
>> p=sym('[r/k a; b*a 0]');
>> c=sym('[r; d]');
>> nhat2=simple(linsolve(p,c))
```

yielding

```
nhat2 = [          1/b/a*d]
        [r/a-r/b/a^2/k*d]
```

where we've labeled the equilibrium with two trophic levels as `nhat2`. These formulas are familiar from the last chapter. Recall also that K must be sufficiently large to support the second level. The minimum K that will support one trophic level above the base level is found by typing

```
>> kmin2=simple(solve(sym(nhat2,2,1),'k'))
```

yielding

```
kmin2 = 1/b/a*d
```

It's perhaps easier if the minimum K that will support two trophic levels is formated as $d/(ab)$, but the form above is what MATLAB returns. The minimum r is simply 0; i.e., for two trophic levels to exist, r must be positive and K greater than $d/(ab)$. This is old ground, and now let's move to new territory. Let's add more trophic levels.

The model having three trophic levels is

$$\begin{aligned}\frac{dN_1}{dt} &= rN_1(1 - N_1/K) - aN_1N_2 \\ \frac{dN_2}{dt} &= baN_1N_2 - aN_2N_3 - dN_2 \\ \frac{dN_3}{dt} &= baN_2N_3 - dN_3\end{aligned}$$

The interior equilibrium is found by our solving

$$\begin{pmatrix} r/K & a & 0 \\ ba & 0 & -a \\ 0 & ba & 0 \end{pmatrix} \begin{pmatrix} N_1 \\ N_2 \\ N_3 \end{pmatrix} = \begin{pmatrix} r \\ d \\ d \end{pmatrix}$$

MATLAB will do this for us when we type

```
>> p=sym('[r/k a 0; b*a 0 -a; 0 b*a 0]');
>> c=sym('[r; d; d]');
>> nhat3=simple(linsolve(p,c))
```

This yields

```
nhat3 = [          k-k/b/r*d]
        [            1/b/a*d]
        [b*k-1/r*k*d-1/a*d]
```

To support three trophic levels, the base trophic level must be even more productive than when it supported two. Specifically, r must be big enough that the equilibrium abundance of the first trophic level is positive. The minimum r is found by our typing

```
>> rmin3=simple(solve(sym(nhat3,1,1),'r'))
```

yielding

```
rmin3 = d/b
```

There is also a minimum K needed to maintain a positive abundance in the third trophic level. This minimum is found by our typing

```
>> kmin3=simple(solve(sym(nhat3,3,1),'k'))
```

yielding

```
kmin3 = r*d/a/(b*r-d)
```

This minimum K also depends on r, so, rather arbitrarily, let's suppose that r is twice the minimum value needed, and see what the minimum K must be for this r. We type

```
>> kmin3_2rmin=simple(subs(kmin3,symop(rmin3,'*','2'),'r'))
```

yielding

```
kmin3_2rmin = 2/b/a*d
```

So, if r is greater than $(2d)/b$ and K greater than $(2d)/(ab)$, then three levels can exist. These requirements mean that to support three levels, the base level has to be more productive than it was when it supported two levels.

We can continue in this vein. The solution for four trophic levels is found by our typing

```
>> p=sym('[r/k a 0 0; b*a 0 -a 0; 0 b*a 0 -a; 0 0 b*a 0]');
>> c=sym('[r; d; d; d]');
>> nhat4=simple(linsolve(p,c))
```

This yields

```
nhat4 = [                           1/b/a*d+d/b^2/a]
        [        -r/b/a^2/k*d-r/b^2/a^2/k*d+r/a]
        [                                  1/b/a*d]
        [-1/a^2/k*r*d-r/b/a^2/k*d+b/a*r-1/a*d]
```

For four trophic levels to exist, the abundance of the second level is positive only if K is big enough. Typing

```
>> kmin4=simple(solve(sym(nhat4,2,1),'k'))
```

yields

```
kmin4 = d*(b+1)/b^2/a
```

For the fourth level to be positive, r must be big enough. This value can be found by our typing

```
>> rmin4=simple(solve(sym(nhat4,4,1),'r'))
```

yielding

```
rmin4 = d*b*a*k/(-b*d-d+b^2*a*k)
```

The minimum r depends on K, so if we arbitrarily assume that K is twice the minimum for four levels, then the minimum r boils down to

```
>> rmin4_2kmin=simple(subs(rmin4,symop(kmin4,'*','2'),'k'))
```

yielding

```
rmin4_2kmin = 2*d/b
```

Thus, for four trophic levels, r should be greater than $(2d)/b$, which is the same requirement as for three levels, and K should be greater than $d(b+1)/(ab^2)$.

Let's do one more level and then stop. Typing

```
>> p=sym('[r/k a 0 0 0; b*a 0 -a 0 0;
          0 b*a 0 -a 0; 0 0 b*a 0 -a; 0 0 0 b*a 0]');
>> c=sym('[r; d; d; d; d]');
>> nhat5=simple(linsolve(p,c))
```

yields[3]

```
nhat5 = [                        -k/b/r*d-k/r/b^2*d+k]
        [                             1/b/a*d+d/b^2/a]
        [              -1/r*k*d-k/b/r*d+b*k-1/a*d]
        [                                     1/b/a*d]
        [-1/r*d*b*k-1/a*d-1/a*b*d+b^2*k-1/r*k*d]
```

The minimum r for five levels is

```
>> rmin5=simple(solve(sym(nhat5,1,1),'r'))
```

yielding

```
rmin5 = d*(b+1)/b^2
```

and the minimum K is

```
>> kmin5=simple(solve(sym(nhat5,5,1),'k'))
```

yielding

```
kmin5 = d*(b+1)*r/a/(-b*d-d+r*b^2)
```

The minimum K depends on r, so when r is arbitrarily taken to be twice the minimum needed, the minimum K reduces to

```
>> kmin5_2rmin=simple(subs(kmin5,symop(rmin5,'*','2'),'r'))
```

which is

```
kmin5_2rmin = 2*d*(b+1)/b^2/a
```

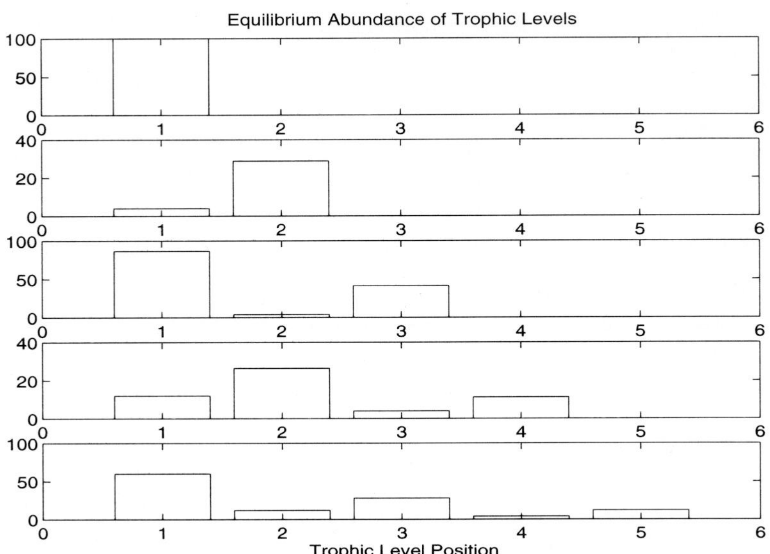

Figure 7.27: The equilibrium abundance of each trophic level, within communities of one to five trophic levels. The base level is level 1. The top panel of the figure is for a community with one trophic level, progressing toward the bottom panel, which is a community with five trophic levels.

Let's put some numbers in these formulas, so we can see how the number of possible trophic levels depends on the productivity of the base level.

To assign some parameters, type

```
>> global r k a b d;
>> a = .5; b = .5; d = 1;
```

Here we've made the parameters `global` because we'll be using them again later in a simulation. To see what the minimum r's and K's are for the maintenance of up to five trophic levels, type

```
>> rmin = [0 0 eval(rmin3) eval(rmin4_2kmin) eval(rmin5)]
>> kmin = [0 eval(kmin2) eval(kmin3_2rmin) eval(kmin4) ...
>>          eval(kmin5_2rmin)]
```

yielding collectively

```
rmin = 0     0     2     4     6
kmin = 0     4     8    12    24
```

[3]The definition of `p` should be typed on one line. The input has been edited to two lines here to avoid overrunning the margin.

The first entry in each vector is the r and K needed to support one trophic level, the next entry is the r and K needed to support two trophic levels, and so forth. So, if r is greater than 6 and K greater than 24, the productivity of the first trophic level should be able to support a total of five levels.

To find out the equilibrium abundance of the trophic levels, let's take values comfortably higher than the minimum needed; type

```
>> r = 15; k = 100;
```

Next, let's evaluate the equilibrium abundances of each trophic level, in communities having one level, two levels, and so on. We type

```
>> nhat1_n = [k 0 0 0 0];
>> nhat2_n = [eval(sym(nhat2,1,1)) eval(sym(nhat2,2,1)) 0 0 0];
>> nhat3_n = [eval(sym(nhat3,1,1)) eval(sym(nhat3,2,1)) ...
>>            eval(sym(nhat3,3,1)) 0 0];
>> nhat4_n = [eval(sym(nhat4,1,1)) eval(sym(nhat4,2,1)) ...
>>            eval(sym(nhat4,3,1)) eval(sym(nhat4,4,1)) 0];
>> nhat5_n = [eval(sym(nhat5,1,1)) eval(sym(nhat5,2,1)) ...
>>            eval(sym(nhat5,3,1)) eval(sym(nhat5,4,1)) ...
>>            eval(sym(nhat5,5,1))];
```

Here `nhat1_n` is the numerical abundance of all the trophic levels if only one level is present, `nhat2_n` is the equilibrium abundance of all the trophic levels if only two levels are present, and so forth. To compare all these, prepare a figure by typing

```
>> figure
>> subplot(5,1,1); bar(nhat1_n)
>> subplot(5,1,2); bar(nhat2_n)
>> subplot(5,1,3); bar(nhat3_n)
>> subplot(5,1,4); bar(nhat4_n)
>> subplot(5,1,5); bar(nhat5_n)
```

This yields Figure 7.27. The striking feature of this figure is that the communities with an odd number of trophic levels have a high abundance in the first level, and the communities with an even number of levels have a low abundance in the first level. Also, the relative abundance of the adjacent levels vary in opposite directions.

To simulate the sequential addition of trophic levels, we integrate the differential equations for a five-level model. We begin with only one level. We'll let it come to equilibrium and then introduce the second level, let it come to equilibrium, and then introduce the second level, and so forth up to all five levels. To integrate the model with five levels, first type in the function below with a text editor and save it as `troph5.m`

```
function ndot=troph5(t,n)
 global r k a b d;
 ndot(1,1)=r*n(1)*(1-n(1)/k)-a*n(1)*n(2);
```

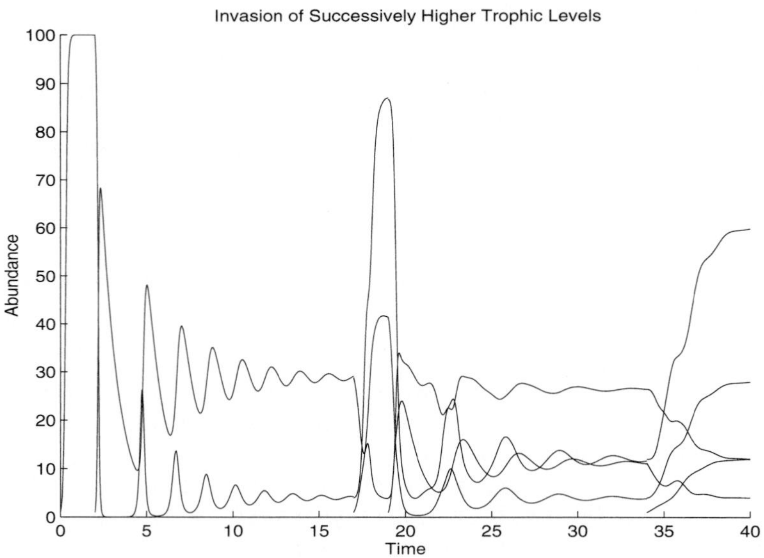

Figure 7.28: The sequential addition of trophic levels. The first curve to leave the origin is the abundance of the first trophic level, which rises to equilibrium at K. The second trophic level is then added when the time is two. It rises in abundance as the first trophic level declines. When the time is 17, a third trophic level is added, and the system quickly approaches a new equilibrium at which the first and third levels are abundant and the second level is low. At time 19 the fourth level is added, and at time 28 the fifth level is added.

```
ndot(2,1)=b*a*n(1)*n(2)-a*n(2)*n(3)-d*n(2);
ndot(3,1)=b*a*n(2)*n(3)-a*n(3)*n(4)-d*n(3);
ndot(4,1)=b*a*n(3)*n(4)-a*n(4)*n(5)-d*n(4);
ndot(5,1)=b*a*n(4)*n(5)-d*n(5);
```

Then we can use it with good ol' `ode45` to produce a sequence of five trajectories. Type

```
>> [ta na] = ode45('troph5',0,2,[1;0;0;0;0]);
>> [tb nb] = ode45('troph5',0,15,[nhat1_n(1);1;0;0;0]);
>> [tc nc] = ode45('troph5',0,2,[nhat2_n(1);nhat2_n(2);1;0;0]);
>> [td nd] = ode45('troph5',0,15,[nhat3_n(1);nhat3_n(2);nhat3_n(3);1;0]);
>> [te ne] = ode45('troph5',0,6,[nhat4_n(1);nhat4_n(2);nhat4_n(3); ...
                   nhat4_n(4);1]);
```

Finally, to graph these trajectories back to back on one graph, we type

```
>> figure
>> hold on
```

```
>> plot(ta,na)
>> plot(tb+ta(length(ta)),nb)
>> plot(tc+tb(length(tb))+ta(length(ta)),nc)
>> plot(td+tc(length(tc))+tb(length(tb))+ta(length(ta)),nd)
>> plot(te+td(length(td))+tc(length(tc))+tb(length(tb))+ta(length(ta)),ne)
```

This yields Figure 7.28. The abundance of the lowest trophic level rises to 100, then drops when a second trophic level is added, then rises again when a third level is added, and so forth.

The conservation implication of this model is realized by our running it in reverse, starting with, say four levels, and removing the top one. A common food chain in lakes is from phytoplankton to zooplankton, to small planktivorous fish, to large picivorous fish. A four-level system will have a small standing stock of phytoplankton, leading to clear water. But if the picivorous fish are somehow lost, the resulting three-level system will feature uninviting pea-soup green water.

Ecologists have long debated why plants are common and conspicuous in the terrestrial biosphere. Some have suggested that animals are not capable of eating most types of vegetation. Others have suggested that the populations of plant-eating insects are being held in check by predatory insects and birds. If plants are sufficiently inedible to herbivores that herbivore populations can only do occasional damage to plants, then the loss of high-trophic-level animals does not undermine the biosphere's plant-dominated structure. Whereas, if plants are edible enough to herbivores that plant populations would be decimated if the top predators were lost, then the loss of birds, wasps, spiders, and other predators on herbivorous insects would pose a serious threat to the biosphere as we know it.

7.2 Biogeochemical Function in Earth Systems

"Earth systems" are systems containing interacting biological and physical components. The atmosphere, for example, originally obtained its oxygen from plant photosynthesis, and the dynamics of CO_2 and O_2 on the earth involve simultaneous exchanges among the atmosphere, the oceans, the solid earth, and the biosphere. This is the grandest earth system of them all, and many smaller ones exist too. Soil forms by the reciprocal interaction of vegetation and its decomposition with the physical substrate. Water runoff from watersheds influences, and is influenced by, the growth of vegetation along the drainage basin of the watershed. In principle, an earth system involves a reciprocal interaction between biological and physical components, but often the biological part is dynamically controlled by the physical part and the system is called an earth system nonetheless. In the marine realm, the population dynamics of coastal organisms is controlled by the circulation pattern of the water that transports marine larvae from the sites where they were born to where they will settle. This is a one-way interaction, because the water's circulation is not itself affected much, if at all, by the larvae it carries. Similarly, the phylogenetic radiation of organisms in large part reflects the pattern of continental drift, and the breakup and collision of tectonic plates. The interdisciplinary study of continental drift and phylogenetic radiation is called "vicariant biogeography." The organisms living on the plates don't affect the rate of continental drift, so again the interaction is one-way only. Regardless, the spirit of the phrase "earth systems" as used in ecology is that ecological processes are interdisciplinary combinations of biology, geology, oceanography, and atmospheric science. Therefore, we'll conclude this book with some topics that require an earth-systems perspective, and we'll look at ecological processes in the context of their interaction with physical processes.

7.2.1 Nonequilibrium Thermodynamics

Unless one is a vitalist, one supposes that ecological systems obey the same laws of physics and chemistry that nonliving physical systems do. Indeed, the chemical reactions of metabolism all make perfect thermodynamic sense. A body's enzymes catalyze chemical reactions that would occur anyway, except a lot slower, if they were not taking place within a living cell. This is what makes the origin of life thermodynamically inevitable. Why not extend this point of view to an entire ecological system?

A thermodynamic vision of an ecosystem has long been attractive to ecologists. One can readily translate between a thermodynamic description of an ecological system and an organism-based description; one simply changes units. One can measure the total number of calories in an organism just as one can measure the number of calories in a dish of ice cream or a slab of beef. By multiplying the number of organisms by the calories per organism, one obtains the amount of calories in a population. By summing over all the populations in a trophic level, one obtains the total number of calories in the trophic level. Moreover, one can also directly convert from rates of herbivory and predation to rates of energy flow between the trophic levels. If an average predator captures x prey per day, then the calories flowing from the prey trophic level to the predator trophic level is x times the calories per average prey individual times the number of predators. So, just by changing units, one can translate an organism-based description of an ecological system to a thermodynamic description based on the energy content of the trophic levels and the flows between them. These flows must be consistent with the laws of thermodynamics.

Why would one want to convert an organism-based description of an ecological system to a thermodynamic one? Two reasons. First, a thermodynamic description is sometimes easier to integrate into an earth-systems model. If there are only a few biological pieces in the overall model, then by majority rule, so to speak, the biological parts can be converted into thermodynamic units to accord with the rest of the model. Second, the laws of thermodynamics might provide some useful constraints on the functioning of an ecological system. For example, suppose you're told that all a certain predator eats is one prey item of its own size per year yet the predator gives birth to ten large offspring per year. Well, you should suspect a mistake here—it sounds like some offspring are being made out of nothing, and that the predator is thereby violating all the laws of thermodynamics at once. The laws of thermodynamics imply, everything else being equal, that energy content and energy flows dwindle from the first trophic level to the last. The thermodynamics of an ecological system determine an envelope within which the system can operate.

Can thermodynamics do more for ecology than tell us what's impossible? Can it tell us

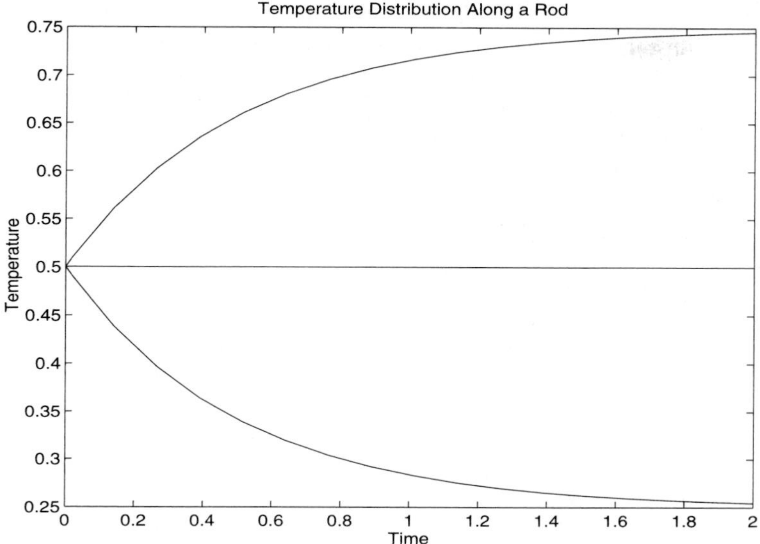

Figure 7.29: Temperature in the top, middle, and bottom sections of a bar through time, starting from a uniform temperature distribution.

what will actually happen? Can it tell us how big the various trophic levels will be, and how fast nutrients and energy will flow between trophic levels? The potential for thermodynamics to predict ecological structure and function took a leap forward in the 1950's when the study of thermodynamics was extended to nonequilibrium systems. Until relatively recently, thermodynamics was limited to predicting the state of a system at equilibrium, such as the equilibrium concentration of various reactants in a test tube. Now, at least metaphorically, thermodynamics can also predict what happens in systems prevented from coming to equilibrium, and that are maintained in a steady state instead.[4] Here's a simple illustration of how nonequilibrium thermodynamics works, and you'll then be able to see, at least metaphorically, how nonequilibrium thermodynamics might be applied to ecological systems in the future.

Imagine that a metal bar is positioned vertically in front of you. Heat at a constant temperature is being applied to the top of the bar, and the bottom of the bar is sitting in a cold bath, which is also being held at a constant temperature. Obviously heat will flow from the top to the bottom down the metal bar. Let's look at the temperature along the bar from top to bottom, and divide the bar into thirds. What is the temperature of the bar

[4] In thermodynamics, an equilibrium is a state in which the state variables are constant through time *and* the macroscopic flows are zero, whereas a steady state is a state in which the state variables are also constant through time, but the flows into and out from the system are nonzero. In ecology, we normally don't bother with this distinction.

at the top third, middle third, and bottom third? Let T_u be the upper temperature being applied to the top of the bar, T_1, T_2, and T_3 be the temperatures in the top, middle, and bottom thirds of the bar, and T_b be the temperature in the cold bath at the base of the bar. As you may recall from way back in Chapter 1, heat flows between two spots at a rate proportional to the temperature difference—this is called Fick's law. So the rate at which heat is flowing from the source above the bar into the top third of the bar is $a(T_u - T_1)$ where a is a constant of proportionality (the conductivity of the bar for heat). Similarly, the rate at which heat is flowing out of the top third of the bar into the middle third is $a(T_1 - T_2)$. Let's further assume that the temperature within a section of the bar increases or decreases by one degree for every net unit of heat that enters or leaves it. (This means that the specific heat of the bar is one, as it is for water.) The change in the temperature of the top third of the bar is therefore the heat flow in minus the heat flow out, and similarly for the other sections of the bar. In symbols, we then have

$$\begin{aligned}
\frac{dT_1}{dt} &= a(T_u - T_1) - a(T_1 - T_2) \\
\frac{dT_2}{dt} &= a(T_1 - T_2) - a(T_2 - T_3) \\
\frac{dT_2}{dt} &= a(T_2 - T_3) - a(T_3 - T_b)
\end{aligned}$$

This system of differential equations can be solved with our now-familiar techniques. In particular, let's check out the steady state. Typing

```
>> t1dot = 'a*(tu-t1)-a*(t1-t2)';
>> t2dot = 'a*(t1-t2)-a*(t2-t3)';
>> t3dot = 'a*(t2-t3)-a*(t3-tb)';
>> [t1hat,t2hat,t3hat] = solve(t1dot,t2dot,t3dot,'t1,t2,t3')
```

will get MATLAB to tell us that the temperature along the bar at steady state is

```
t1hat = 3/4*tu+1/4*tb
t2hat = 1/2*tu+1/2*tb
t3hat = 1/4*tu+3/4*tb
```

The steady-state pattern is a linear decline in temperature along the bar from top to bottom. To find the approach to this steady state from an initial condition, we integrate these differential equations. Because these equations are linear differential equations, we could do the integration symbolically, but it's a bit quicker here to use our trusty ol' `ode23` routine instead. So, with a text editor, type in the following function and save it as `therm.m`,

```
function tdot=therm(time,t)
 global a tu tb;
 tdot(1,1) = a*(tu  -t(1))-a*(t(1)-t(2));
 tdot(2,1) = a*(t(1)-t(2))-a*(t(2)-t(3));
 tdot(3,1) = a*(t(2)-t(3))-a*(t(3)  -tb);
```

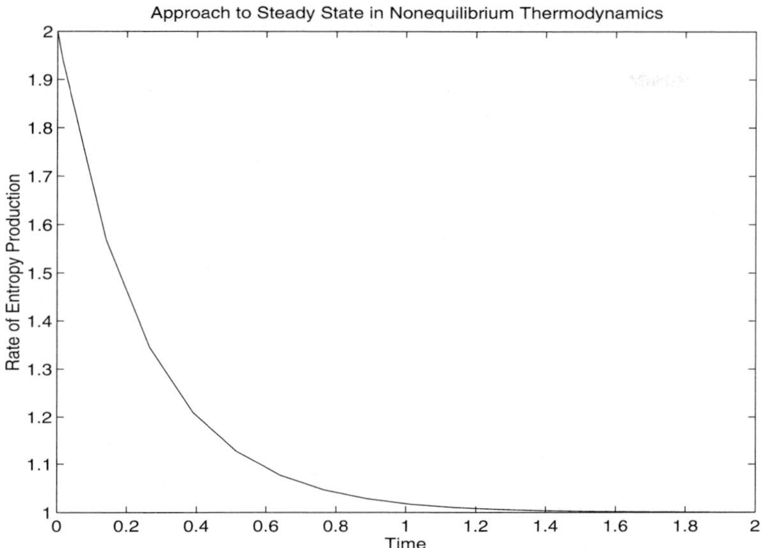

Figure 7.30: The rate of entropy production as heat flows through a bar. The rate of entropy production declines as the temperature within the bar approaches its steady-state distribution.

To generate an illustration, let's use as an initial condition a bar whose temperature is the same throughout, namely $(T_u + T_b)/2$. Typing

```
>> global a tu tb;
>> a = 1; tu = 1; tb = 0;
>> [time temperature] = ode23('therm',0,2,[(tu+tb)/2,(tu+tb)/2,(tu+tb)/2]);
```

generates a trajectory of temperature vs. time, which we can plot by typing

```
>> figure
>> plot(time,temperature)
```

This yields Figure 7.29. The curves show how the temperature along the bar approaches the steady state distribution.

So where's the thermodynamics? We've done all this so far without any mention of energy or entropy. So now the thermodynamics comes in, and we can state the main result of nonequilibrium thermodynamics: as the bar approaches its steady state temperature distribution, the rate of entropy production along the bar declines to a minimum.

Well, I hope you've got your favorite physical chemistry text lying around, and if so, open it to the definition of entropy. You'll probably find two definitions. One refers to how much pattern the system exhibits, and the other is a formula for calculating entropy during heat exchange. Both definitions are relevant here, and they are equivalent as your physical chemistry text undoubtedly points out at great length.

The state of highest entropy for the rod is that in which the temperature is uniform across its entire length—this is a state with the least amount of pattern. If the rod were put in an insulated box and allowed to come to equilibrium, it would come to this state. According to the second law of thermodynamics, the entropy of an isolated system always increases until equilibrium, and indeed the medal rod would eventually equilibrate with a uniform temperature across it, in accordance with the second law. In contrast, if heat is flowing through the rod in a steady state, the temperature declines linearly from its hot side to its cold side. This nonuniform pattern is *not* the state of highest entropy. The system comes to this pattern because it is in a steady state and not in equilibrium.

Yet, as we will see, the linear pattern of temperature does have a thermodynamic interpretation: it is the pattern that causes the flow of heat through the rod to generate entropy at the slowest possible rate. To see how this idea works, we'll have to find out how fast entropy is being generated as heat flows through the rod and see how this speed depends on the temperature distribution across the rod. Now we need the other definition of entropy, the one involving a formula for computing the entropy from the heat flow. The change in entropy, dS, generated when a system absorbs an amount of heat, dQ, at temperature, T, for a differential change in state at constant pressure, is

$$dS = \frac{dQ}{T}$$

Suppose heat is flowing from one spot to another, say from a high temperature at T_l on the left to a low temperature at T_r on the right. The rate at which the heat is flowing, dQ/dt, is, by Fick's law, $-a(T_l - T_r)$. For a unit amount of heat flow from the spot at the left to that at the right, the rate at which the entropy changes is the change in the reciprocal of the temperature, $1/T$, in going from the left to the right. The change in $1/T$ in going from left to right is approximately $-(T_l - T_r)/((T_l + T_r)/2)^2$ because the derivative of T^{-1} is T^{-2}. So if we combine the amount of the heat flow with the change in entropy per unit heat flow, we get that the overall rate at which entropy is being produced is $dS/dt = a(T_l - T_r)^2/((T_l + T_r)/2)^2$. Returning to the rod, we have four exchanges of heat in all. We have the flow from the source above the rod into the top third of the rod, the flow from the top third to the middle, the flow from the middle to the bottom third, and the flow from the bottom third into the cold bath. So the total rate of entropy production as heat is flowing down the rod is

$$\frac{dS}{dt} = a\frac{(T_u - T_1)^2}{((T_u + T_b)/2)^2} + a\frac{(T_1 - T_2)^2}{((T_u + T_b)/2)^2} + a\frac{(T_2 - T_3)^2}{((T_u + T_b)/2)^2} + a\frac{(T_3 - T_b)^2}{((T_u + T_b)/2)^2}$$

This overall rate of entropy production is a function of three variables, T_1, T_2 and T_3, which describe the temperature distribution across the rod. There is a particular set of these T's that minimizes dS/dt.

MATLAB will solve for the T's that minimize the rate of entropy production. First we have to tell MATLAB about the formula for dS/dt, which has to be stretched over two lines because the formula is so long. We type

```
>> sdot = 'a*(tu-t1)^2/((tu+tb)/2)^2+a*(t1-t2)^2/((tu+tb)/2)^2';
>> sdot = [sdot '+a*(t2-t3)^2/((tu+tb)/2)^2+a*(t3-tb)^2/((tu+tb)/2)^2'];
```

Next, we differentiate `sdot` with respect to each of the three variables and set these derivatives equal to zero, yielding a set of three simultaneous equations for `t1`, `t2`, and `t3`. Solving these equations for `t1`, `t2`, and `t3` yields the temperature distribution that minimizes the rate of entropy production. All this is done in one line, by our typing

```
>> [t1min,t2min,t3min]= ...
>>        solve(diff(sdot,'t1'),diff(sdot,'t2'),diff(sdot,'t3'),'t1,t2,t3')
```

yielding

```
t1min = 3/4*tu+1/4*tb
t2min = 1/2*tu+1/2*tb
t3min = 1/4*tu+3/4*tb
```

These temperatures that minimize dS/dt are exactly the same temperatures that describe the steady-state temperature distribution across the rod, as we've already illustrated in Figure 7.29. But we now see that this steady-state temperature distribution is also interpreted thermodynamically as the state that minimizes the rate of entropy production. Indeed, we can also illustrate how the rate of entropy production starts out high and progressively slows down until the steady state is attained. Typing

```
>> t1=temperature(:,1);
>> t2=temperature(:,2);
>> t3=temperature(:,3);
>> figure
>> plot(time,eval(sym2ara(sdot)))
```

yields figure 7.30. The figure shows that entropy is being generated very quickly at first, but its rate of production progressively slows down through time, until it bottoms out when the steady state is attained.

So what's all this have to do with ecology, you're undoubtedly wondering by now. Well, no one knows for certain yet, but how the temperature along a bar is organized as a result of heat flowing through it might be a metaphor for how ecosystems organize. As energy flows through the trophic levels of an ecosystem, perhaps the amount of energy that accumulates in each level is that which brings about the lowest overall rate of entropy production by the ecosystem.

Before you toss away your bird books or your field guides to wildflower identification for texts on physical chemistry, recall that nonequilibrium thermodynamics requires the kinetics to be specified at the beginning. Our heat-flow example illustrates this point. The formula for the rate of entropy production, dS/dt, included Fick's law in it. You can't write down a formula for dS/dt unless you know the kinetics of exchange between the different components of the system. And somebody like Mr. Fick has to find out what those kinetics are[5]. A thermodynamic approach to ecosystems doesn't bypass the determining of the kinetics of exchange between ecosystem compartments, and that brings us right back to predator-prey and other population interactions. Furthermore, the kinetics completely determine the system's trajectory anyway. Figure 7.29 told the whole story of heat flow in the rod and didn't use thermodynamics. Figure 7.30, which does have thermodynamics in it, offers an attractive covering principle, but doesn't do the hard work—it piggybacks on knowledge of the kinetics. Because a thermodynamic approach to ecosystems doesn't save any effort we don't already have to expend, it is superfluous, in a sense. But the study of non-equilibrium thermodynamics may remain attractive. It promises generalizations that apply to all thermodynamic systems and may help us to understand the workings of ecosystems in a broader context.

What thermodynamics does do is settle the age-old question of whether we eat to live, or live to eat. According to thermodynamics, we live to eat. Now let's see what the entire biosphere eats.

7.2.2 Carbon Exchange in the Biosphere

Suppose humanity wants to continue burning fossil fuels at the rate it has been. This action will increase the atmosphere's CO_2 and continue the global warming trend now in effect. Will the earth somehow come and bail us out of trouble, and let us continue in our ways without any adverse effect? Well, the biosphere has bailed us out to some extent as it is. The increasing CO_2 has led to increasing photosynthesis around the globe, which has apparently led to higher plant biomass worldwide. The extra carbon that now resides in the world's biomass is therefore not in the atmosphere anymore as CO_2, and thus, the biosphere has partly ameliorated the trend to global warming. So, how long can the biosphere keep this up, and how effective is the amelioration?

To answer these questions, models have been developed to describe the biosphere's total uptake and release of carbon with the atmosphere. We offer here an illustration of such a model, one introduced recently by Sarmiento and colleagues. The setup is diagramed in Figure 7.31. The biosphere is envisioned to be a system through which a stream of carbon flows, not unlike the way heat flows through the metal rod of the last section. There are five state variables: x_1 is the amount of metabolic plant material on the earth, which you can think of as the sum of all the leaves on the earth; x_2 is the amount of structural plant material on the earth, which is approximately the same as all the wood on the planet; x_3 is the amount of litter from decomposing metabolic material, which is approximately the same as all the leaf litter in the world; x_4 is the amount of litter from decomposing structural material, which is approximately the same as all the decaying wood on earth; and x_5 is the amount of carbon in the earth's soil. These five compartments, called "flavors of carbon" by one of the model's authors, are hypothesized to be appropriate to describe the aggregate carbon flow through the entire biosphere[6].

[5]Incidentally, so that dS/dt is minimized in a steady state, the kinetics must be linear and symmetric in the state variables. This condition is true of physical systems at a steady state that happens to be very close to the equilibrium state.

[6]The units for the carbon pools, the x's, are really big. A PgC is a "petagram" of carbon, which is also

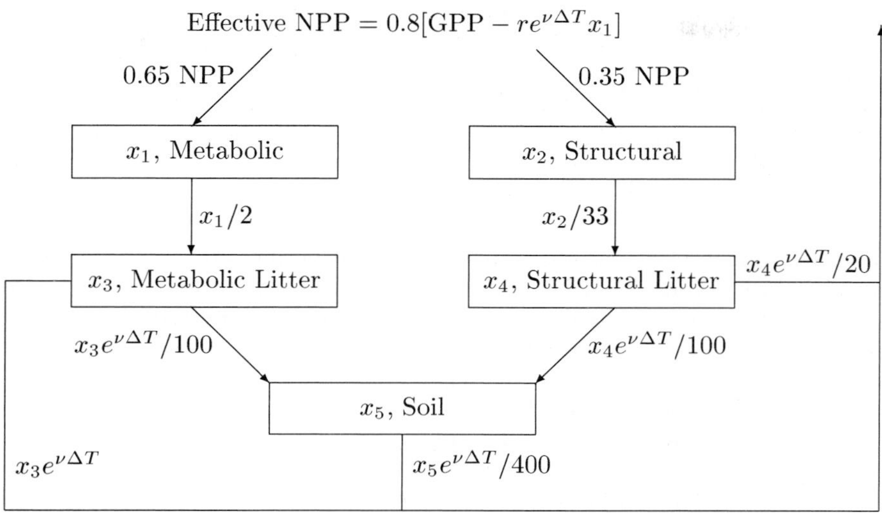

Figure 7.31: A model of carbon exchange between the biosphere and the atmosphere.

The overall input rate of carbon into the biosphere is called the gross primary production (GPP). This parameter's value is determined by the amount of CO_2 in the atmosphere and the global environmental temperature. The temperature is denoted as ΔT, which stands for the difference between the average annual global temperature in a particular year and the long-term average. A positive ΔT stands for a global temperature higher than the long-term average. We'll have more to say about the GPP shortly. Some of the GPP that flows into the biosphere is immediately returned to the atmosphere as respiration, which occurs at the rate $re^{\nu\Delta T}$ per unit of metabolic plant material. The remainder is called the net primary production (NPP), and of this only 80% flows into the biosphere proper, the other 20% being overhead used during the construction of the plants' structural material. The effective NPP (i.e., the 80% of the NPP left after construction costs are removed) is then allocated into the compartments as shown in the figure. Decomposition in the litter categories and in the soil releases CO_2 that is returned to the atmosphere. The transfer rates vary greatly from box to box in the diagram. The transition from x_1 to x_2 is fastest, taking two years on the average, whereas the decomposition rate of carbon in the soil, x_5, takes 400 years on the average. The other transfer rates are between these extremes. A more complete system diagram using SIMULINK appears later in the section. For now, Figure 7.31 should be viewed as a sketch of the model.

An interesting aspect of the model is how it can reconcile two competing effects of global

identical to a GtC, which is "gigatonne" of carbon. A petagram and a gigatonne are 10^{15} grams.

warming. On one hand, the increase in CO_2 makes plants photosynthesize faster and CO_2 is removed from the atmosphere. On the other hand, increasing the global temperature makes plants respire faster and speeds up decomposition, so that CO_2 is released back into the atmosphere. Whether the biosphere will continue to ameliorate global change to the atmosphere depends on whether the increased photosynthesis exceeds the increased respiration and decomposition, when all the compartments are taken into account.

Gross Primary Production

The GPP has a submodel of its own. The GPP could be measured by satellite using remote sensing techniques, or could be assumed to depend on the fraction of the earth that is covered with vegetation. Instead, we'll use a submodel for the GPP that represents a perturbation scenario; that is, a scenario in which the atmospheric CO_2 and global temperature are somewhat different from today's, but otherwise the biosphere remains much the same as it presently is. If so, the GPP for the earth presumably follows formulas that mimic how a single leaf performs, and how the canopy of a woods works. This is extrapolates relatively small-scale information to the entire world, but when it is fitted to global data, this extrapolation seems appropriate nonetheless. Specifically, the GPP increases according to a Michaelis-Menton formula[7] with respect to the partial pressure of CO_2 in the atmosphere. The GPP also increases according to a Michaelis-Menton formula with respect to the number of canopy layers. The number of canopy layers is correlated with the amount of global vegetation, indexed by the variable, B. The number of canopy layers is bounded by the compensation point, labeled $B^\star$, which is the level of vegetation at which photosynthesis balances respiration. All of this is motivation for the GPP submodel, as defined by the following formula:

$$\text{GPP} = c_1 \int_0^{B^\star} \frac{e^{-\alpha B}}{K_l + e^{-\alpha B}} dB$$

where

$$c_1 = P_{\text{max}} e^{\lambda \Delta t} \frac{p_{\text{CO}_2}}{K_{\text{CO}_2} + p_{\text{CO}_2}}$$

and $B^\star$ is the root of the equation

$$\frac{e^{-\alpha B}}{K_l + e^{-\alpha B}} = c_2$$

where

$$c_2 = r \frac{e^{\nu \Delta T}}{P_{\text{max}} e^{\lambda \Delta T} \frac{p_{\text{CO}_2}}{K_{\text{CO}_2} + p_{\text{CO}_2}}}$$

The first equation represents an integral over all canopy levels up to the compensation point. The second equation gives the dependency of photosynthesis on the partial pressure of CO_2

[7] A Michaelis-Menton equation is $dx/dt = \frac{V_{\text{max}} x}{K_m + x}$ where V_{max} is called the maximal velocity because $dx/dt = V_{\text{max}}$ when $x \to \infty$, and K_m is called the half-saturation constant because if $x = K_m$ then $dx/dt = V_{\text{max}}/2$.

and on the global temperature. The third and fourth equations are for the value of B that yields a canopy thickness such that the bottom layer is at the compensation point. The various coefficients have been measured from plant-physiological studies as reported in the literature or are fitted to time series on global atmospheric CO_2 and temperature, and are explicitly defined below.

All this looks complicated, but with MATLAB we can boil it down to an explicit formula for GPP, using p_{CO_2} and ΔT. First, let's define the several global parameters. `pco2_gw` and `deltat_gw` will hold our scenarios of how the p_{CO_2} and ΔT change under global warming, `gpp` will hold the formula for GPP that we are about to derive with MATLAB's help, `r` is the respiration rate, and `nu` is the coefficient relating the respiration and decomposition rates to global temperature.

```
>> global pco2_gw deltat_gw gpp r nu
```

To get the formula for GPP, we start with the formula for $B^\star$. We type

```
>> bstar = simple(solve('exp(-alpha*b)/(kl+exp(-alpha*b))=c2','b'));
>> const2 = 'r*exp(nu*deltat)/(pmax*exp(lam*deltat)*(pco2/(kco2+pco2)))';
>> bstar = simple(subs(bstar,const2,'c2'))
```

At this point, MATLAB replies with a messy, but correct, answer. Next, to find the explicit formula for the GPP, we type

```
>> gpp = simple(int('c1*exp(-alpha*b)/(kl+exp(-alpha*b))','b','0','bstar'))
>> const1 = 'pmax*exp(lam*deltat)*(pco2/(kco2+pco2))';
>> gpp = simple(subs(gpp,const1,'c1'));
>> gpp = simple(subs(gpp,bstar,'bstar'))
```

and again MATLAB replies with a messy, but correct, answer. Finally, let's hardwire the formula with the numerical values of some of the coefficients that are not changing in any of the scenarios. We'll type

```
>> gpp = subs(gpp,'5.9','pmax');
>> gpp = subs(gpp,'383','kco2');
>> gpp = subs(gpp,'0.25','kl');
>> gpp = subs(gpp,'0.039','alpha');
>> gpp = subs(gpp,'0.043','lam')
```

So we're done, and just for the record, here is the formula for the GPP as a function of p_{CO_2} and ΔT, where I've edited the output so that it spreads over four lines:

```
gpp = 151.2820512820513*exp(4.3e-2*deltat)*
      pco2*(-log(-1.475*pco2*exp(4.3e-2*deltat)/
      (-5.9*exp(4.3e-2*deltat)*pco2+383*r*exp(nu*deltat)+
      r*exp(nu*deltat)*pco2))+.2231435513142098)/(383+pco2)
```

To get a feel for what this messy formula says, let's plot it out. We'll be using respiration parameters of

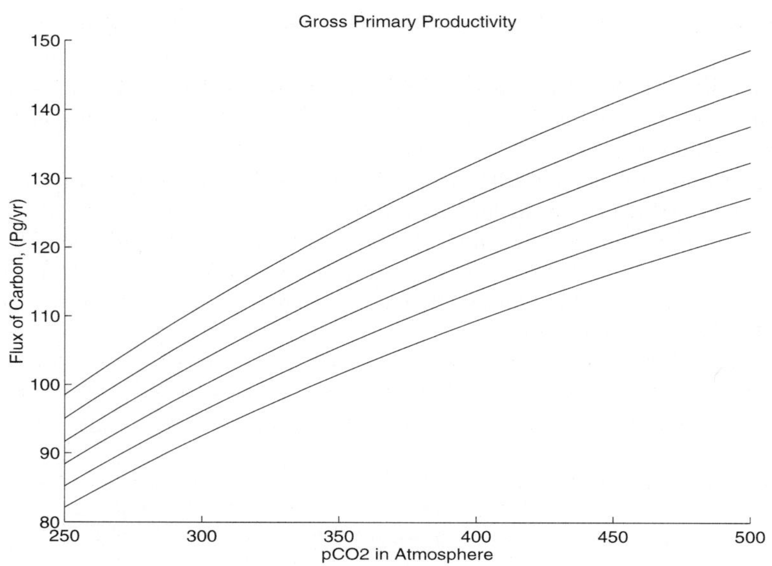

Figure 7.32: The gross primary production (GPP) as a function of atmospheric p_{CO_2} for various levels of global warming, ΔT. The highest curve is for $\Delta T = 4$ and successively lower curves are for $\Delta T = 3, 2, 1, 0, -1$.

```
>> r = 0.39;
>> nu = 0.086;
```

Now typing

```
>> figure
>> hold on
>> pco2 = 250:10:500;
>> deltat = -1;
>> plot(pco2,eval(sym2ara(gpp)),'y')
>> deltat = 0;
>> plot(pco2,eval(sym2ara(gpp)),'m')
>> deltat = 1;
>> plot(pco2,eval(sym2ara(gpp)),'c')
>> deltat = 2;
>> plot(pco2,eval(sym2ara(gpp)),'r')
>> deltat = 3;
>> plot(pco2,eval(sym2ara(gpp)),'g')
>> deltat = 4;
>> plot(pco2,eval(sym2ara(gpp)),'w')
```

yields Figure 7.32. The figure shows that the GPP for the entire biosphere increases with both p_{CO_2} and ΔT.

This then is the submodel for GPP used as the input to the biosphere model. It's important to realize that this submodel for GPP is logically independent of the biosphere model, and it could be replaced with other models. Because of the simplicity of the curves in Figure 7.32, I would replace this model with a linear model, a bivariate regression of GPP against both both p_{CO_2} and ΔT, rather than bother with a complicated justification involving canopy levels, compensation points, and so forth. More fundamentally, this GPP model does not depend dynamically on the state of the biosphere. If all the plants in the world suddenly disappeared, this model predicts that photosynthesis would occur anyway, according to the present p_{CO_2} and ΔT. Thus, this model is of a passive biosphere that simply receives its GPP from on high for distribution into five compartments. Nonetheless, let's see what the biosphere does with the carbon that's handed to it, and postpone using another GPP submodel to extend the model to the future.

Carbon Dynamics in the Biosphere

To see how the model for carbon flow through the biosphere works, we'll start with the case in which both p_{CO_2} and ΔT don't change through time. Then we'll look into the case where both of these rise through time in accord with global warming. In the equations for the model, the notation can be condensed somewhat by our letting $d = e^{\nu \Delta T}$ and $g = \text{GPP}$. To acquaint MATLAB with the model, we can type

```
>> dx1 = '0.65*0.8*(g-r*d*x1)-x1/2'
>> dx2 = '0.35*0.8*(g-r*d*x1)-x2/33'
>> dx3 = 'x1/2-d*(1+1/100)*x3'
>> dx4 = 'x2/33-d*(1/20+1/100)*x4'
>> dx5 = 'd*((x3+x4)/100-x5/400)'
```

These equations are read directly off of Figure 7.31. They form a system of linear differential equations with constant coefficients. The formulas for the equilibrium distribution of carbon throughout the biosphere are found by our typing

```
>> [x1hat,x2hat,x3hat,x4hat,x5hat] = ...
>> solve(dx1,dx2,dx3,dx4,dx5,'x1,x2,x3,x4,x5')
```

yielding

```
x1hat = 26.*g/(26.*r*d+25.)
x2hat = 231.0000000000000*g/(26.*r*d+25.)
x3hat = 12.87128712871287*g/d/(26.*r*d+25.)
x4hat = 116.6666666666667*g/d/(26.*r*d+25.)
x5hat = 518.1518151815182*g/d/(26.*r*d+25.)
```

The equilibrium is always positive, and the elements are directly proportional to the gross primary production, g. To find the Jacobian at this equilibrium, we type

```
>> jacob = jacobian(sym([dx1 '; ' dx2 '; ' dx3 '; ' dx4 '; ' dx5]), ...
>>                   sym('[x1,x2,x3,x4,x5]'))
```

yielding

```
jacob = [-.520*r*d-1/2,      0,            0,          0,          0]
        [     -.280*r*d, -1/33,            0,          0,          0]
        [           1/2,     0, -101/100*d,          0,          0]
        [             0,  1/33,            0,   -3/50*d,          0]
        [             0,     0,     1/100*d,   1/100*d, -1/400*d]
```

The most interesting aspect of the Jacobian is that it depends only on r and d, the respiration and decomposition rates. Therefore, the speed with which the equilibrium is approached doesn't depend on the gross primary production, g. Also, the matrix is triangular (all entries above and to the right of the main diagonal are zero) and its eigenvalues are simply the elements of the main diagonal.

To illustrate, let's use the 1990 values of p_{CO_2} and ΔT in a scenario for the future in the absence of global change. Type

```
>> pco2_gw = '350'
>> deltat_gw = '0.364'
```

With this scenario, g and d are evaluated by our typing

```
>> pco2=eval(pco2_gw)
>> deltat=eval(deltat_gw)
>> g=eval(gpp)
>> d=exp(nu*deltat)
```

yielding collectively

```
pco2 = 350
deltat = 0.3640
g = 106.9732
d = 1.0318
```

The equilibrium is evaluated by our typing

```
>> x1hat_n=eval(x1hat)
>> x2hat_n=eval(x2hat)
>> x3hat_n=eval(x3hat)
>> x4hat_n=eval(x4hat)
>> x5hat_n=eval(x5hat)
```

yielding

```
x1hat_n = 78.4296
x2hat_n = 696.8164
x3hat_n = 37.6299
x4hat_n = 341.0814
x5hat_n = 1.5148e+03
```

This shows that the biosphere's carbon is located, in rank order, in the soil, in tree trunks, in decaying wood, in leaves, and in leaf litter. To find the speed with which this equilibrium is approached, evaluate the eigenvalues of the Jacobian, which are simply the elements along the main diagonal. Type

```
>> for i=1:5
>>   lambda(i,1)=eval(sym(jacob,i,i));
>> end
>> lambda
```

yielding

```
lambda = -0.7092
         -0.0303
         -1.0421
         -0.0619
         -0.0026
```

All the eigenvalues are negative, indicating that the equilibrium is stable. Clearly the soil compartment equilibrates about 100 times slower than the leaf compartment.

We can integrate the differential equations to produce some trajectories. With a text editor, type in the following function, and save it as `biospher.m`.

```
function xdot=biospher(t,x)
 global pco2_gw deltat_gw gpp r nu
 pco2=eval(pco2_gw);
 deltat=eval(deltat_gw);
 g=eval(gpp);
 d=exp(nu*deltat);
 xdot(1,1) = 0.65*0.8*(g-r*d*x(1))-x(1)/2;
 xdot(2,1) = 0.35*0.8*(g-r*d*x(1))-x(2)/33;
 xdot(3,1) = x(1)/2-d*1.01*x(3);
 xdot(4,1) = x(2)/33-d*0.06*x(4);
 xdot(5,1) = d*((x(3)+x(4))/100-x(5)/400);
```

In this function, the potentially time-varying parameters, g and d, are evaluated anew each time, so the function can be used with global-change scenarios as well as global-constancy scenarios. Then, trajectories starting with 1990 values of the state variables and running for 200 years are produced by our typing

```
>> [time,x] = ode45('biospher',0,200,[77;639;30;313;1217]);
```

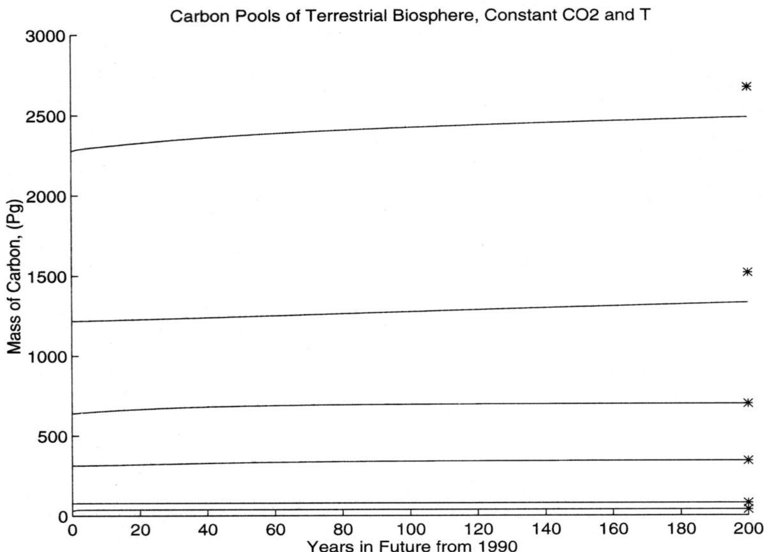

Figure 7.33: Carbon pools in the biosphere based on a scenario in which p_{CO_2} and ΔT remain constant at their 1990 values. The initial condition for the carbon pools is their state in 1990. Curves from top to bottom are for total carbon pool, x_5, x_2, x_4, x_1, and x_3.

To synthesize a record of the total carbon in the biosphere, we type

```
>> x_total = x(:,1) + x(:,2) + x(:,3) + x(:,4) + x(:,5);
```

and total carbon at equilibrium is defined similarly as

```
>> xhat_total = x1hat_n + x2hat_n + x3hat_n + x4hat_n + x5hat_n;
```

A figure presenting the trajectory, together with the equilibrium points marked as asterisks, is then made by our typing

```
>> figure
>> hold on
>> plot(time,x)
>> plot(time,x_total,'w')
>> plot(time(length(time)),x1hat_n,'y*')
>> plot(time(length(time)),x2hat_n,'m*')
>> plot(time(length(time)),x3hat_n,'c*')
>> plot(time(length(time)),x4hat_n,'r*')
>> plot(time(length(time)),x5hat_n,'g*')
>> plot(time(length(time)),xhat_total,'w*')
```

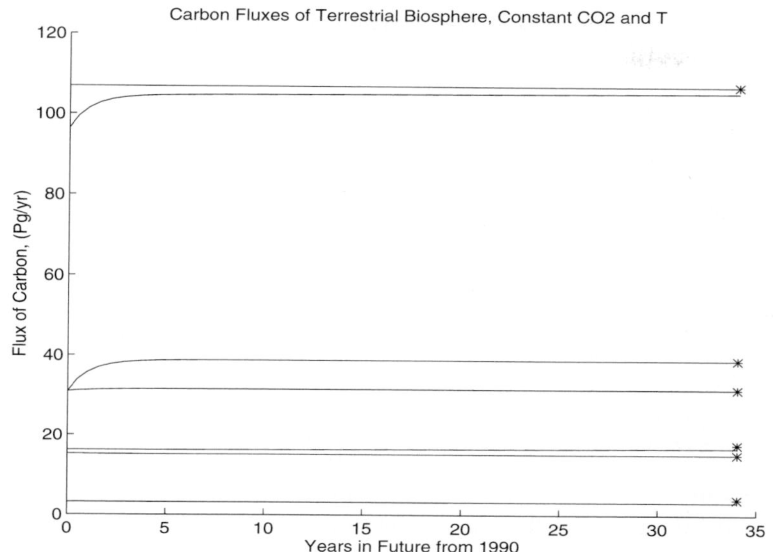

Figure 7.34: The biosphere's carbon fluxes that are based on p_{CO_2} and ΔT remaining constant at their 1990 values. The curves from top to bottom are for GPP (the flux into the biosphere), the total flux out from the biosphere to the atmosphere, the decomposition of x_3, respiration, the decomposition of x_4, the construction costs, and the decomposition of x_5.

This appears as Figure 7.33.

The record of the fluxes can also be synthesized from the trajectory of the carbon pools. Typing

```
>> gpp_projected = g*ones(size(time));
>> flux(:,1) = r*d*x(:,1);
>> flux(:,2) = 0.2*(gpp_projected-r*d*x(:,1));
>> flux(:,3) = d*x(:,3);
>> flux(:,4) = d*x(:,4)/20;
>> flux(:,5) = d*x(:,5)/400;
>> flux_total = flux(:,1) + flux(:,2) + flux(:,3) + flux(:,4) + flux(:,5);
>> flux1hat = r*d*x1hat_n;
>> flux2hat = 0.2*(g-r*d*x1hat_n);
>> flux3hat = d*x3hat_n;
>> flux4hat = d*x4hat_n/20;
>> flux5hat = d*x5hat_n/400;
>> fluxhat_total = flux1hat+flux2hat+flux3hat+flux4hat+flux5hat;
```

defines the record of fluxes as well as the fluxes expected at the equilibrium. The first 35 years of this record, together with asterisks for the equilibrium, are plotted by our typing

```
>> subtime=find(time<35);
>> figure
>> hold on
>> plot(time(subtime),gpp_projected(subtime),'b')
>> plot(time(subtime),flux(subtime,:))
>> plot(time(subtime),flux_total(subtime),'w')
>> plot(time(length(subtime)),flux1hat,'y*')
>> plot(time(length(subtime)),flux2hat,'m*')
>> plot(time(length(subtime)),flux3hat,'c*')
>> plot(time(length(subtime)),flux4hat,'r*')
>> plot(time(length(subtime)),flux5hat,'g*')
>> plot(time(length(subtime)),fluxhat_total,'w*')
```

This results in Figure 7.34. At equilibrium, the sum of the fluxes whereby carbon is returned to the atmosphere equals the GPP, as must be so for a steady state.

Biosphere Dynamics and Global Change

To model how the biosphere will respond to global change, we need a scenario for how p_{CO_2} and ΔT are anticipated to change in the future. To illustrate, let's imagine that the trends of increasing p_{CO_2} and ΔT from the past are continued into the future. The p_{CO_2} in the atmosphere has been increasing approximately exponentially in the last hundred years or so. We'll mark time from 1990, and according to the data, we want an exponential function that equals 350 for 1990, and equals 280 for the year 1800. Typing

```
>> pco2_gw='350*exp(a*t)'
>> a=solve('280=350*exp(a*(1800-1990))','a');
>> pco2_gw=subs(pco2_gw,a,'a')
```

yields

```
pco2_gw = 350*exp(-1/190*log(4/5)*t)
```

which is an exponential function that fits the past data nicely. Similarly, the ΔT has been increasing approximately linearly in the past. It equaled about -0.5 in the year 1870 and about 0.4 in 1995. We still want to mark time from 1990. So, typing

```
>> deltat_gw='b+c*t'
>> [b,c]=solve('-0.5=b+c*(1870-1990)','0.4=b+c*(1995-1990)','b,c');
>> deltat_gw=subs(deltat_gw,b,'b');
>> deltat_gw=subs(deltat_gw,c,'c')
```

yields

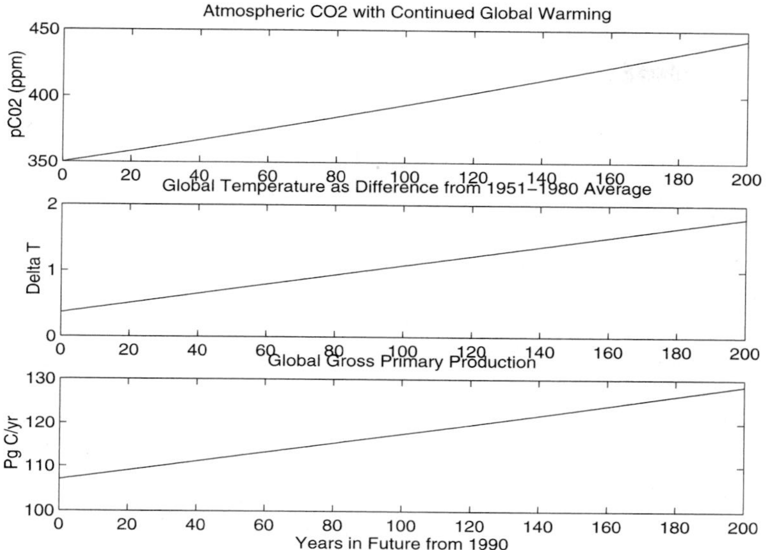

Figure 7.35: A scenario of global change. The top panel illustrates the p_{CO_2} anticipated during the next 200 years if present trends continue. The middle panel shows the projected ΔT based on present trends. The bottom panel shows the GPP projected with the GPP submodel.

```
deltat_gw = .364+7.2e-3*t
```

which is a linear function that fits the past data nicely. To make a figure of the global-change scenario we will be using, type

```
>> t=0:200;
>> pco2=eval(pco2_gw);
>> deltat=eval(deltat_gw);
>> g=eval(sym2ara(gpp));
>> figure
>> subplot(3,1,1)
>> plot(t,pco2)
>> subplot(3,1,2)
>> plot(t,deltat)
>> subplot(3,1,3)
>> plot(t,g)
```

This results in Figure 7.35. Generating trajectories for the biosphere during a global-change scenario is done exactly the same way as for the global-constancy scenario considered previously. Specifically, type

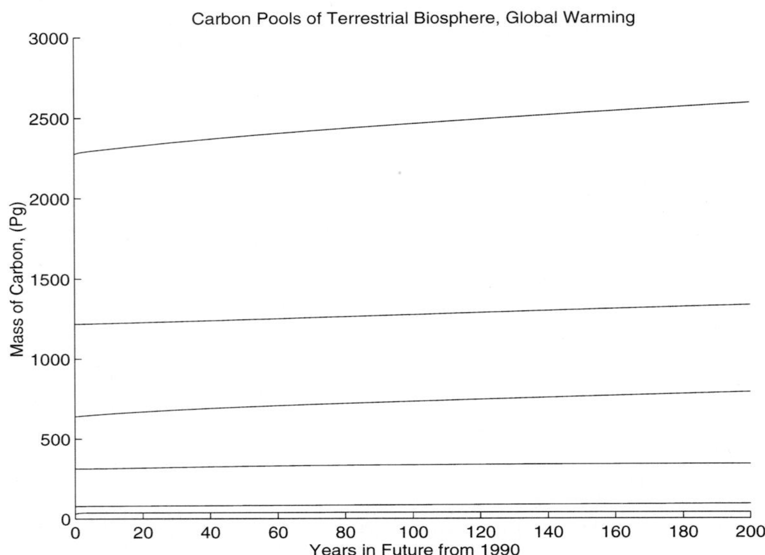

Figure 7.36: The biosphere's carbon pools based on a global-change scenario in which p_{CO_2} and ΔT increase according to past trends. The initial condition for the carbon pools is their state in 1990. Curves from top to bottom are for total carbon pool, x_5, x_2, x_4, x_1, and x_3.

```
>> [time,x] = ode45('biospher',0,200,[77;639;30;313;1217]);
>> x_total = x(:,1) + x(:,2) + x(:,3) + x(:,4) + x(:,5);
```

and we have the trajectories for each of the biosphere's compartments plus the record of the total of the carbon pools. Typing

```
>> figure
>> hold on
>> plot(time,x)
>> plot(time,x_total,'w')
```

yields Figure 7.36 which shows these trajectories. Clearly the biosphere is expanding as the p_{CO_2} increases through time. This expansion shows that the biosphere is continuing in part to ameliorate the projected increase of carbon dioxide in the atmosphere.

To check out the magnitude of the carbon fluxes between the biosphere and the atmosphere, we again calculate all the fluxes by typing

```
>> t=time;
>> pco2=eval(pco2_gw);
>> deltat=eval(deltat_gw);
>> d=eval('exp(nu*deltat)');
```

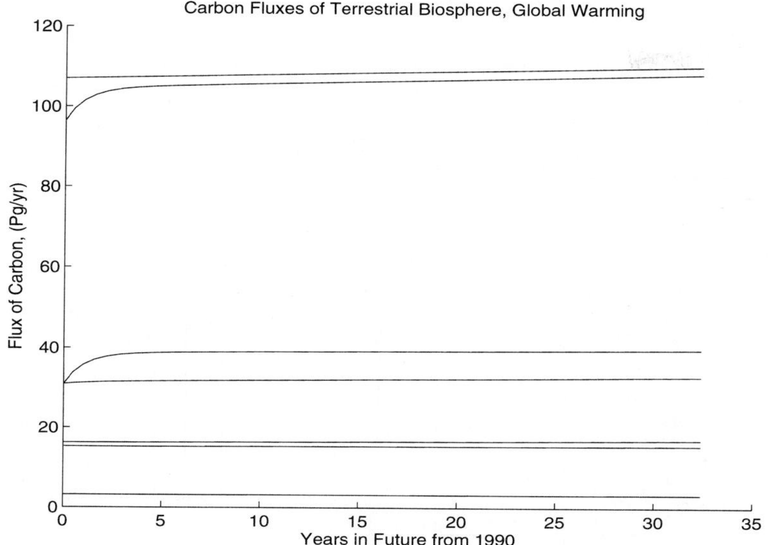

Figure 7.37: The biosphere's carbon fluxes based on a global-change scenario in which p_{CO_2} and ΔT increase according to past trends. The curves from top to bottom are for GPP (the flux into the biosphere), total flux out from biosphere to atmosphere, decomposition of x_3, respiration, decomposition of x_4, construction costs, decomposition of x_5.

```
>> gpp_projected=eval(sym2ara(gpp));
>> flux=[];
>> flux(:,1) = r.*d.*x(:,1);
>> flux(:,2) = 0.2.*(gpp_projected-r.*d.*x(:,1));
>> flux(:,3) = d.*x(:,3);
>> flux(:,4) = d.*x(:,4)/20;
>> flux(:,5) = d.*x(:,5)/400;
>> flux_total = flux(:,1) + flux(:,2) + flux(:,3) + flux(:,4) + flux(:,5);
```

To then graph the first 35 years, we type

```
>> subtime=find(time<35);
>> figure
>> hold on
>> plot(time(subtime),gpp_projected(subtime),'b')
>> plot(time(subtime),flux(subtime,:))
>> plot(time(subtime),flux_total(subtime),'w')
```

This results in Figure 7.37. The gap between the curve for the GPP into the biosphere and the total flux out of the biosphere indicates that the biosphere is accepting more carbon

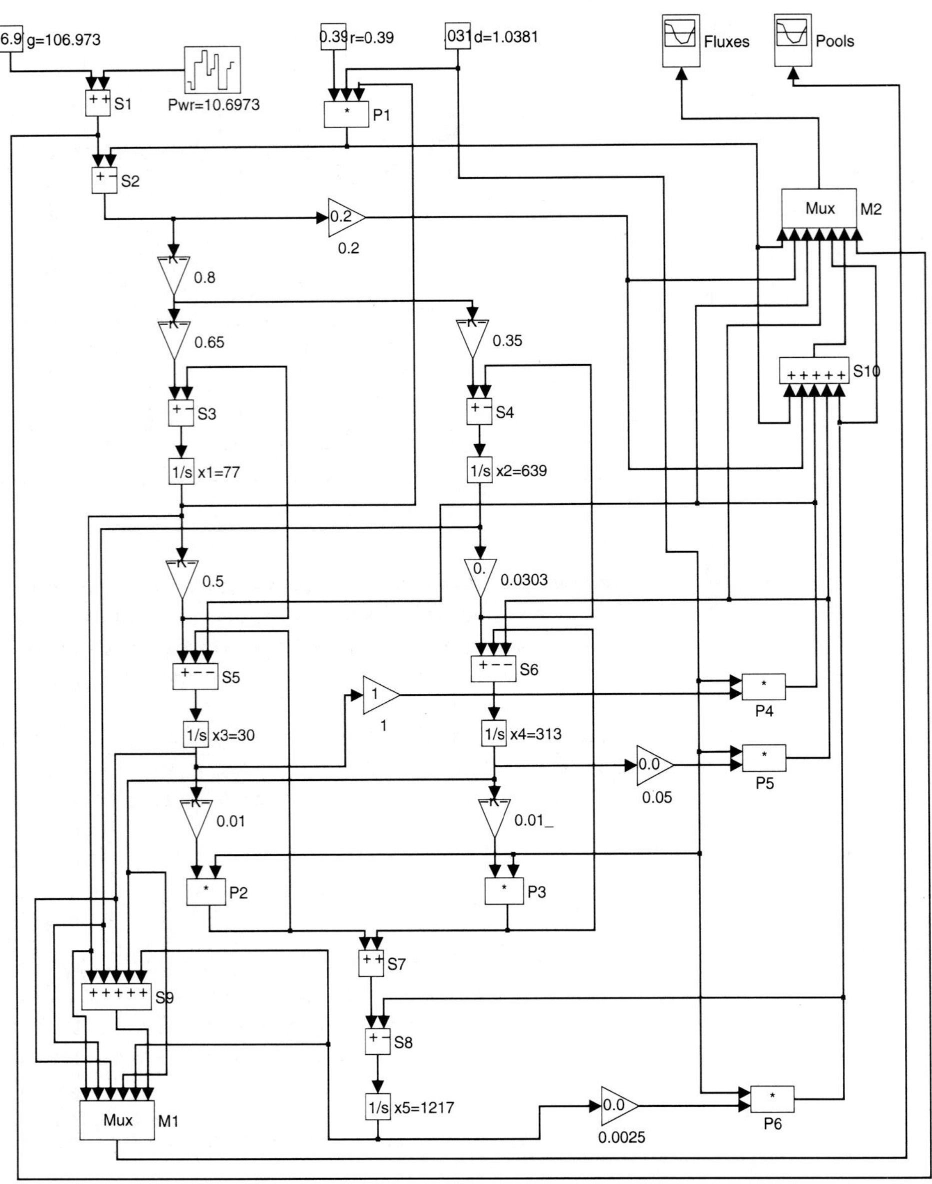

Figure 7.38: The system diagram for a model of carbon exchange between the biosphere and the atmosphere; the model uses SIMULINK.

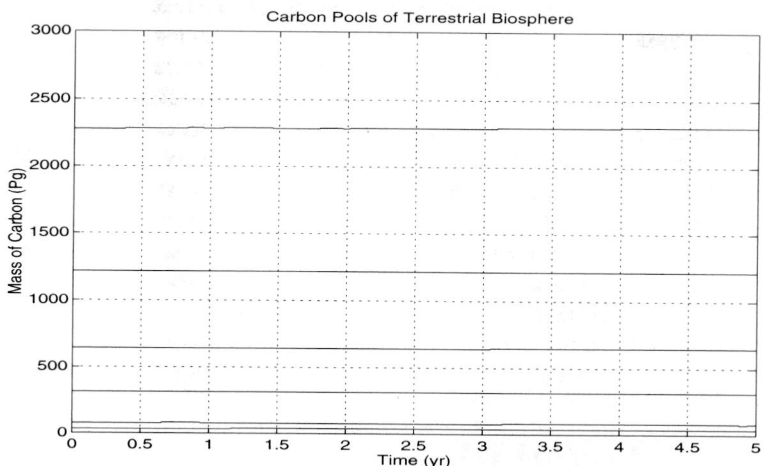

Figure 7.39: Carbon pools in the biosphere with random fluctuations in the GPP. The initial condition for the carbon pools is their state in 1990. The curves from top to bottom are for total carbon pool, x_5, x_2, x_4, x_1, and x_3.

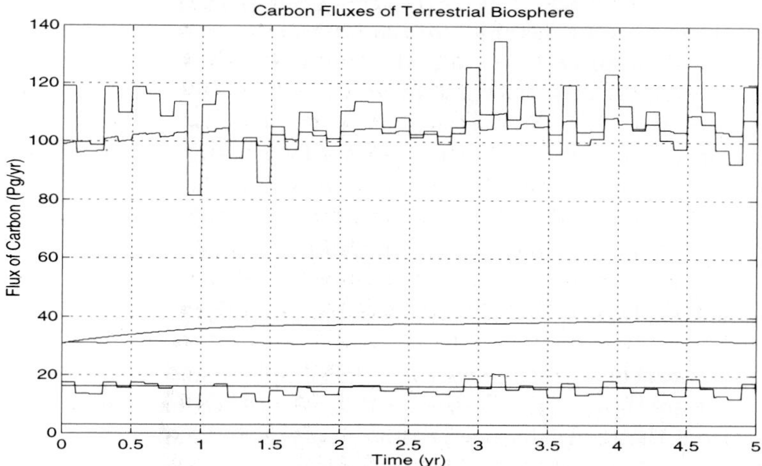

Figure 7.40: Carbon fluxes in the biosphere with random fluctuations in the GPP. The curves from top to bottom are for GPP (the flux into the biosphere), total flux out from biosphere to atmosphere, decomposition of x_3, respiration, decomposition of x_4, construction costs, decomposition of x_5.

than it is releasing. This is a biogeochemical ecosystem service that helps buffer us from global warming.

It's easy to investigate other scenarios for biosphere response to global change too. Just code the formula for the scenario as a function of `t` into `pco2_gw` and `deltat_gw`. You may also wish to code the formula for an alternative model of GPP into `gpp`. Then have `ode45` integrate `'biospher'` over your desired time interval and from your desired initial condition. The hardest part is making the graph.

Still another approach to modifying the biosphere model is using the SIMULINK toolkit, as we now illustrate.

Systems Approach to Global Change

The response of the biosphere to global change is an excellent subject for a systems approach. One may enlarge on the number of compartments if one adds herbivores and their predators. One may also add space: one could define a set of compartments for say, each square degree of latitude and longitude, and thereby cover the globe with boxes, using parameters within each box that are right for its locale. To get you started, I've used SIMULINK to draw up the system diagram for the biosphere model that appears as Figure 7.38.

The model takes as input signals the gross primary production, `g`, the respiration rate, `r`, and the temperature-sensitivity coefficient for respiration and decomposition, `nu`. It produces two graphs as outputs, the carbon pools through time and the carbon fluxes through time.

To make the system diagram, drag in the components and hook them up as illustrated in Figure 7.38. The various constants are then initialized as shown on the diagram, the integration boxes are initialized to the initial conditions for the carbon-pool sizes, and the graph boxes are initialized with the appropriate time intervals, vertical scales, and graph colors.

To illustrate an easy application of the SIMULINK system diagram, I've attached a band-limited white-noise generator to the input signal for the gross primary production. This allows us to see how fluctuations in the atmosphere are damped somewhat by the biosphere's response. Figure 7.39 shows the output graph of system carbon pools through time, and Figure 7.40 shows the output graph of carbon fluxes through time. Clearly, the fluctuations in the biospheric pool sizes and fluxes are smoother than the fluctuations in the GPP.

In this model of carbon exchange between the biosphere and the atmosphere, the biosphere simply processes the carbon that's handed to it by the GPP submodel. The possibility of a connection from the biosphere back to the atmosphere is not considered. So, to conclude the book, we turn to another model in which the biosphere and geosphere do strongly interact to form the epitome of a coupled biological-physical earth system, called Gaia.

7.2.3 The Gaia Hypothesis

Gaia is the ancient Greek word for the earth goddess. The Gaia hypothesis views the whole earth as a living organism, and was introduced scientifically by Lovelock in the 1960's. In part, it marks a return to the superorganism view of ecological systems that dates to the 1920's. What's new is the spiritual interpretation given to the entire planet as an integrated and self-regulating system. The spiritual dimension of the Gaia hypothesis harmonizes with the Native American tradition of the planet as a woman, a theme eloquently expressed by the ecofeminist writer, Paula Gunn Allen. As we will see, there is a definite sense in which a coupled biosphere and geosphere do form a self-regulating system. Most people would not consider this enough to declare the earth a living organism[8]. Still, the implications of the Gaia hypothesis for how we think about the world are profound.

Lovelock introduced a simple model, called "Daisyworld," as a metaphor for how the biosphere and geosphere interact in a self-regulating manner. Focus on the earth's temperature. The earth's temperature attains the value whereby the input of heat from the sun balances the loss of heat to space via black-body radiation. The rate at which heat is absorbed from the sun depends on the reflectivity of the planet's surface, which is called its "albedo." The albedo varies between zero, which is dark and absorbs light, and one, which is shiny and reflects light. The albedo of the planet's surface is assumed to be 0.4. Here's the main idea. Imagine one or more species of vegetation (daisies) can cover the globe. One species of vegetation is lighter than the ground, and has an albedo of 0.65—think of silvery leaves of desert plants; the other is darker than the ground, and has an albedo of 0.25—think of the deep green leaves of understory plants in a forest. As the vegetation spreads across the planet, it changes the reflectivity of the planet, and thereby changes the planet's temperature. So an interaction develops between the spread of the biosphere and the global temperature of the planet. Let's see how this will play out. First we look at the biosphere-geosphere system with one kind of vegetation, which can be either light or dark, and then look at the earth system with both kinds of vegetation.

One-Component Biosphere

Let a_g be the albedo of the ground and a_1 be the albedo of the vegetation type, either dark or light. Let p be the fraction of the planet that is presently covered by vegetation. The present albedo of the world is then the average of the two albedos, which may be written in MATLAB's notation, as

```
>> aw = '((1-p)*ag+p*a1)'
```

The luminosity of the sun is assumed to be capable of varying over geological time. The sun's present luminosity is 1.0, by convention. A younger, cooler sun has a luminosity of say, 0.6 and an older, hotter sun a luminosity of 1.2. The temperature (in °C) of the world with its albedo and the sun's luminosity works out to be

```
>> tw = '(((1.7e+10)*l*(1-aw))^(1/4)-273)'
```

This formula results when solar input balances black-body radiation, once all the correct physical constants are used. Meanwhile, the temperature of the vegetation is

[8]Is your house a living organism just because it has a thermostatically controlled central heating system?

```
>> t1 = '(u*(aw-a1)+tw)'
```

which means that if the vegetation is lighter than the average albedo of the world, (aw-a1) is negative. This relation implies that the plants are cooler than the global temperature, whereas if the vegetation is darker than the world's albedo, the plants are hotter than the global temperature. The parameter u is set to 20. It's further assumed that the birth rate of the vegetation depends on the temperature of the plants, with 22.5°C being optimal for all vegetation, regardless of whether they are light or dark. Therefore, the birth rate of the vegetation may be taken as a downward opening quadratic whose peak is at 22.5. To alert MATLAB of this, type

```
>> r1 = '(1-v*(22.5-t1)^2)'
```

v is a paramter that will be set to 0.00032. Although r1 is called a birth rate, it's really more like a colonization rate, to vacant substrate from occupied substrate. The rate at which the vegetation spreads over the planet is assumed to be

```
>> pdot = 'p*((1-p)*r1-d1)'
```

Vacant substrate, measured by (1-p), is colonized from occupied substrate, measured by p, at rate r1. Occupied patches also die out at rate d1. The idea of occupied patches colonizing vacant patches at rate r1 is reminiscent of the model for a metapopulation, the logistic weed, considered in Chapter 4. pdot is the differential equation we have to integrate to see how the vegetation spreads across the planet. pdot contains r1, which depends on t1, which depends on aw, which depends on p. So, to determine the full expression for pdot as a function of p, we need to substitute all these relations. To do this, we type

```
>> tw = subs(tw,aw,'aw')
>> t1 = subs(t1,aw,'aw')
>> t1 = subs(t1,tw,'tw')
>> r1 = subs(r1,t1,'t1')
>> pdot = subs(pdot,r1,'r1')
```

yielding

```
pdot = p*((1-p)*(1-v*(295.5-u*((1-p)*ag+p*a1-a1)
       -361.0873136847277*l^(1/4)*(1-(1-p)*ag-p*a1)^(1/4))^2)-d1)
```

where I've spread the formula out over two lines. Although its derivation has seemed complicated, Daisyworld for a single type of vegetation boils down to a single differential equation for the variable, p, which is the fraction of the planet occupied by the biosphere.

The right-hand-side of pdot is the production function for planetary coverage. To see its graph, assign values to the parameters and type

```
>> global ag l u v a1 d1;
>> ag = 0.4;
>> l = 1;
```

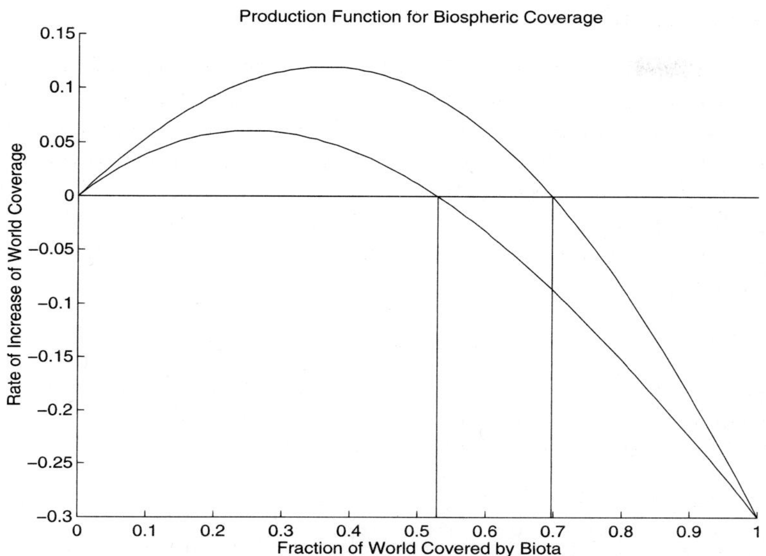

Figure 7.41: The production function for the spread of a single type of vegetation across the planet. The top curve is for a light vegetation and the bottom curve is for dark vegetation. Equilibria occur where the production functions cross the horizontal axis, and are marked with vertical lines.

```
>> u = 20;
>> v = 3.2e-4;
>> d1 = 0.3;
>> a_dark = 0.25;
>> a_light = 0.65;
```

Then, to generate a figure, type

```
>> figure
>> hold on
>> plot([0 1],[0 0],'w')
>> p = 0:.01:1;
>> a1 = a_dark;
>> plot(p,eval(sym2ara(pdot)),'b')
>> a1 = a_light;
>> plot(p,eval(sym2ara(pdot)),'c')
```

This yields Figure 7.41. The production function for dark vegetation is graphed in blue and for light vegetation in cyan. What jumps out from the figure is that the production functions very closely resemble a logistic model. The production functions look like downward opening

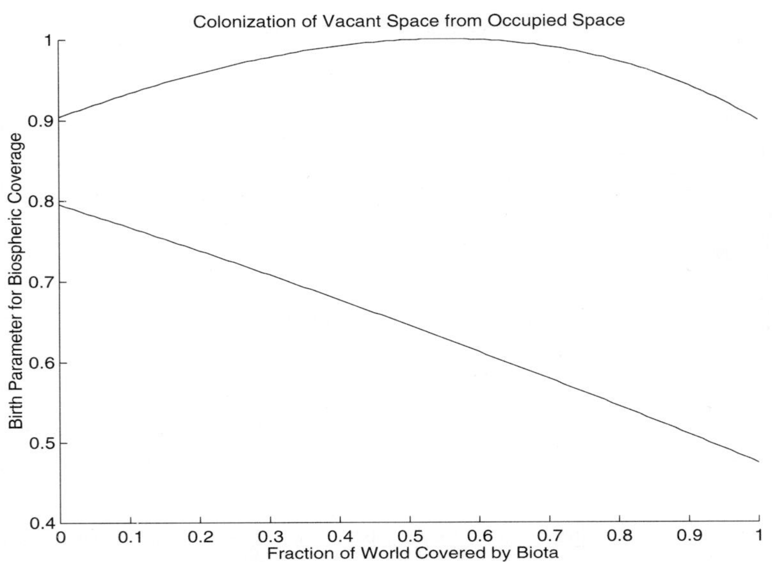

Figure 7.42: The growth parameter, r_1, as a function of the fraction of the planet covered by vegetation, p. Top curve is for a light vegetation and bottom curve for dark vegetation.

parabolas, just like the production function for the logistic equation discussed way back in Chapter 4.

The equilibrium level of planetary coverage, `phat`, is found numerically by MATLAB's `fzero` function. With a text editor, type in the production function, and save it as `gaia_eq.m`,

```
function pdot=gaia_eq(p)
 global ag l u v a1 d1;
 pdot = p*((1-p)*(1-v*(295.5-u*((1-p)*ag+p*a1-a1) ...
       -361.0873136847277*l^(1/4)*(1-(1-p)*ag-p*a1)^(1/4))^2)-d1);
```

Then the equilibrium for the dark vegetation is found by our typing

```
>> a1 = a_dark;
>> phat_dark = fzero('gaia_eq',1)
```

yielding

```
phat_dark = 0.5284
```

Similarly, for the light vegetation, we type

```
>> a1 = a_light;
>> phat_light = fzero('gaia_eq',1)
```

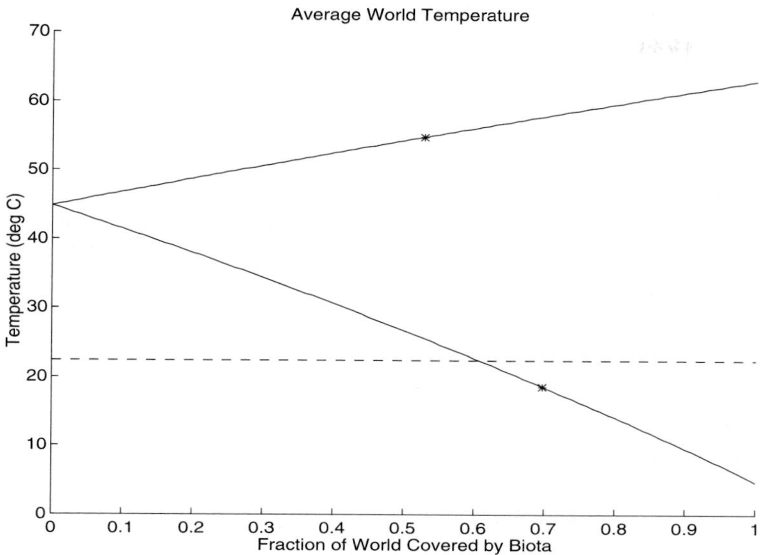

Figure 7.43: Effect of planetary vegetation coverage on world temperature. Top curve is for dark vegetation and bottom curve for light vegetation. Dashed horizontal line is the biologically optimal temperature. Asterisks mark equilibrium points. In the absence of vegetation, world temperature is 45°C. The dark vegetation comes to an equilibrium planetary coverage that makes the earth's temperature much higher than optimum, and the light vegetation comes to an equilibrium that makes temperature lower than optimum.

which yields

```
phat_light = 0.6972
```

The locations of these equilibria can be added as vertical lines to Figure 7.41. Type

```
>> p = 1;
>> plot([phat_dark phat_dark],[0 eval(pdot)],'b')
>> plot([phat_light phat_light],[0 eval(pdot)],'c')
```

The reason that Daisyworld with one vegetation type is very close to a logistic model is that in the formula for `pdot` the coefficient, `r1`, doesn't change very much with `p`. If `r1` is independent of `p`, then the model is exactly a logistic model, as discussed in Chapter 4 for the logistic weed. To see how `r1` depends on `p`, type

```
>> figure
>> hold on
>> p = 0:.01:1;
```

```
>> a1 = a_dark;
>> plot(p,eval(sym2ara(r1)),'b')
>> a1 = a_light;
>> plot(p,eval(sym2ara(r1)),'c')
```

This results in Figure 7.42. The rather slight variation of `r1` with `p` is not enough to make the model much different from the logistic model.

Because it changes the world's albedo, the coverage of vegetation affects the world's temperature. The world's temperature as a function of the fraction of the planet covered by vegetation is illustrated by our typing

```
>> figure
>> hold on
>> p = 0:.01:1;
>> a1 = a_dark;
>> plot(p,eval(sym2ara(tw)),'b')
>> a1 = a_light;
>> plot(p,eval(sym2ara(tw)),'c')
>> p = phat_dark;
>> a1 = a_dark;
>> plot(p,eval(tw),'b*')
>> p = phat_light;
>> a1 = a_light;
>> plot(p,eval(tw),'c*')
>> plot([0 1],[22.5 22.5],'w--')
```

This results in Figure 7.43. The world temperatures when the vegetation's planetary coverage is at equilibrium are indicated with asterisks, and the biologically optimal temperature of 22.5°C is shown for reference as a dashed horizontal line. If dark vegetation covers the planet, it raises the world's temperature above the biological optimum, whereas light vegetation depresses the world's temperature below the biological optimum.

To see the world's temperature change dynamically as the vegetation spreads across the planet, we can integrate the equation for `pdot`. We'll use the familiar `ode23` command. So, with a text editor, type the following function and save it as `gaia_dy.m`

```
function pdot=gaia_dy(t,p)
 global ag l u v a1 d1;
 pdot = p*((1-p)*(1-v*(295.5-u*((1-p)*ag+p*a1-a1) ...
       -361.0873136847277*l^(1/4)*(1-(1-p)*ag-p*a1)^(1/4))^2)-d1);
```

Then the trajectories for the spread of dark and light vegetation are generated by typing

```
>> a1 = a_dark;
>> [t_dark p_dark] = ode23('gaia_dy',0,20,0.01);
>> a1 = a_light;
>> [t_light p_light] = ode23('gaia_dy',0,20,0.01);
```

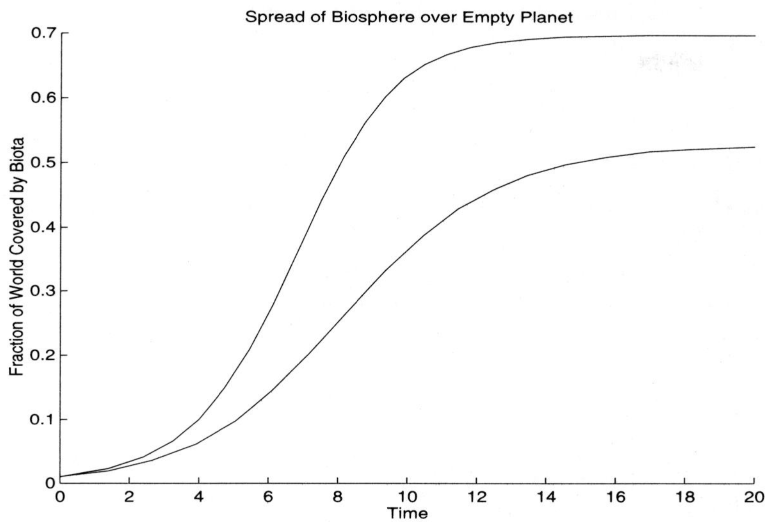

Figure 7.44: Spread of the biosphere across the planet. The top curve is for light vegetation and the bottom curve for dark vegetation.

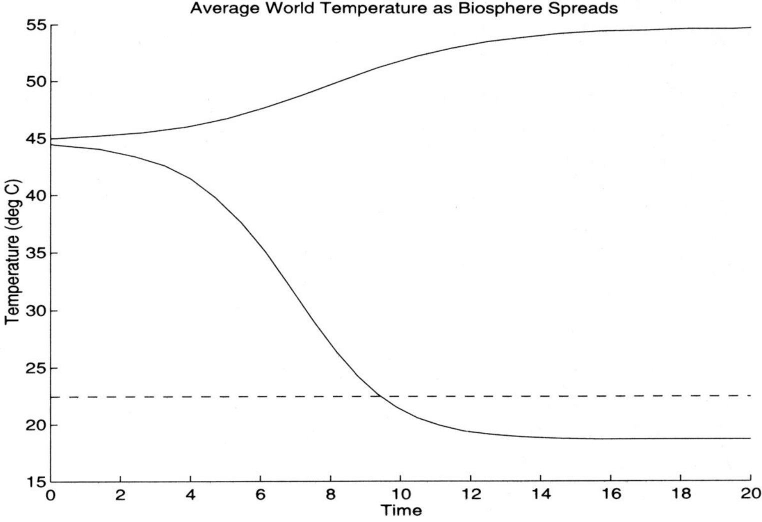

Figure 7.45: The world's temperature as the biosphere spreads across the planet. The top curve is for dark vegetation and the bottom curve for light vegetation. The dashed horizontal line is the biologically optimal temperature.

A graph of these trajectories is then made by our typing

```
>> figure
>> hold on
>> plot(t_dark,p_dark,'b')
>> plot(t_light,p_light,'c')
```

This appears as Figure 7.44. The curves show sigmoid growth of the vegetation, as expected of a logistic-like model.

As the vegetation spreads, the world's temperature changes too. The trajectory of temperature change brought about by the spread of vegetation is graphed with

```
>> figure
>> hold on
>> a1 = a_dark;
>> p = p_dark;
>> plot(t_dark,eval(sym2ara(tw)),'b')
>> a1 = a_light;
>> p = p_light;
>> plot(t_light,eval(sym2ara(tw)),'c')
>> plot([0 t_light(length(t_light))], [22.5 22.5],'w--')
```

as shown in Figure 7.45.

Next, suppose the plant's biosphere has come to its equilibrium level of coverage. What happens as the sun ages? The equilibrium biosphere will track the changes in the sun. To model the aging of the sun, let the luminosity parameter, `l`, vary from 0.60 to 1.20. The following code generates two vectors for each type of vegetation. One vector holds the equilibrium biosphere coverage, and the other the world's temperature, for each luminosity. To generate these vectors, type

```
>> lrange=0.60:0.025:1.20;
>>
>> phat_dark=[];
>> twhat_dark=[];
>> a1=a_dark;
>> for l=lrange
>>  p = fzero('gaia_eq',1);
>>  phat_dark = [phat_dark p];
>>  twhat_dark = [twhat_dark eval(tw)];
>> end
>>
>> phat_light=[];
>> twhat_light=[];
>> a1=a_light;
>> for l=lrange
```

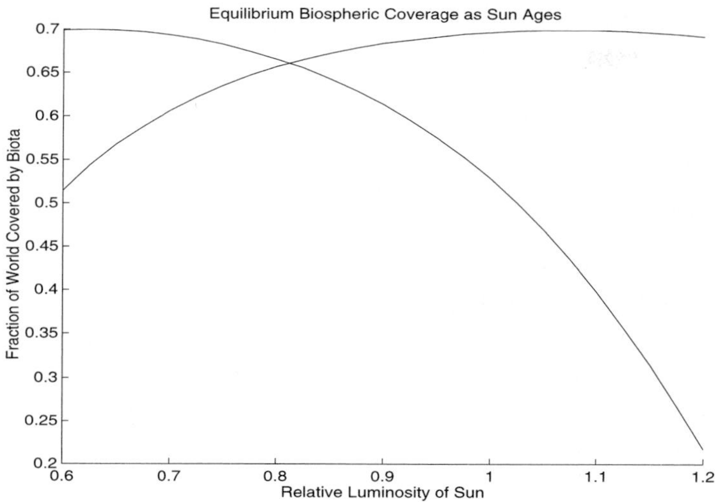

Figure 7.46: The equilibrium coverage of the planet by biota, as a function of the sun's relative luminosity. The top curve at a luminosity of 0.60 is for dark vegetation, and the other curve for light vegetation.

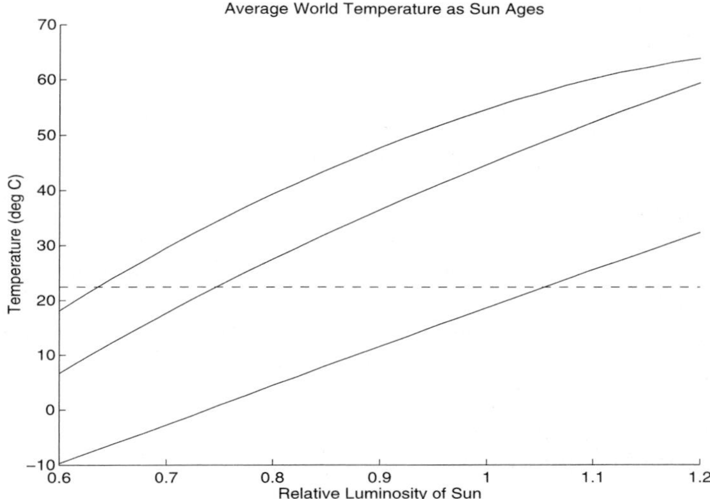

Figure 7.47: The world's temperature when the biota has come to its equilibrium coverage, as a function of the sun's relative luminosity. The top curve is for dark vegetation, the bottom curve for light vegetation. For reference, the middle curve is the world temperature without any biota. The dashed horizontal line is the biologically optimal temperature.

```
>>  p = fzero('gaia_eq',1);
>>  phat_light = [phat_light p];
>>  twhat_light = [twhat_light eval(tw)];
>> end
```

To graph the equilibrium biosphere-coverage vectors, type

```
>> figure
>> hold on
>> plot(lrange,phat_dark,'b')
>> plot(lrange,phat_light,'c')
```

This results in Figure 7.46. When the sun is young, the dark vegetation does best, but as the sun ages and gets brighter, the earth is increasingly better for the light vegetation.

To graph the world temperatures that occur with these equilibrium levels of coverage, type

```
>> figure
>> hold on
>> plot(lrange,twhat_dark,'b')
>> plot(lrange,twhat_light,'c')
```

For reference, let's also put a curve for the temperature of a world without any biota on it. Type

```
>> p = 0; l = lrange;
>> plot(lrange,eval(sym2ara(tw)),'w')
```

Let's also add the dashed horizontal line for the biologically optimal temperature:

```
>> plot([lrange(1) lrange(length(lrange))],[22.5 22.5],'w--')
```

This all results in Figure 7.47. Clearly the planet is always hotter if covered with dark plants and cooler if covered with light plants, regardless of how bright the sun is.

What Daisyworld shows us so far is an *interaction* between a biota's spreading across the globe and global temperature. But it doesn't show anything that might be called *self-regulation.* What makes this model interesting is how matters change when the biosphere has two coexisting components, when the light and dark types of vegetation both share the planet at the same time.

Two-Component Biosphere

When there are two distinct vegetation components in the biosphere, two state variables are needed, `p1` for the fraction of the world covered by the first type of vegetation (dark) and `p2` for the coverage of the second type of vegetation (light). The model with two types of vegetation is the straight-forward extension of that with one type. The world's albedo is defined as the spatial average of the albedo of the ground, the albedo of plant-1, and that of plant-2, according to

```
>> aw = '((1-p1-p2)*ag+p1*a1+p2*a2)'
```

The formula for the world's temperature, given the overall albedo, is the same as before

```
>> tw = '(((1.7e+10)*l*(1-aw))^(1/4)-273)'
```

The temperature of each plant type depends on its albedo relative to the world's albedo according to the same formulas as before:

```
>> t1 = '(u*(aw-a1)+tw)'
>> t2 = '(u*(aw-a2)+tw)'
```

Also, the birth rates of the two vegetation types are the same as before, with 22.5° as the optimal temperature for both.

```
>> r1 = '(1-v*(22.5-t1)^2)'
>> r2 = '(1-v*(22.5-t2)^2)'
```

Finally, the differential equations for each of the two vegetation types are also the same, in that each vegetation type colonizes vacant substrate from the substrate it already occupies:

```
>> p1dot = 'p1*((1-p1-p2)*r1-d1)'
>> p2dot = 'p2*((1-p1-p2)*r2-d2)'
```

If `r1` and `r2` were both independent of `p1` and `p2`, then this system would reduce to a special case of the Lotka-Volterra competition equations, as discussed in the section on Lotka-Volterra weeds in Chapter 6. But the `r`'s do depend on the `p`'s and to develop the differential equations explicitly, we need to go through the following chain of substitutions:

```
>> tw = subs(tw,aw,'aw')
>> t1 = subs(t1,aw,'aw')
>> t1 = subs(t1,tw,'tw')
>> t2 = subs(t2,aw,'aw')
>> t2 = subs(t2,tw,'tw')
>> r1 = subs(r1,t1,'t1')
>> r2 = subs(r2,t2,'t2')
>> p1dot = subs(p1dot,r1,'r1')
>> p2dot = subs(p2dot,r2,'r2')
```

yielding

```
p1dot = p1*((1-p1-p2)*(1-v*(295.5-u*((1-p1-p2)*ag+p1*a1+p2*a2-a1)
       -361.0873136847277*l^(1/4)*(1-(1-p1-p2)*ag-p1*a1-p2*a2)^(1/4))^2)-d1)

p2dot = p2*((1-p1-p2)*(1-v*(295.5-u*((1-p1-p2)*ag+p1*a1+p2*a2-a2)
       -361.0873136847277*l^(1/4)*(1-(1-p1-p2)*ag-p1*a1-p2*a2)^(1/4))^2)-d2)
```

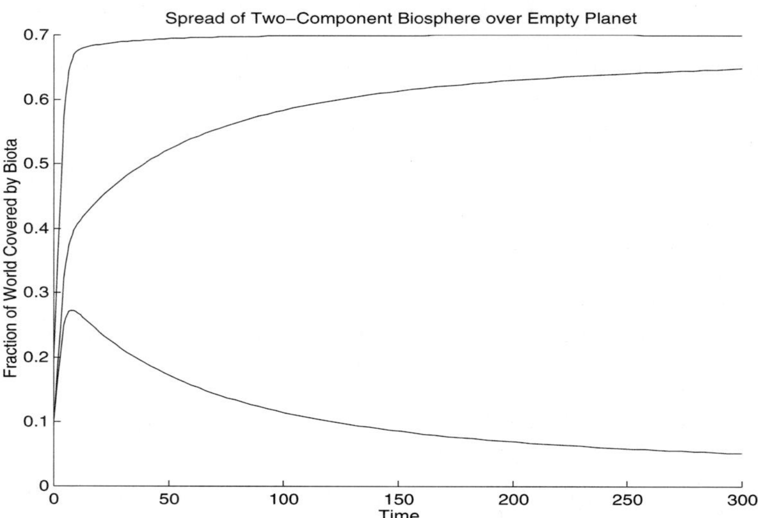

Figure 7.48: The spread of a two-component biosphere across the planet. The top curve is for the sum of both vegetation types, the middle curve is for light vegetation and the bottom curve for dark vegetation.

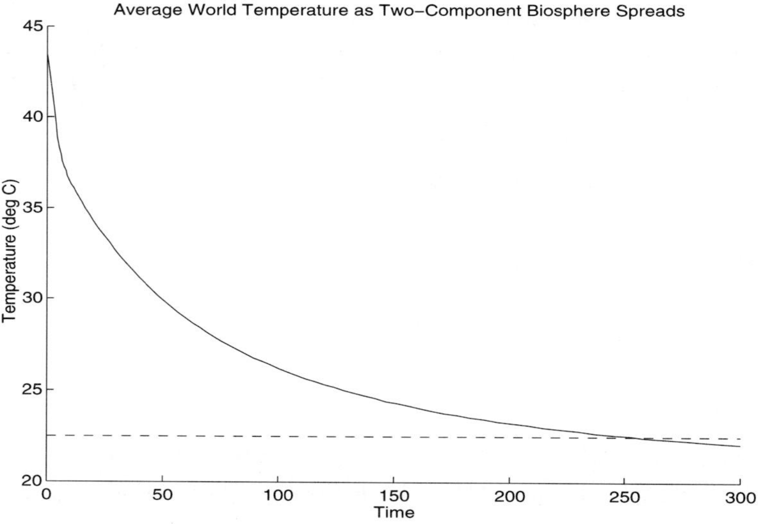

Figure 7.49: The world's temperature as a two-component biosphere spreads across the planet. The dashed horizontal line indicates the biologically optimum temperature.

So, Daisyworld with a two-component biosphere boils down to a pair of competition equations.

To see what happens as the competition between the two vegetation types unfolds, we'll transcribe the formulas for `p1dot` and `p2dot` into a function that we can integrate numerically with good ol' `ode45`. So, with a text editor, type in the following function, and save it as `gaia2_dy.m`.

```
function pdot=gaia2_dy(t,p)
 global ag l u v a1 a2 d1 d2;
 pdot(1,1) = p(1)*((1-p(1)-p(2))*(1-v*(295.5 ...
             -u*((1-p(1)-p(2))*ag+p(1)*a1+p(2)*a2-a1) ...
             -361.0873136847277*l^(1/4)*(1-(1-p(1)-p(2))*ag ...
             -p(1)*a1-p(2)*a2)^(1/4))^2)-d1);
 pdot(2,1) = p(2)*((1-p(1)-p(2))*(1-v*(295.5 ...
             -u*((1-p(1)-p(2))*ag+p(1)*a1+p(2)*a2-a2) ...
             -361.0873136847277*l^(1/4)*(1-(1-p(1)-p(2))*ag ...
             -p(1)*a1-p(2)*a2)^(1/4))^2)-d2);
```

Next, to define the parameters, type

```
>> global ag l u v a1 a2 d1 d2;
>> d2 = d1;
>> a1 = a_dark;
>> a2 = a_light;
>> l = 1;
```

A set of trajectories for the spread of the two-component biosphere will start with a 10% cover by each component. To produce these trajectories, type

```
>> [t p] = ode45('gaia2_dy',0,300,[0.1 0.1]);
```

The results are then graphed by our typing

```
>> p1 = p(:,1);
>> p2 = p(:,2);
>> figure
>> hold on
>> plot(t,p1,'b')
>> plot(t,p2,'c')
>> plot(t,p1+p2,'g')
```

This yields Figure 7.48. The spread of the first component (dark vegetation) is shown in blue, of the second component (light vegetation) in cyan, and the summed coverage by both types combined is shown in green. The biosphere quickly spreads to near its equilibrium total level of coverage, and thereafter the relative proportions of the two vegetation types slowly change. The two types coexist; the light plants are more abundant than the dark plants.

As the two-component biosphere spreads, the world's temperature changes, and to graph its change through time, type

```
>> figure
>> hold on
>> plot(t,eval(sym2ara(tw)),'g')
>> plot([0 t(length(t))], [22.5 22.5],'w--')
```

This results in Figure 7.49. The world's temperature approaches the vicinity of the biologically optimal temperature, indicated by the dashed horizontal line. Contrast this with Figure 7.45, in which the trajectories for the world's temperature with a one-component biosphere proceeded with indifference to the biologically optimal temperature. The two-component model offers us the first glimpse of biotic regulation of the earth's temperature.

To see if the appearance of regulation in Figure 7.49 is just a coincidence, let's consider how the biosphere affects the earth's temperature for a range of luminosities. As before, we'll consider a range of luminosities from a young dim star to a bright old star. We'll generate vectors holding the equilibrium coverage levels for each vegetation type, and a vector for the world temperature at the biospheric equilibrium. We type

```
>> p1hat=[];
>> p2hat=[];
>> twhat=[];
>> for l=lrange
>>   [t_both p_both] = ode45('gaia2_dy',0,300,[0.1 0.1]);
>>   p1 = p_both(length(t_both),1);
>>   p2 = p_both(length(t_both),2);
>>   p1hat = [p1hat p1];
>>   p2hat = [p2hat p2];
>>   twhat = [twhat eval(tw)];
>> end
```

The results are now easily graphed. The equilibrium biospheric coverage for both vegetation types and for their sum is graphed by our typing

```
>> figure
>> hold on
>> plot(lrange,p1hat,'b')
>> plot(lrange,p2hat,'c')
>> plot(lrange,p1hat+p2hat,'g')
```

yielding Figure 7.50. The dark plants predominate when the sun is young and dim, the light plants when the sun is old and bright. The sum of the coverage remains constant across the range of solar luminosities. So as the sun ages, the relative proportions of the two components in the biosphere change, but their sum remains constant. Contrast this with Figure 7.46, where a biosphere with just one component either shrinks or expands as a whole when the sun's luminosity changes.

The most interesting figure is the last. Typing

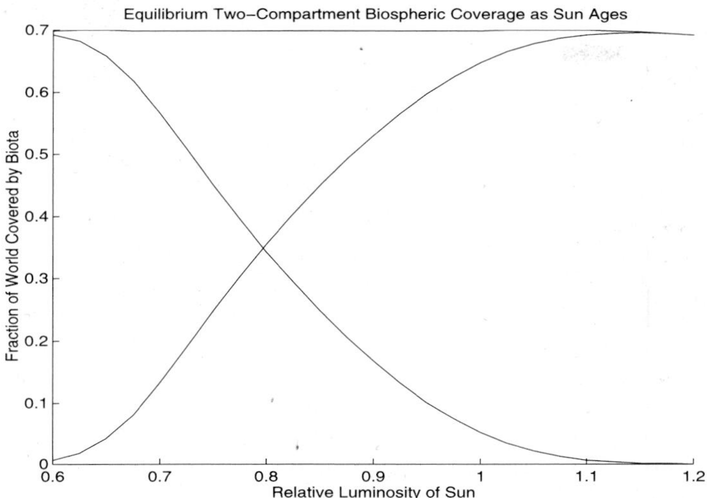

Figure 7.50: Equilibrium coverage of planet by a two-component biosphere as a function of the sun's relative luminosity. The top curve is for the sum of the two components, middle curve at a luminosity of 0.60 is for dark vegetation, and the other curve for light vegetation.

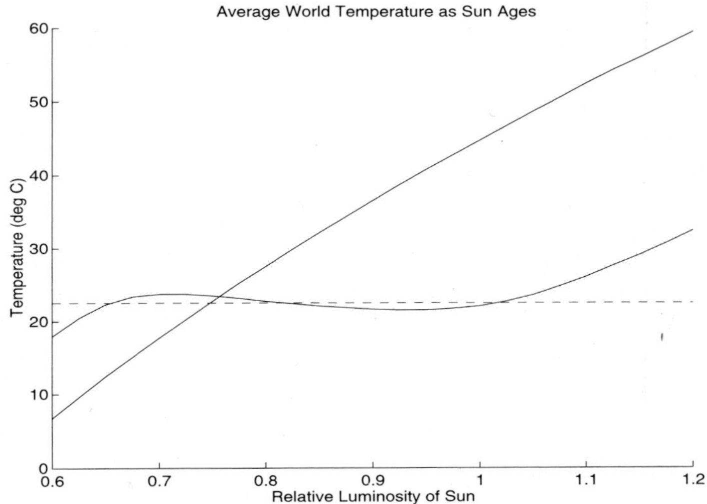

Figure 7.51: The world's temperature occurring when a two-component biosphere has come to its equilibrium coverage as a function of the sun's relative luminosity. The dashed horizontal line is the biologically optimal temperature. For reference, the world temperature without any biota is the monotonically increasing curve.

```
>> figure
>> hold on
>> plot(lrange,twhat,'g')
>> p1 = 0; p2=0; l = lrange;
>> plot(lrange,eval(sym2ara(tw)),'w')
>> plot([lrange(1) lrange(length(lrange))],[22.5 22.5],'w--')
```

yields Figure 7.51 which shows the world temperature being maintained near the biologically optimal temperature during nearly the entire transition from a dim young star to a bright old star. Contrast this with Figure 7.47, where a biosphere with just one component simply makes the planet as a whole either hotter or colder, with seeming indifference to where the biologically optimum temperature is.

Figure 7.51 shows what can validly be called self-regulation of the planet's temperature by the biosphere. Here's what's happening and why it works. For any given luminosity, if the temperature should increase, the light plants expand and the dark plants shrink, so that the temperature decreases. Conversely, if the temperature should decrease, the light plants shrink and the dark plants expand, so that the temperature increases. Competitive equilibrium is attained between the two biosphere components only when the temperature is at their common optimum, 22.5°C. It's the competitive adjustments between the two components of the biosphere that, in the aggregate, regulate the world's temperature at the biological optimum.

Daisyworld offers a valuable introduction to the many surprising findings that lie ahead in our exploring the dynamic interactions between the biosphere and the geosphere. Daisyworld itself may not feature the most important of the biosphere-geosphere interactions. The albedo of the earth is largely set by the extent of the oceans and by the area covered with ice, and the opportunity for vegetation to affect the earth's overall albedo seems limited. In contrast, the biosphere is well known to have huge effects on the atmosphere, and a model along the lines of Daisyworld could be framed around geochemical processes. Perhaps the comparatively realistic but non-interactive model of the last section could be merged with the approach used in the Daisyworld of this section.

We have come to the end of the book. We started with a model for a tiny lizard sunning itself on a rock, and trying to stay comfortable outdoors. We conclude with a model for the entire planet, with the biosphere keeping the planet comfortable for its inhabitants. Along the way we have spanned the full range of scales that ecology is about. I hope you've enjoyed the trip.

7.3 Further Readings

Bazzaz, F. (1990). The response of ecosystems to rising global CO_2 levels. *Ann. Rev. Ecol. Syst.*, 21:167–196.

Carpenter, S. and Kitchell, J. (1988). Consumer control of lake productivity. *BioScience*, 38:764–769.

Case, T. J. (1990). Invasion resistance arises in strongly interacting species-rich model competition communities. *Proceedings of the National Academy of Sciences (USA)*, 87:9610–9614.

Clements, F. E. (1916). *Plant Succession: An Analysis of the Development of Vegetation.* Carnegie Institute Publication No. 242, Washington D. C.

Cohen, J. E. (1989). Food webs and community structure. In Roughgarden, J., May, R. M., and Levin, S. A., editors, *Perspectives in Theoretical Ecology*, pages 181–202. Princeton University Press.

D'Ancona, U. (1954). The struggle for existence. *Biblioteca Biotheoretica*, VI:1–274.

Dial, R. and Roughgarden, J. (1995). Experimental removal of insectivores from rain forest canopy: direct and indirect effects. *Ecology*, 76:1821–1834.

Diamond, J. M. (1973). Distributional ecology of New Guinea birds. *Science*, 179:759–769.

Diamond, J. M. (1975). Assembly of species communities. In Cody, M. L. and Diamond, J. M., editors, *Ecology and Evolution of Communities*, pages 342–444. Belknap, Harvard University Press.

Ehrlich, P. R. and Ehrlich, A. H. (1992). The value of biodiversity. *Ambio*, 21:219–226.

Ehrlich, P. R. and Mooney, H. A. (1983). Extinction, substitution and ecosystem services. *BioScience*, 33:248–254.

Field, C., Chapin, S., Matson, P., and Mooney, H. (1992). Responses of terrestrial ecosystems to the changing atmosphere: a resource-based approach. *Ann. Rev. Ecol. Syst.*, 23:201–235.

Gilpin, M. and Diamond, J. (1976). Calculation of immigration and extinction curves from the species-area-distance relation. *Prod. Nat. Acad. Sci. (USA)*, 73:4130–4134.

Gleason, H. A. (1926). The individualistic concept of the plant association. *Bull. Torrey Bot. Club*, 53:1–20.

Goldwasser, L. and Roughgarden, J. (1993). Construction and analysis of a large Caribbean food web. *Ecology*, 74:1216–1233.

Golley, F. B. (1960). Energy dynamics of a food chain of an old-field community. *Ecol. Monogr.*, 30:187–206.

Holling, C. S. (1992). Cross-scale morphology, geometry and dynamics of ecosystems. *Ecol. Monogr.*, 62:447–502.

Lovelock, J. E. (1992). A numerical model for biodiversity. *Phil. Trans. R. Soc. Lond. B*, 338:383–391.

MacArthur, R. (1972). *Geographical Ecology: Patterns in the Distribution of Species.* Harper and Row.

MacArthur, R. H. and Wilson, E. O. (1963). An equilibrium theory of insular zoogeography. *Evolution*, 17:373–387.

May, R. (1973). *Stability and Complexity in Model Ecosystems.* Princeton University Press.

Merchant, C. (1994). *Key Concepts in Critical Theory: Ecology.* Humanities Press.

Odum, E. P. (1971a). *Fundamentals of Ecology.* Saunders.

Odum, H. T. (1971b). *Environment, Power and Society.* Wiley.

O'Neill, R. V., DeAngelis, D. L., Waide, J. B., and Allen, T. F. H. (1986). *A Hierarchical Concept of Ecosystems.* Princeton University Press.

Persson, L., Andersson, G., Hamrin, S., and Johansson, L. (1988). Predator regulation and primary productivity along the productivity gradient of temperate lake ecosystems. In Carpenter, S., editor, *Complex Interactions in Lake Communities*, pages 45–65. Springer-Verlag.

Pimm, S. (1982). *Food Webs.* Chapman and Hall.

Potter, C., Randerson, J., Field, C., Matson, P., Vitousek, P., Mooney, H., and Klooster, S. (1993). Terrestrial ecosystem production: a process model based on global satellite and surface data. *Global Biogeochemical Cycles*, 7:811–841.

Power, M. (1990). Effects of fish in river food webs. *Science*, 250:811–814.

Power, M., Parker, M., and Wootton, J. (1996). Disturbance and food chain length in rivers. In Polis, G. and Winemiller, K., editors, *Food Webs*, pages 286–297. Chapman and Hall.

Preston, F. (1948). The commonness and rarity of species. *Ecology*, 29:254–283.

Prigogine, I. (1961). *Introduction to the Thermodynamics of Irreversible Processes.* Wiley.

Roughgarden, J. (1979). *Theory of Population Genetics and Evolutionary Ecology: An Introduction.* Macmillan (Reprinted 1996, Prentice Hall).

Roughgarden, J. (1995). Anolis *Lizards of the Caribbean: Ecology, Evolution, and Plate Tectonics.* Oxford University Press.

Sarmiento, J., Quéré, C. L., and Pacala, S. (1995). Limiting future atmospheric carbon dioxide. *Global Biogeochemical Cycles*, 9:121–137.

Schimel, D., Kittel, T., and Parton, W. (1991). Terrestrial biogeochemical cycles: global interactions with the atmosphere and hydrology. *Tellus, Ser. A*, 43A:188–203.

Schoener, T. W. (1983). Rate of species turnover decreases from lower to higher organisms: a review of data. *Oikos*, 41:372–377.

Schoener, T. W. (1989). Food webs from the small to the large. *Ecology*, 70:1559–1589.

Schoener, T. W. (1993). On the relative importance of direct versus indirect effects in ecological communities. In Kawanabe, H., Cohen, J., and Iwasaki, K., editors, *Mutualism and Community Organization: Behavioral, Theoretical, and Food-Web Approaches*, pages 365–411. (Publisher Not Recorded).

Simberloff, D. and Wilson, E. O. (1969). Experimental zoogeography of islands: the colonization of empty islands. *Ecology*, 50:278–296.

Spiller, D. and Schoener, T. W. (1994). Effects of top and intermediate predators in a terrestrial food web. *Ecology*, 75:182–196.

Williamson, M. (1983). The land-bird community of Skokholm: ordination and turnover. *Oikos*, 41:378–384.

7.4 Application: Many Small or One Large Nature Reserve?

The species-area relationships introduced in Section 7.1.3 of this chapter formalize the well-known tendency of large habitat islands to contain more species than do small habitat islands[9]. Soon after island biogeography was introduced as a theoretical framework, conservation organizations appreciated its relevance to their efforts to preserve biodiversity. Unfortunately, occasionally the enthusiasm for this theory outstripped careful thinking. For instance, the International Union for the Conservation of Nature and Natural Resources published guidelines for park design that claimed it was always best to have a fixed amount of area in one large parcel as opposed to several small parcels. While this claim may sometimes be true, it does not follow from island biogeographic theory.

To see this, we can examine the total number of species expected for areas divided into one, two, three, and so on pieces. We develop a MATLAB function called `seven1.m` that has as its input total area (`a`), the number of equal-sized pieces the area is divided into (`n`), and the number of species in the species pool (`p`). Its output predicts the total number of species for all of the pieces summed together. The way this function works is that it generates an expected number of species for each piece of habitat and then multiplies this times the number of pieces of habitat, and subtracts the number of species that are expected to be redundant when a random draw from the species pool is assumed. As written below, the function uses an exponent of 0.263 and an intercept of 2.256. (The parameter values will change some according to the taxa and habitats being examined, but the shape of the function remains the same for a wide variety of cases.)

[9]This section is contributed by Peter Kareiva.

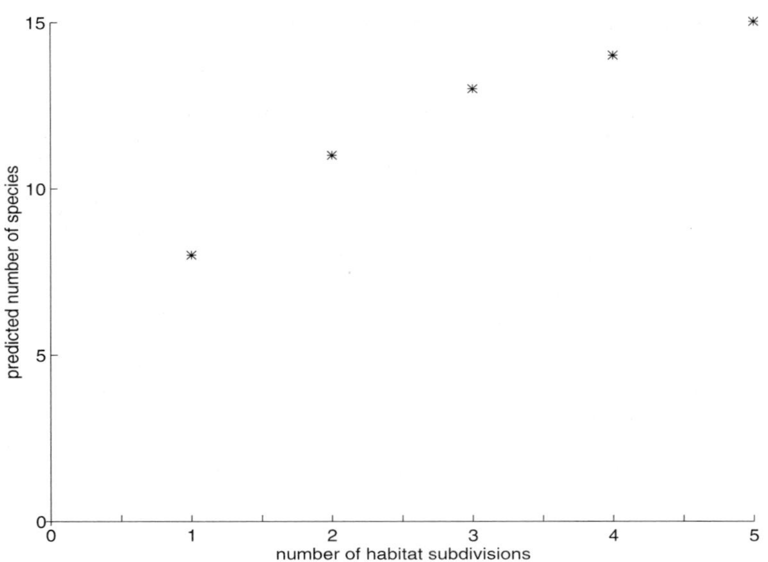

Figure 7.52: Using the species-area relationship given by $S = kA^z$, with $k = 2.256$ and $z = 0.263$, the graph shows the predicted number of species in a habitat parceled into 1 to 5 pieces. While the total area remains the same, the number of species increases with an increasing number of subdivisions.

```
function stot = seven1(a,n,p)
 s = 2..256*(a/n)^.263;
 suma = 0;
 for k = 1:n
   coef = seven2(n)/(seven2(k)*seven2(n-k));
   suma = suma + coef*(p*(-s/p)^k);
 end;
 st = round(-suma);
 stot = (st);
```

where `stot` is the total number of species, and `seven2` is an additional MATLAB file that calculates factorials as shown below.

```
function prod = seven2(int)
 prod=1;
 for i = 2:int
   prod = prod*i
 end;
```

By substituting in different values of `n` for the same total area, we can generate graphs that show how the total number of species is expected to change with subdivision. Using a total area of 100, and dividing it into two parcels, three parcels, four parcels, and five parcels, yields a total number of expected species that increases as the parcel size declines (noting that if 100 is divided into two pieces each piece has area 50, if into three pieces each piece has area 33.33, and so on).

The results are graphed in Figure 7.52. The important lesson is that the first impression of many conservation biologists was entirely wrong—the species-area curve of island biogeography does not predict that it is better to have one large park than to have many small parks that sum to the same area. (If a nested-subset species-area relationship applies, however, one large area does preserve more species than many small areas that sum to the same total area.)

Simberloff, D. and L. Abele. 1976. Island biogeography theory and conservation practice. *Science* 191:285–286.

Simberloff, D. 1988. The contribution of population and community biology to conservation science. *Ann. Rev. Ecol. Systematics* 19: 473–511.

7.5 Application: Disturbance Affects Food Webs

The simple food-web models discussed in section 7.1.5 can provide useful tools for the consideration of human alterations of ecosystems[10]. A nice example of this use involves analyses of multitrophic dynamics in rivers of California.

Many California rivers have as key ingredients, algae (A), predator-resistant herbivores (D) that consume algae, predator-susceptible herbivores (H) that consume algae, and predators that attack the herbivores (P). The simplest food web for such an ecosystem is:

$$\begin{aligned}
dA/dt &= b_a A L e^{-caA} - c_h HA - c_d DA - m_a A \\
dH/dt &= b_h c_h AH - c_p PH - m_h H \\
dP/dt &= b_p c_d HP - m_p P \\
dD/dt &= b_d c_d AD - m_d D
\end{aligned}$$

where L is incident light; the b_x is the conversion efficiency of the consumed resource, x; and m_x is the density-independent loss rate of species-x due to disturbance or pollution.

The simplest possibility is a system with all of the m_x's equal to zero, in which case there is no disturbance or pollution loss. When we do this and substitute in parameters that a team of ecologists have estimated for rivers in northern California, we find that there is no way all four species can coexist. To perform this analysis, we construct a MATLAB function that numerically solves the above ordinary differential equations, with the file called `seven3.m`:

[10]This section is contributed by Peter Kareiva.

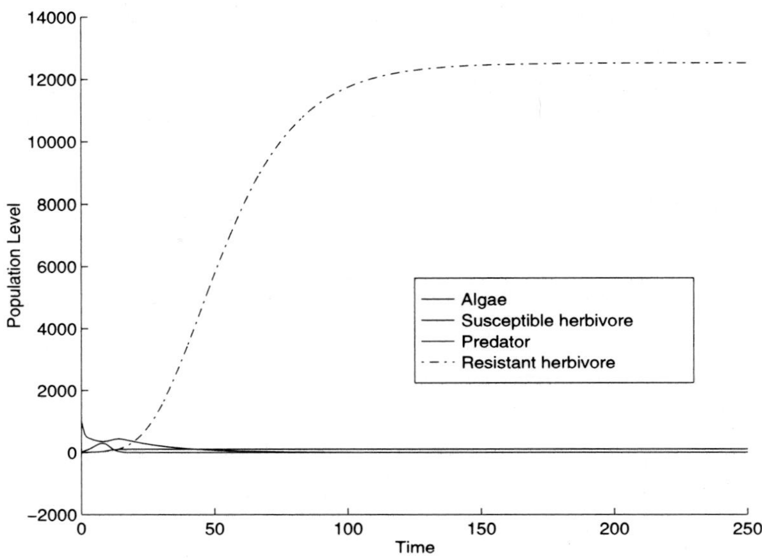

Figure 7.53: Population trajectories of the algae, predator-susceptible herbivore, predator-resistant herbivore, and predator, over time, when there is no disturbance in the system.

```
function ndot = seven3(t,n)
 global ba L ca ch cd ma bh cp mh bp mp bd md;
 ndot(1,1) = ba*n(1)*L*exp(-ca*n(1))-ch*n(2)*n(1)-cd*n(4)*n(1)-ma*n(1);
 ndot(2,1) = bh*ch*n(1)*n(2)-cp*n(3)*n(2)-mh*n(2);
 ndot(3,1) = bp*cp*n(2)*n(3)-mp*n(3);
 ndot(4,1) = bd*cd*n(1)*n(4)-md*n(4);
```

Now we can use this function, with no disturbance and the estimates for parameters derived from field experiments, to see the expected time course of the system over 50 years. We type

```
>> global Ao Ho Po Ro ba bh bp bd ca ch cp cd ma mh mp md L
>> Ao=1000; Ho=25; Po=10; Ro=5; L=1250;
>> ca=.01; ba=.1;
>> bh=.1; ch=.01;
>> cp=.01; bp=.1;
>> bd=.05; cd=.01;
>> ma=0; mh=0; mp=0; md=0;
>> [t a]=ode45('seven3',0,250,[Ao;Ho;Po;Ro]);
>> h = figure; hold on
>> plot(t,a(:,1),'g-')
```

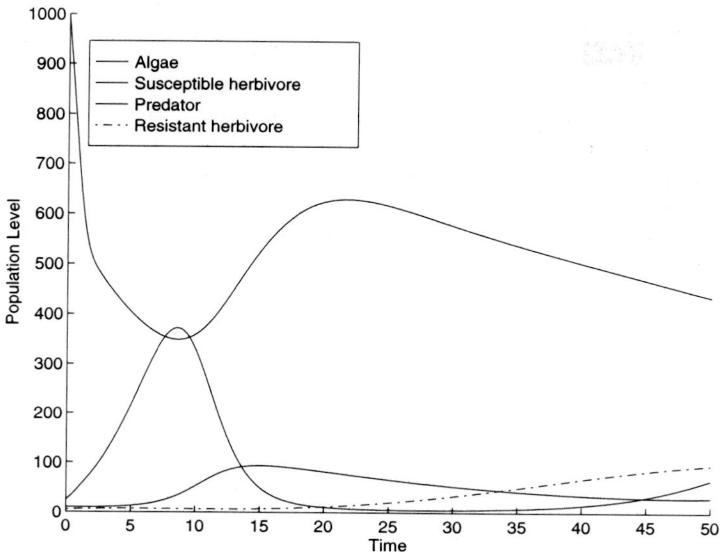

Figure 7.54: Population trajectories of the algae, predator-susceptible herbivore, predator-resistant herbivore, and predator, over time with disturbance in the system which affects the predator-resistant herbivores 20 times more than the predator-susceptible herbivores.

```
>> plot(t,a(:,2),'y-')
>> plot(t,a(:,3),'r-')
>> plot(t,a(:,4),'b-.')
```

We see from Figure 7.53 that the resistant herbivore levels off to an equilibrium after 100 years, with the algae at very low densities and the other species extinct. Now we can add disturbance, and track the dynamics:

```
>> ma=.01; mh=.01; mp=.05; md=.2;
>> [t a]=ode45('seven3',0,50,[Ao;Ho;Po;Ro]);
>> h=figure; hold on
>> plot(t,a(:,1),'g-')
>> plot(t,a(:,2),'y-')
>> plot(t,a(:,3),'r-')
>> plot(t,a(:,4),'b-.')
```

In the above simulation, shown in Figure 7.54, disturbance took a greater toll on the resistant herbivore than on the susceptible herbivore by a ratio of 20:1 (`md:mh`). This variation allows all four species to coexist in the system.

The important lesson is that in food webs of many interacting species, changes in disturbance rates can alter patterns of abundance and ultimately affect biodiversity. The unfortunate reality is that we humans, through the building of dams or flood-control channels, or the dumping of pollutants, alter disturbance rates in aquatic systems all over the world—it is hard to find a river system that has not been largely altered by human activity. Even when these activities do not seem that dangerous in terms of direct toxicity, they can drive species extinct. We should not be surprised that hundreds of fish and aquatic invertebrates are on the brink of extinction in U.S. waterways alone.

Wootton, T., Parker, M. and Power, M. 1996. Effects of disturbance on river food webs. *Science* 273: 1558–1561.

Programming in MATLAB

A computer is a machine that follows instructions. A sequence of instructions is a computer program. A set of allowed instructions is a computer "language." Any particular computer has one underlying basic set of instructions that it can obey, called the machine language of the computer—this is the language that the computer's hardware understands and it is called a "low-level" language. Nobody bothers to program in the computer's machine language because it's too tedious. Instead, people write their programs in a language of their choice, called a "high-level" language. The high-level program is then translated into machine language so that it can actually be executed. A basic difference between a Macintosh and a Windows PC is that each has a different machine language. Yet one can write the same high-level program to run on both machines because the high-level program is translated into the appropriate low-level language when needed.

Today, one has many choices of high-level languages that are convenient to different tasks. Here we'll be using the MATLAB language. In the tutorial below, we'll start with concepts common to every high-level language, and then mention some of the commands special to MATLAB. The MATLAB manual and `help` pages are the definitive reference. The purpose of this tutorial is to assist in getting you oriented, so that you can use the MATLAB reference material.

A MATLAB Program

A program in MATLAB is created with a text editor and saved as a text file. It must be saved as a text-only file. It must have a name that ends with `.m` such as `myname.m`, and as such, may be called an "m-file." To write a program to add 2 + 2, create a file with one line, `2 + 2`, and save it as `add2.m` Then to execute the program, at MATLAB's command prompt, `>>`, simply type `add2` This makes MATLAB search your directory (or folder) for a file whose name is the command you've just typed (`add2` in this case) with the suffix `.m` appended. MATLAB then executes the file `add2.m` just as though you had typed it at the command line. So, that's all there is to writing and running programs in the MATLAB language. Create the program with a text editor, save it as a text file with the `.m` suffix, start up MATLAB, and from its command line, type the name of the program you've previously saved (not including the `.m` suffix.)

Often you may compose a program in advance, and enter it into the computer with your text editor. A neat trick though, if you're not sure whether a command is going to work, or if you're developing your program by interacting with MATLAB at its command line, is to save a transcript of your activity with the `diary` command. For example, typing `diary myfile` will save the transcript under the file name `myfile`. When you later exit from MATLAB, the transcript is available for including or pasting into the program you're developing. And don't forget, the manual page for each of MATLAB's commands is always available if you type `help` at the command line. For example, to learn all about the `diary` command, type `help diary`

Now let's see what the ingredients are in a MATLAB program.

Data Types, Variable Names, and Assignments

A datum in MATLAB is a two-dimensional matrix. The matrix can be filled with either numbers or letters. A single number or single letter is a special case of a matrix whose dimensions are simply 1×1. A variable name is a sequence of characters that begins with a letter. MATLAB pays attention to only the first 19 characters in the name, and allows the underscore character. So, a valid name for a variable is `my_first_variable`

A variable is assigned to a datum with the = (equal) sign. For example, the variable `a` is assigned to the number, 7, by our typing `a = 7` This is a command, namely, the assignment command. A command means that an action is carried out, and the assignment `a = 7` means that the number 7 is stuffed into the memory location reserved for the variable `a`. MATLAB uses either the end-of-line character or the semicolon to discern where a command ends. When a command ends without a semicolon, MATLAB automatically prints out the result of the command. In this case, after executing `a = 7`, MATLAB replies that `a = 7`, which is obvious. But if you had written that `a = 7 + 7`, MATLAB would reply that `a = 14` which is perhaps less obvious. To prevent MATLAB from echoing the result of each command it carries out, use a semicolon to end the command. Thus, the command `a = 7;` would be executed silently. To confirm that `a` does indeed now equal 7, type `a` all by itself on a line, and MATLAB will reply with 7. If a command is too long to fit on one line, then type `...` (three periods) at the end of a line to get MATLAB to continue reading on to the next line without prematurely concluding that the command is complete.

If you've programmed in other languages, you'll note that input and output occur automatically and naturally. A huge amount of most programs is involved in gathering input from the user and formating output back to the user. Here, you can inspect what a variable's value is just by typing its name on a line by itself.

To assign a variable to a letter rather than to a number, use single quotation marks around the letter. To assign the variable `b` to the letter, `z`, type `b = 'z'`. If you had typed `b = z` then MATLAB would say either that `z` is undefined because it expects `z` to be the name of a variable, or if `z` has in fact been previously defined, it assigns `b` to the same value that `z` had been previously defined as. Please be sure you understand the difference between using `z` as a meaningless letter and as the name of a variable.

A vector of length n is a matrix whose dimensions are $1 \times n$. A vector of numbers, say 1, 2, and 3, is assigned by typing `a = [1 2 3]` or `a = [1,2,3]`. The space and comma are interchangable. A vector of letters is assigned in any of the following ways: b = ['x' 'y' 'z'] or b = ['x','y','z'] or b = 'xyz'. A vector of letters is called a "string" in computerese.

A matrix is entered row by row, with each row separated by semicolons. For example a = [1 2 3; 4 5 6; 7 8 9] defines a 3×3 matrix. Instead of spaces, one could use commas within the rows. Similarly, a 3×3 matrix of letters could be entered as b = ['pqr';'stu';'vwx'].

The variable name, such as a above, refers to the entire matrix. To access a particular entry in the matrix, you have to know the row and column position of the entry. The number 4 in the matrix a above is accessed as a(2,1) and the letter 's' in the matrix b above is accessed as b(2,1). The convention is that the first index counts down starting at one from top to bottom, and the second index counts across starting at one from left to right. If you want to know what's in the $(3,2)^{th}$ position of a matrix a, type a(3,2), and MATLAB will output the value at that location, and if you want to change the value at that location to something else, say 8, type a(3,2) = 8.

In MATLAB you don't have to declare variables in advance of using them. If you want to let x = 7 just do it. In other languages you often have to declare in advance that x will be the name of a variable before you're allowed to assign it a number to it. Once a variable has been assigned a value, it is known throughout the program (except in external subroutines as discussed later). To find out what variable names are defined at any particular time, type who or for more detail, type whos. To undefine a variable, say x, type clear x, and to undefine all the variables, and restart with a clean slate, type clear all.

Numeric Operations

MATLAB allows numbers and the variables that refer to numbers to be combined with the usual operations of + (addition), - (subtraction), * (an asterisk, multiplication), / (division), and ^ (caret, shift-6 on the keyboard, for exponentiation). For example x=7^7 assigns x to 7^7.

MATLAB also has a convenient scheme for doing calculations in bulk. If a = [1,2;3,4] and b=[5,6;7,8] then the matrices can be added and subtracted as a whole with the + and - signs. Thus a+b equals [6,8;10,12], a-b equals [-4,-4;-4,-4]. Perhaps more interestingly, if the *, /, and ^ symbols are preceded by a . (period), matrices are multiplied, divided, and exponentiated element by element. For example, a.*b equals [5,12;21,32], a./b equals [.2,.3333;.4286,.5] and a.^b equals [1,64;2187,65536]. MATLAB calls these bulk calculations "array operations."

MATLAB has a bunch of built-in functions, such as sin, cos, log, sqrt, and exp for the sine, cosine, natural logarithm, square root, and exponential functions. It has a lot more too that you can find when you type help elfun and go from there. (Incidentally, if the manual page is too long to fit on your screen, type more on to make MATLAB scroll page by page. Press the space bar to display each screen full of text.) Most of the elementary functions perform their operations in bulk on all the elements of a matrix. For example, if a=[1,2;3,4] then cos(a) returns [.8415,.9093;.1411,-.7568], which is the cosine of each element of a.

String Operations

Some special commands are available for strings. To list them, type `help strfun`, and the details of each can then be determined with the use of `help` for each one. In particular, one command is simply to tell whether a variable refers to a string or not. If `a='xyz'` then `isstr(a)` returns 1, otherwise it returns 0. Another command allows substitutions within strings. For example, if `a='My coffers are filled with lead.'` then the command `strrep(a,'lead','gold')` turns lead into gold, changing your life forever. The pattern to be replaced (lead) doesn't have to be the same length as the pattern to be substituted (gold). Finally, the command `eval` is used to convert a string into a numerical answer. For example, if `a='2+2'` then `eval(a)` yields the number, 4. Obviously, the string that is evaluated must represent a valid formula, one whose variables are fully defined when the evaluation is carried out. For example, `eval('a+b')` will work only if `a` and `b` have already been defined at this point as having values for which addition is legitimate.

Symbolic Operations

MATLAB can rearrange algebraic formulas for you. The steps used to rearrange a formula are called symbolic operations. MATLAB didn't have this capability until rather recently. MATLAB has joined forces with another software product, called Maple, and a merged product is sold as the academic (student) edition. If you're buying a professional edition of MATLAB, you'll have to purchase the symbolic operations separately, which are packaged as the Symbolic Math Toolbox. Either way, if you type `help symbolic` you'll get a list of the commands that are available for symbolic operations.

To interface with the Symbolic Toolbox, MATLAB uses strings. One creates a formula as a string, processes it with one or more symbolic operations, and winds up with another string, which hopefully is more useful than the formula in the original string. Thus, from the standpoint of MATLAB, the entire Symbolic Toolbox is just a bunch of commands that rearrange strings. From our point of view, these rearrangements are important because they are carried out according to algebraic rules. Let's now visit some of the more important of the symbolic operations.

The command `symop` is used to construct complicated formulas from simple ones. You can call `symop` with up to sixteen arguments at a time, and `symop` puts all the arguments together and tries to make some sense of the result. The arguments can be strings themselves, or the characters `+ - * / ^ ( )` These must always be surrounded by single quotation marks. They are used for addition, subtraction, multiplication, exponentiation, and grouping respectively. For example, the formula `a='(p+q)/(x+y)'`, which is a string, could also be assembled by typing `a=symop('(','p','+','q',')','/','(','x','+','y',')')` which is the hard way, of course, but you get the idea. So, `symop` simply combines small strings into big strings. But it also tries to make some sense out what it gets. For example, `symop('2','+','2')` returns 4, not just 2+2. When `symop` returns `4` in this case, the "4" is a character, not the number 4 itself. To convert the `4` from a character into a number, you must use the `eval` command. For example, `eval('4')` returns the number 4, which you can then use in further calculations. As a further example, suppose `a='x^2'` and `b='x^3'` Next, type `c=symop(a,'/',b)` and MATLAB replies that `c` is `1/x`. Now the `c` is another string, which you can confirm if you type `isstr(c)` and see that the answer is 1. The `c` produced by `symop` in this way is identical to what you'd get by typing `c='1/x'`. At this point you could assign `x` to a number, say by typing `x=4`, and you could now evaluate the formula represented by `c`. You'd type `eval(c)` and MATLAB would reply with .25. Thus, `symop` is used to assemble a formula, and `eval` is used to evaluate the formula, once the variables in the formula have been assigned numerical values.

Now that a formula is assembled, how do we rearrange it? One useful command is to see what a formula looks like after it's been assembled. A relatively attractive printout of a formula is obtained by our typing `pretty(a)` where `a` is a formula, such as `a='(p+q)/(x+y)'` To display it correctly, the `pretty` function has to understand the formula. If you made a mistake in typing in the formula for `a`, say by forgetting a closing parenthesis, then attempting to display it with `pretty` would return an error. So, `pretty` can be used to check if a formula is well formed, which is an important side benefit of its main purpose.

Now that we've looked at the formula, we may want to change one of the variables in it. To replace one variable with another, say a variable named `x` with `y` in the formula `a`, type `subs(a,'y','x')` For example, `subs('x^2','y','x')` yields the string `y^2` Notice that the symbolic operation, `subs`, and string operation, `strrep`, do about the same thing, but unfortunately have a reverse syntax. With `subs` the order of the arguments is (formula, new variable, old variable) whereas with `strrep` the order of arguments is (string, old pattern, new pattern).

Further rearranging of formulas is done with the commands `simplify`, `expand`, `factor`, and `collect`. Also, the command `simple` combines several of the formula rearrangement commands into one overall package, and tries to find the best way of arranging a formula. You'll have to check these out individually, and by trial and error see which works best for the formula you're working with. Another useful command is `numden` which splits up the formula for a fraction into its numerator and denominator; it returns one string for the numerator and other string for the denominator, so that you can work with these separately.

The main command to solve algebraic equations is `solve`. It has an extensive syntax to cover many possibilities. An example of solving a single linear equation is `x = solve( 'a*x-b','x')` which means to solve $ax - b = 0$ for the variable, x, and which yields the string, `b/a`. This string, in turn, is assigned to the MATLAB variable, `x`. An example of solving a quadratic equation is `x=solve('a*x^2-b','x')` which means to solve $ax^2 - b = 0$ for x. This yields a matrix of characters, the top row being the first root, and the second row being the second root, equivalent to `[' 1/a*(a*b)^(1/2)';'-1/a*(a*b)^(1/2)']`. This matrix of characters is then assigned to the MATLAB variable, `x`. Finally, an example of solving a pair of equations for two unknowns is `[x,y]=solve('a*x+b*y-c','d*x+e*y-f','x,y')`, which means to solve $ax + by - c = 0$ and $dx + ey - f = 0$ for x and y. This yields two strings, `(-b*f+c*e)/(a*e-d*b)` which is assigned to the MATLAB variable, `x`, and `(a*f-d*c)/(a*e-d*b)` which is assigned to the MATLAB variable `y`. Overall, `solve` is extremely powerful and is one of the most used of all the MATLAB commands.

For calculus, the command to take derivatives with respect to a single variable is `diff`.

(For multiple variables, also see the `jacobian` command.) For example[11], `diff('x^2','x')` yields the string, `2*x`. Integration with respect to one variable is done with the `int` command. An example of an indefinite integral is `int('x^2','x')` which yields the string, `1/3*x^3`. An example of a definite integral of x^2 with respect to x from a to b is `int('x^2', 'x','a','b')` which yields the string `1/3*b^3-1/3*a^3`.

Another symbolic operation is to do calculations to arbitrary precision. You want π to 100 decimal places? You've come to the right place. Check out the `vpa` and `digits` commands.

What's interesting about a language that merges both symbolic and numerical operations is that a computer program can be developed that does both. The program can start out with simple formulas, rearrange and combine them, and then evaluate the symbolic result numerically. A typical numerics-only computer language starts with the final formula, and requires that all the derivation be done in advance with paper and pencil.

Rearranging Matrices

In most computer languages, matrices have a fixed size. If the matrix `a` is declared as a 2×2 matrix to begin with, it stays that way forever. In MATLAB a matrix can grow or shrink and have columns added or removed, so they become flexible vessels for holding data. To start, an empty matrix is created by typing `a=[]`. This means that the name `a` is now reserved (as can be confirmed by typing `whos`). To append an element, say 7, to this matrix, type `a=[a 7]`. The matrix now consists of the single number, 7. To append another element, say 8, type `a=[a 8]`, and the matrix now becomes `[7 8]`. And so forth, to as large as you wish. The matrix now is a 1×2 row vector.

To convert a row vector into a column vector, type `a=a'`, and `a` now is `[7; 8]` and its dimension is 2×1. The `'` symbol is the single quote mark, and the operation of producing a column vector from a row vector is called the "transpose" operation. The transposition means the movement of each element at position i, j to position j, i, so that `a(i,j)` becomes `a(j,i)`. The transposition operation works for matrices of any size, not just vectors. (Because the transposition operation is a single quote mark, if you're entering a string, which is a sequence of characters enclosed by single quote marks, and forget to type one of the quote marks, the single quote mark that you did type is interpreted as a transposition operation, and will generate an error of some sort.)

Now consider a matrix `a=[1 2; 3 4]`. A third column can be added to the end by your typing `a(:,3)=[5; 6]`, which yields the matrix `[1 2 5; 3 4 6]`. Notice the `:` symbol (a colon). It means "all". In this case, all rows at column 3 are being assigned a value. To delete the second column, type `a(:,2)=[]` yielding `[1 5; 3 6]`. Similarly, to add a row to the bottom, type `a(3,:)=[7 8]` yielding `[1 5; 3 6; 7 8]`. Thus all columns in row 3 are assigned a value. To delete the second row type `a(2,:)=[]` yielding `[1 5; 7 8]`. Major cosmetic surgery is accomplished with `a=a(:)`, (just one colon) which turns `a` into a long

[11]The `diff` command has two different meanings in MATLAB. If called with a string, it attempts to find the derivative of the formula represented by the string. If called with a vector of numbers, it finds the differences between adjacent elements of the vector.

column vector. It tacks all the columns after each other, yielding `[1; 7; 5; 8]`. (See also the `reshape`, `fliplr`, `flipud` and `rot90` commands.)

If there are matrices changing size all the time, you may want to know the current size of a matrix. Consider `a=[1 2; 3 4; 5 6]`. `size(a)` returns `[3 2]`, which is a row vector itself whose first element is the number of rows in `a` and second element is the number of columns. `length(a)` returns 3, a number, which is the size of the largest dimension in `a`. Also, `size(a,1)` returns specifically the number of rows, and `size(a,2)` the number of columns, in `a`, which are 3 and 2, respectively.

Premade matrices with any desired number of rows and columns are generated with the `zeros`, `ones`, `eye`, and `rand` commands. For example, `zeros(3,2)` is `[0 0; 0 0; 0 0]`, `ones(3,2)` is `[1 1; 1 1; 1 1]`, and `eye(3,2)` is `[1 0; 0 1; 0 0]`. `eye` is the identity matrix, with 1's on the main diagonal and 0's elsewhere. `rand(3,2)` is a 3×2 matrix of random numbers, each distributed uniformly between 0 and 1. For example, the first call to `rand(3,2)` yields `[.2190 .6793; .0470 .9347; .6789 .3835]` and the second call to `rand(3,2)` yields another set of random numbers. The random number generator is initialized to the state it has when MATLAB starts up, by your typing `rand('seed',0)`. Also, typing `rand` by itself without any arguments returns one random number. Also `randn` can be used for normally distributed random numbers.

Another family of premade matrices is generated by use of the colon between bottom and top limits. `a=1:3` generates the vector `[1 2 3]`. The idea is that using limits qualifies the colon from meaning "all" into "almost all." That is, `1:3` means all the integers between 1 and 3. Similarly, `a=[1:3;4:6]` generates the matrix `[1 2 3; 4 5 6]`. Moreover, the "almost all" doesn't have to refer to only integers, but can use any step size. For example, `a=1.0:0.1:1.2` generates the vector `[1.0 1.1 1.2]`. The syntax for generating premade matrices by use of the colon symbol is `bottom_limit:step:upper_limit`, and means all the numbers from the bottom limit, in steps of step, up to and including the upper limit. You can even use negative steps, in which case the bottom limit becomes effectively the top limit, and the top limit becomes the bottom limit because you're counting down from the left limit to the right limit.

In most computer languages, you have to pick out the elements of a matrix one at a time, such as `a(1,2)`, which is the element in the first row and second column. In MATLAB, you can also select a whole bunch of elements at once. Let `a=[1:4;5:8;9:12;13:16]`, i.e., `[1 2 3 4; 5 6 7 8; 9 10 11 12; 13 14 15 16]`. All the elements in the first or second rows that are also in the second or fourth columns are selected together by typing `a([1 2],[2 4])`, yielding `[2 4; 6 8]`. The 3×3 matrix in the top left corner of `a` is selected when you type `a(1:3,1:3)`, yielding `[1 2 3; 5 6 7; 9 10 11]`. Thus, by enumerating the numbers of the rows and columns you want, you can select pieces of a matrix for further use.

Another way to select a whole bunch of elements at once from a particular matrix is with another matrix of the same size that consists only of 0's and 1's. The 1's are in the locations where an element is to be selected and the 0's in locations where an element is to be ignored. For example, if `a=[1 2 3 4]` then `a([0 1 0 1])` picks out the second and fourth elements yielding `[2 4]`.

All of the points above apply only to regular MATLAB matrices filled with numbers or characters. Symbolic matrices are another matter altogether. A symbolic matrix is made with the sym command. For example, a=sym('[u,v,w;x,y,z]') makes a symbolic matrix. If instead you had typed a=[u,v,w;x,y,z] MATLAB would have tried to put the numerical values of u, v, w, x, y and z into the matrix instead of the symbols themselves. The size of a symbolic matrix is reported with the symsize command. For example, symsize(a), where a=sym('[u,v,w;x,y,z]') returns the numeric vector [2 3], to indicate two rows and three columns.

sym is also used to extract components of a symbolic vector. Recall solving for the roots of a quadratic equation, as in x=solve('a*x^2-b','x'). The answer, x, is a symbolic column vector, [' 1/a*(a*b)^(1/2)';'-1/a*(a*b)^(1/2)']. Typing symsize(x) yields [2 1] indicating two rows and one column. To assign the formula for first root to the variable, x1, type x1=sym(x,1,1) and for the second root to the variable, x2, type x2=sym(x,2,1). The syntax is (symbolic matrix, row position, column position). If numbers have already been assigned to a and b, you can get the numerical value of these roots if you type eval(x1) and eval(x2). eval unfortunately evaluates only one formula at a time, not a matrix of formulas, so you have to use sym to extract each formula from the symbolic matrix of solutions and evaluate them one by one.

A third use of sym is to change an element within a symbolic matrix that has already been defined. For example, if a=sym('[w,x;y,z]') then sym(a,2,2,'z^2') returns the symbolic matrix, [w,x;w,z^2]. The syntax for sym here is (symbolic matrix, row position, column position, expression to be placed at the position).

Because formulas and symbolic matrices are, in the final analysis, just matrices of characters, you can fake the action of sym with the string operations. For example, b=['[w,x]'; '[y,z]'] yields a matrix that is identical to a=sym('[w,x;y,z]'). On occasion when the sym function wasn't up to the job, I've had to resort to the string operations to produce a formula, but try not to do this. Using string operations on symbolic matrices is a "hack," which means getting the right result the wrong way.

For-Loops

To make the computer repeat a sequence of instructions over and over again, you place the instructions in what is called a "for-loop." The idea is to define a counter, say i, which goes from a lower limit to an upper limit, say from 1 to 3 in steps of 1. For each value of the counter a bunch of commands are carried out, and when the counter is all done, the computer moves on to whatever is next in the program. The syntax of a for-loop in MATLAB is

```
for i=[1,2,3]
 Put your commands here.
end
```

The loop begins with a line in which the counter is identified, which in this example is i, and the counter is told to equal 1, then 2, and finally 3. The commands are carried out for each step of the counter, which in this case is three times. The set of commands to be executed are those up to the line with `end`. Thus, a for-loop is consists of the `for` statement itself, the set of commands to be done over and over again, and the `end` statement that marks the bottom of the loop. For example, suppose the commands are to print the value of the counter squared, and the square root of the counter. Then the for-loop to do this is

```
for i=[1,2,3]
 i^2
 i^(1/2)
end
```

The result is `1 1 4 1.4142 9 1.7321`. As you can see, the first time through, `i` is 1, so both `i^2` and `i^(1/2)` equal 1, the next time `i` is 2, so `i^2` is 4 and `i^(1/2)` is 1.4142, and so forth.

It's considered good style to indent the commands within a for-loop. Also, you don't need to enumerate each value of the counter, as in `for i=[1,2,3]`, because you can take advantage of the : (colon) symbol. An equivalent way to start the for-loop is `for i=1:3`. Indeed, using the colon is the only way to go if you want the counter to run from 0 to 1 in steps of 0.01, which you should write as `for i=0:0.01:1` instead of explicitly writing out a vector with 101 components.

Another way to make loops is with the `while` command—type `help while` for all the details.

If-Expressions

Often you want a command to be performed only in certain circumstances. For example, it's considered bad form to divide a number by zero. Here's the way to calculate the reciprocal of x, provided x is not zero:

```
if x~=0
 1/x
end
```

The first line begins with `if` followed by the condition, which in this case is `x~=0` which means "x not equal to zero." Then the next line is a command to be carried out if the condition is true. There can more than one command here. All the commands up until the next `end` statement are executed if the condition is true.

The kinds of conditions that can be checked are: `x<y`, `x>y`, `x<=y`, `x>=y`, `x==y`, `x~=y`. The last two may not seem obvious. The `~=` means "not equal to" and is constructed with the tilde character followed by the equal sign. The == means "equal to" and is constructed with two equal signs. The reason two equal signs are used is that one equal sign by itself means an assignment command (`x=y` means "make x equal to y") whereas the two equal signs indicate a state of affairs (`x==y` means "is x equal to y?"), and as such, is either true (represented as a 1) or false (represented as a 0).

The relations, `< > <= >= == ~=`, can be used to compare matrices of the same size, not just single numbers. The elements of the matrices at corresponding positions are compared and a matrix of 1's and 0's is returned to summarize all the comparisons. For example, `[5 5]>[4 6]` returns `[1 0]` to indicate the relation is true for the first element but not the second.

Another situation arises when you want to do one set of commands if a condition is true and another set if the condition is false. The way to accomplish this is

```
if condition
 Commands to do if all elements of the condition are true.
else
 Commands to do if at least one element of the condition is false.
end
```

For even more complicated decision sequences there is an `elseif` that can be used too. Consult `help if` for more detail.

Subroutines and Functions

When you are computer programming, you may come up with a way to do something that really works well, and want to use the same sequence of instructions again some other time. A "subroutine" is a package of instructions that can be used again in future applications. A "function" is a special case of a subroutine that accepts information from a program, processes it, and returns information back to the program that called on it in the first place. What you send to a function are called its "arguments." The arguments are put in parentheses after the function's name. (If a function is mentioned in the text or index without explicitly listing its arguments, then a pair of parentheses indicates that it is a function, as in my_function().) MATLAB has many built-in functions, such as `cos(x)`. Here `x` is the argument with which the cosine function is called. When you want the cosine of a number, say 2, type `cos(2)` and the 2 is passed to a subroutine which determines the cosine, and then returns the answer to the program that asked for it. Now you might think that we're limited to the functions that originally come with MATLAB—these are the functions documented in the manual and in the `help` pages. But in fact, we can add as many functions to MATLAB as we wish, and thereby extend the capability of MATLAB. Here's how to do this.

Suppose you've come up with a groovy way to calculate the length of the hypotenuse of a triangle. If a is the length of one side, b the length of the other, and c the length of the hypotenuse, then $c = \sqrt{a^2 + b^2}$. Let's extend MATLAB by making this remarkable discovery available to all. Let's define a function to be called `hypot` that we'll be able to use with any pair of numbers for the two sides, a, and b. If a and b are 1 and 2 respectively, then we want to be able to type `hypot(1,2)` and have MATLAB give us the length of the hypotenuse in this case. Similarly, if the sides are 3 and 4, then we want to type `hypot(3,4)` and have the MATLAB respond with the hypotenuse in this case too. Here's how to construct the function `hypot`. With a text editor we make a separate file, yes, a separate file that is not part of any other program, that is named `hypot.m` and that contains the following lines:

```
function c=hypot(a,b)
 c=(a^2+b^2)^(1/2);
```

This tiny file has only two lines in it. Once this tiny file has been saved in your working directory or folder as `hypot.m`, you can type `hypot(1,2)` and MATLAB responds with `2.2361` and you can type `hypot(3,4)` and MATLAB responds with `5`. That's all there is to it. To add a function to MATLAB, simply create a separate file with the instructions for the function, where the file's name must be the name of the function with a `.m` suffix appended to it.

Now look at the function in more detail. The first line says that the function's name is `hypot`, that it expects to receive two pieces of information to work with (these are called the function's arguments), and to return one piece of information. What the function names these pieces of information is private to it. The function will name the information it receives as `a` and `b` and will name what it returns as `c`. The program that requests `hypot` to do its thing may have different names for these pieces of information. The program that requests `hypot` to do something is referred to as the "calling program." The only name shared by both the calling program and the function is the function's own name, `hypot`. Thus, the variable names used in the function are separate from any names used in the calling program. How does the function know whether it's using the information given it as intended? It's the calling program's responsibility to call the function with the information in the correct order—the function will assume it's OK to assign the first argument to what it calls a and the second argument to what it calls b. It's also the responsibility of the calling program to know what to do with the answer that the function supplies. `hypot` assumes it's OK to return the square root of a^2 plus b^2, which it calls c, and the calling program has to be prepared to accept this information.

Suppose you don't like keeping secrets. By default, the names of variables used in the calling program and the functions it calls are private. There is a way to publicize the names of variables so that both the calling program and the functions it calls share the same names. Names of variables are publicized with the `global` command. Suppose you want `x`, `y`, and `z` to be shared by both a calling program and a function. Then *both* the calling program *and* the function must have the statement `global x y z` in them before these variables are used. This device turns out to be necessary when you want to get information to a function in a way that bypasses its calling convention. (It's needed with the `ode23` and `ode45` commands, among others.)

A function doesn't have to return only a single number, it can return a vector or a matrix to the calling program. Also, a function can call other functions, just as long as every function is either built-in or in a separate file by itself. A function cannot call itself (function calls are not recursive). See `help function` for more detail.

Input and Output

MATLAB allows for both textual and graphical input and output. The textual output (i.e., the output that MATLAB types out to you in the command window) is primarily to echo each command after it is executed, provided you didn't terminate the command with a semicolon. For more complex formated output, use the `fprintf` function. MATLAB has a set of functions, more or less fashioned after their counterparts in the C programming language, to deal with input and output to and from a file. A special case of a file is the screen itself, and `fprintf` can be used to format output to the screen as well as to an external file. Typically, you would use `fopen` to open an external file, `fprintf` to write to the file, `fscanf` to read from the file, and `fclose` to close the file. The screen, however, is automatically open to begin with, and is designated as file number 1 (the so-called standard output). To find out more about these functions type `help iofun` and go from there.

The graphical input and output facilities of MATLAB are extensive; they are listed for everything when you type `help graphics`, or `help plotxy` and `help plotxyz` are used specifically for the two- and three-dimensional graphs respectively. Key commands include `figure` that brings up a graphical window, and `hold on` that makes sure successive drawings to the window do not erase previous drawings. Two-dimensional graphs are usually drawn with the `plot` function. You really owe it to yourself to study the manual page for this function; type `help plot`. Basically, `plot` expects as its arguments a vector of x coordinates (the horizontal axis) and a vector of y coordinates for the vertical axis. The first element of the x-vector makes a point together with the first element of the y-vector, and so forth. The x and y vectors must be the same length. A third argument to `plot` controls the line style and color. If only one vector is given instead of two, the vector that is given is assumed to be a y-vector and a default x vector is assumed to be a vector of indices; i.e., the default x is `[1:length(y)]`. The default line type is a solid line, which is yellow on a black background on the computer screen. Many other colors and line styles are available. Semi-log plots are made with the `semilogx` and `semilogy` commands and log-log plots with `loglog`. Areas are filled in the the `fill` command. Axes are labeled with the `xlabel` and `ylabel` commands, and a title placed with the `title` command. Labels within the graph are placed with the `text` command. The axis characteristics are controlled with `axis` command. Input can also be obtained from a graph when clicks of the mouse are used with the `ginput` command. Special-purpose graphs include `bar`, `compass`, `errorbar` and `hist`. For three-dimensional graphs, consult `plot3`. Finally, a graph may be saved as a file or printed out in hard copy with the `print` command.

Spread-Sheet Operations

MATLAB has commands that let you use a matrix as though it were a spread sheet, with rows and columns for data. These are listed when you type `help datafun`. These functions include `max`, `min`, `mean`, `std`, `median`, `sum`, `product`, `cumsum`, and `sort`. Suppose one counts the number of dots on the left and right rear legs of a frog. Each frog is a row in the data matrix, and the first column is for the left leg and the second column the right leg. With two frogs we might have `a=[5 6; 4 5]`. Therefore, the average number of dots on the left leg is `mean(a(:,1))`, and on the right leg is `mean(a(:,2))`. The total number of dots on the first frog is `sum(a(1,:))` and on the second frog is `sum(a(2,:))`. The maximum number of dots on any leg is `max(max(a))`. Two calls to `max` are needed because the first call returns a vector with the maximum number from each column, and the second call returns the maximum of these. `std` returns the standard deviation, and `sort` rearranges a vector in ascending order. `cumsum` returns the cumulative sum, `cumsum([1 2])` returns `[1 3]`.

The `find` command is also useful in spread-sheet operations. Find returns the positions of a vector that are nonzero. This isn't particularly useful by itself, but is useful when it's combined with the relational operations of `< > <= >= == ~=`. Suppose the snout-vent length of four lizards in mm is `a=[60 62 59 61]`. The mean, `mean(a)` is 60.5 mm. Which lizards are larger than the mean? `a>mean(a)` returns `[0 1 0 1]`. The identification numbers of the lucky lizards who are larger than the mean are found with `find(a>mean(a))`, which returns `[2 4]`. And just how big are the lucky lizards? Their sizes can be found in either of two ways—your choice. The vector of 0's and 1's from the comparison operation can index directly into the matrix, as in `a(a>mean(a))` which returns `[62 61]`. Alternatively, the vector of locations from `find` can also be used, as in `a(find(a>mean(a)))`, which also yields `[62 61]`.

Linear Algebra

MATLAB began as an interactive frontend for some famous subroutines in numerical linear algebra called LINPACK and EISPACK, which were written in FORTRAN. Today MATLAB retains the ability for numerical computations with matrices, and can now do symbolic manipulation of matrices with the Symbolic Toolkit from Maple. Let's start with the symbolic commands.

Consider the 2×2 symbolic matrix `a=sym('[a11,a12;a21,a22]')` which has `'a11'` and `'a12'` on the top row and `'a21'` and `'a22'` on the bottom row. Similarly, let `b=sym('[b11, b12;b21,b22]')`. To add, subtract and multiply matrices, use `symop` together with the appropriate symbol. For example, `symop(a,'-',b)` yields `sym('[a11-b11,a12-b12;a21-b21, a22-b22]')`. Also, `symop(a,'*',b)` yields `sym('[a11*b11+a12*b21, a11*b12+a12*b22;a21*b11+a22*b21,a21*b12+a22*b22]')`. The determinant of the matrix is `determ(a)` which yields `'a11*a22-a12*a21'`. The transpose of the matrix is `transpose(a)` yielding `sym('[a11,a21;a12,a22]')`. The inverse of the matrix is `inverse(a)` yielding `sym('[a22/(a11*a22-a12*a21), -a12/(a11*a22-a12*a21); -a21/(a11*a22-a12*a21), a11/(a11*a22-a12*a21)]'`. The solution to a system of linear equations can be obtained directly, without bothering with the explicit inverse, with `linsolve`. If `c=sym('[c1;c2]')` then the solution to `'a*x=c'` is found as `x=linsolve(a,c)` which yields `x=sym('[(-a12*c2+c1*a22)/(a11*a22-a12*a21); (-(-a11*c2+a21*c1)/(a11*a22-a12*a21)]')`. `linsolve` works only with symbolic linear equations and contrasts with `solve` that works with symbolic nonlinear equations too. The characteristic polynomial of `a` with `'x'` as the variable in the polynomial is found as `charpoly(a,'x')` yielding `'x^2-x*a22-a11*x+a11*a22-a12*a21'`. (The terms in `'x'` can be grouped together with `collect`.) Finally, the eigenvalues, which are the roots of the characteristic polynomial, are `e=eigensys(a)` yielding a 2×1 symbolic column vector `sym('[1/2*a22+1/2*a11+1/2*(a22^2-2*a11*a22+a11^2+4*a12*a21)^(1/2), 1/2*a22+ 1/2*a11-1/2*(a22^2-2*a11*a22+a11^2+4*a12*a21)^(1/2)]')`. The individual eigenvalues are accessed with `sym`. The first is `sym(e,1,1)` and the second is `sym(e,2,1)`. The `eigensys` command will also return formulas for the eigenvectors. Clearly, the symbolic approach to linear algebra produces very messy formulas unless the matrices happen to have an especially simple structure to begin with. Therefore, in practice, linear algebra requires numerical methods.

MATLAB's numerical matrix commands correspond to the symbolic ones above. Let `a=[1,2;3;4]` and `b=[0.1,0.2;0.3,0.4]`. Addition, subtraction, and multiplication occur naturally with matrices without any additional special notation. For example, `a-b` yields `[.9,1.8;2.7,3.6]` and `a*b` yields `[.7,1;1.5,2.2]` (This and other numerical answers may be confirmed by use of the formulas in the paragraph above.) The determinant of a matrix is found with `det`. For example, `det(a)` yields -2. The transpose is obtained with a single backwards quote `'`, as in `a'` which yields `[1,3;2,4]`. The numerical inverse is found with `inv` as in `inv(a)`, which yields `[-2,1;1.5,-.5]`. The solution to a set of linear equations can be found without bothering with an explicit inverse matrix when you use the \ (backslash) symbol. For example, if `c=[1;1]`, the solution to `a*x=c` is found as `x=a\c` which yields `[-1;1]`. The coefficients of the characteristic polynomial of `a` is obtained with `poly(a)`, yielding `[1,-5,-2]`, which means that the characteristic polynomial is `'x^2-5*x-2'` in the symbolic variable, `'x'`. Finally, the eigenvalues, which are the roots of the characteristic polynomial, are found as `eig(a)`, which returns `[-0.3723;5.3723]`. `eig` will also find the eigenvectors corresponding to these eigenvalues.

Differential Equations

Differential equations are solved symbolically with `dsolve`. The calling syntax is to specify the equation as a string, with `D` to indicate a derivative, the initial condition as another string, and the symbol for the independent variable. For example, `dsolve('Dn=r*n', 'n(0)=a','t')` yields `'exp(r*t)*a'`. The calling syntax for more complicated situations, including systems of equations, is documented in `help dsolve`. It seems that `dsolve` prefers its symbolic variables to be single letters, such as `'n'`, and not `'n1'`, for example.

Differential equations are solved numerically with `ode23` and `ode34`. Both use so-called Runge-Kutta formulas, and are called in the same way. The difference is in the step size and number of points returned. `ode23` uses smaller steps and returns more data points than `ode45`. To call either, you must define the differential equation to be solved in a separate file as a function. To pass parameters to the differential equation, declare them as `global`. To solve the equation, `Dn=r*n` numerically, compose the following file with three lines and save it as `expgrow.m`

```
function ndot=expgrow(t,n)
 global r;
 ndot=r*n;
```

The `r` is declared as `global` because it is a parameter to be defined in the the calling program. Notice that the function `expgrow` must have the independent variable, `t`, as the first argument (even though it's not used), and the dependent (or state) variables as the second argument. The function returns the derivative of the state variable, which is privately labeled within the function as `ndot`. The function to be integrated *must* have this format. Once the function is defined as a separate file, it is integrated from the calling program with the following statements, assuming that `r=0.5` and the initial condition is 2.

```
global r;
r = 0.5;
[t,n]=ode23('expgrow',0,1,2);
```

This means that `ode23` should integrate the function `'expgrow'` with the independent variable (which is `t`) running from 0 to 1, and with an initial condition for the dependent variable (which is `n`) of 2. `ode23` returns two vectors, and you must provide variables to accept both of them. The first vector is a set of points for the independent variable, and the second vector contains the dependent variable at these points. In this example, `ode23` returns a column vector of 18 elements for time from 0 to 1, and a column vector also of 18 elements for `n` at each time in the `t` vector. The first element of `n` is 2 and the last turns out to be 3.2974. You can then plot the solution with `plot(t,n)`. If you use `[t,n]=ode45('expgrow',0,1,2);` instead of `ode23` you again wind up with 18 points in the solution, but `n` differs slightly—in the seventh decimal place. In either case, if you want the solution for a time point not in the `t` vector you have to interpolate with the `interp1` command.

A visit to MATLAB's web page will allow one to download still more packages for solving differential equations numerically.

Optimization

Numerical optimization is done with `fmins`. Actually, this command finds values that minimize a function, and to do optimization, which involves maximizing a function, multiply what you want to maximize by -1 and then minimize that. To minimize a function, you must first define it in a separate file. Let's minimize `y=x*(x-a)`. This is a parabola with roots at 0 and `a`, so the minimum is at `a/2`. With `fmins` there are two ways to pass parameters to the function to be minimized. One way is to declare the parameters as `global`. The function to be minimized then accepts the variable as its argument, and returns the value of the function. For example, type in and save the following three lines as the file `parab.m`.

```
function y=parab(x)
 global a;
 y=x*(x-a);
```

Then the following three lines in the calling program will find the minimum of this function assuming `a=1`

```
global a;
a = 1;
fmins('parab',0)
```

The call to `fmins` is to pass a string with the name of the function to be minimized, and a best guess for where the minimum is. Here I guessed that the minimum is at 0 (a bit of a white lie because I know the minimum is at $1/2$, but want `fmins` to do a little work). `fmins` returns with `0.5`, which is correct.

The second way to pass parameters to the function being minimized is to define the function such that the parameters are arguments following the variable. For example, prepare the following two-line file and save it as `parab.m`

```
function y=parab(x,a)
 y=x*(x-a);
```

Now the function as defined in this way is minimized with only one line in the calling program

```
fmins('parab',0,[],[],1)
```

Again `fmins` returns the answer of 0.5. The call to `fmins` has as arguments a string for the name of the function to be minimized, the initial guess of where the minimum is, then two dummy arguments `[]` and `[]`, and finally the value of the first parameter to be passed to the function being minimized (i.e., the value of `a` that is then passed on to `parab`). `fmins` has more elaborate uses as well, which are documented in `help fmins`, and its calling syntax, which seems a little awkward here, is really intended to be upwards compatible with the Optimization Toolbox.

Index